Cyclophanes

Volume II

This is Volume 45 of
ORGANIC CHEMISTRY
A series of monographs
Editor: HARRY H. WASSERMAN

A complete list of the books in this series appears at the end of the volume.

Cyclophanes

Edited by

PHILIP M. KEEHN

Department of Chemistry
Brandeis University
Waltham, Massachusetts

STUART M. ROSENFELD

Department of Chemistry
Smith College
Northampton, Massachusetts

II

1983

ACADEMIC PRESS

A Subsidiary of Harcourt Brace Jovanovich, Publishers
New York   London
Paris   San Diego   San Francisco   São Paulo   Sydney   Tokyo   Toronto

ACADEMIC PRESS, INC.
111 Fifth Avenue, New York, New York 10003

United Kingdom Edition published by
ACADEMIC PRESS, INC. (LONDON) LTD.
24/28 Oval Road, London NW1   7DX

Library of Congress Cataloging in Publication Data

Main entry under title:

Cyclophanes.

  (Organic chemistry, a series of monographs ; v. 45)
  Includes index.
  1. Cyclophanes. I. Keehn, Philip M. II. Rosenfeld,
Stuart M. III. Series.
QD400.C93 1983   547'.59   82-25307
ISBN 0-12-403002-5 (v. 2)

PRINTED IN THE UNITED STATES OF AMERICA

83 84 85 86   9 8 7 6 5 4 3 2 1

Contents

Contributors to Volumes I and II

Numbers in parentheses indicate the pages on which the authors' contributions begin.

MASSIMO D. BEZOARI (359), Dow Chemical U.S.A., Plaquemine, Louisiana 70816.

K. ANN CHOE* (311), Department of Chemistry, Smith College, Northampton, Massachusetts 01063

DONALD J. CRAM (1), Department of Chemistry, University of California, Los Angeles, Los Angeles, California 90024

YUTAKA FUJISE (485), Department of Chemistry, Faculty of Science, Tohoku University, Sendai 980, Japan

YOSHIMASA FUKAZAWA (485), Department of Chemistry, Faculty of Science, Tohoku University, Sendai 980, Japan

HENNING HOPF (521), Technische Universität Braunschweig, Institute of Organic Chemistry, D-3300 Braunschweig, Federal Republic of Germany

SHÔ ITÔ (485), Department of Chemistry, Faculty of Science, Tohoku University, Sendai 980, Japan

PHILIP M. KEEHN (69), Department of Chemistry, Brandeis University, Waltham, Massachusetts 02254

KENJI KOGA (629), Faculty of Pharmaceutical Sciences, University of Tokyo, Tokyo 113, Japan

JOEL F. LIEBMAN (23), Department of Chemistry, University of Maryland Baltimore County, Catonsville, Maryland 21228

SOICHI MISUMI (573), The Institute of Scientific and Industrial Research, Osaka University, Osaka 567, Japan

*A portion of the work was done at the Department of Chemistry, Wellesley College, Wellesley, Massachusetts 02181.

REGINALD H. MITCHELL (239), Department of Chemistry, University of Victoria, Victoria, British Columbia V8W 2Y2, Canada

KAZUNORI ODASHIMA (629), Faculty of Pharmaceutical Sciences, University of Tokyo, Tokyo 113, Japan

WILLIAM W. PAUDLER (359), Department of Chemistry, Portland State University, Portland, Oregon 97207

JAMES A. REISS (443), Department of Organic Chemistry, La Trobe University, Bundoora, Victoria, 3083 Australia

STUART M. ROSENFELD (311), Department of Chemistry, Smith College, Northampton, Massachusetts 01063

IAN SUTHERLAND (679), Department of Organic Chemistry, University of Liverpool, Liverpool L69 3BX, England

Preface

These volumes are intended to provide a comprehensive review of the field of cyclophane chemistry for the period between the earlier volume in this series (*Bridged Aromatic Compounds* by B. H. Smith, 1964) and the present (generally through 1981). In reconciling the enormous growth in this field with our desire to produce a work of manageable size, certain topics have necessarily received less attention than is desirable. Within this limitation we have tried to provide a selection of topics that delineate the past and present of cyclophane chemistry and (it is hoped) point toward some of its future directions.

Bridged Aromatic Compounds provided an excellent evaluation of the field in an "adolescent" stage. Synthetic methods in general have grown in sophistication and power, making the number of potentially available cyclophanes much larger. At the time of writing of the book by Smith, a principle rationale for work in this area was to extend the understanding of molecular systems incorporating unusual structural features (e.g., strain and aryl ring proximity). These earlier results often proved useful in developing, confirming, and refining the theoretical underpinnings of the science. Cyclophanes have provided insight into the ways in which molecules distribute strain, the effects of strain on reactivity, transannular effects on chemistry and spectroscopy, and the criteria for aromatic stabilization. In the present "mature" stage of cyclophane chemistry, we are better able to intelligently define the scope of available compounds and to assess the structural and conformational accommodations made by these systems. This is a propitious time to step back and examine our progress.

Although in a field this large we cannot claim exhaustive coverage, we have tried to touch the areas of major interest by assembling a diverse (both chemically and geographically) group of authors with the shared purpose of critically reviewing our progress at this time of vibrant maturity. The ordering of chapters has been chosen to present background, theory, structure, and spectroscopy followed by a somewhat arbitrary division of cyclophanes into subgroups, roughly in order of increasing structural complexity. We are especially pleased that Professor Donald Cram's introductory description of the development of his thinking in this area reminds us of the human element in scientific research and of the joys and frustrations that we all share. The concluding two chapters point toward two of many future directions, underscoring the view that this field remains a vital and evolving area of chemistry.

Individual chapters have been written so that they may be read with little or no direct reference to other chapters. Each stands alone as a review of a particular area of cyclophane chemistry and therefore some overlap between chapters will be apparent. It has been left to each author to design an organizational approach most suited to the particular material being covered. It is hoped that the advantages of autonomy in this regard will outweigh any compromise in readability of the work as a whole. In any case, shortcomings in this area are the responsibility of the editors and not of the individual authors.

Philip M. Keehn
Stuart M. Rosenfeld

Contents of Volume I

CHAPTER **6**

Synthesis and Properties of Heterophanes

WILLIAM W. PAUDLER

Department of Chemistry
Portland State University
Portland, Oregon

MASSIMO D. BEZOARI

Dow Chemical U.S.A.
Plaquemine, Louisiana

I. INTRODUCTION

Since the original synthesis[1] of paracyclophane in 1949 this type of structure has attracted much attention. The literature on cyclophanes

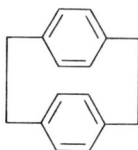

Paracyclophane

through 1964 has been reviewed.[2] This chapter is confined to a discussion of phanes that have at least one heterocyclic ring as part of the cyclophane structure. Such molecules include the *π-excessive* systems, for example, furan, pyrrole, thiophene, and related five-membered rings, and the *π-deficient* systems, such as pyridine and pyridazine. In addition, mixed heterophanes that may contain both π-excessive and π-deficient rings or a heterocyclic and a benzenoid aromatic moiety are considered. The only other concession to brevity is the lack of consideration of the porphine ring system. The reader is referred to other reviews for information on porphines and their congeners.[3,4]

II. NOMENCLATURE

In naming phane compounds, *Chemical Abstracts* uses the *polycyclo system,* which, although unambiguous, is frequently cumbersome. This has prompted some authors to suggest the *cyclophane nomenclature system,* which was originally applied only to carbocyclic molecules. However, because of the large number of new molecules synthesized in recent years, especially those containing heteroatoms, the cyclophane system has evolved into one that attempts to encompass all classes of phanes. This system has been explained in some detail[2] and has been submitted for consideration to the Committee on Organic Nomenclature of the Division of Organic Chemistry of the American Chemical Society. It is this nomenclature system that is used throughout this chapter.

The word *phane* is used as the root for all bridged aromatic molecules. Prefixed to the fragment *-phane* is the class name of the aromatic ring that is bridged. Preceding the full name are two sets of numbers. Those appearing in brackets indicate the size of the bridge (or bridges), and those in parentheses indicate the positions of substitution on the aromatic nucleus. Thus, compound **1** is called [10](2,5)thiophenophane. If more than one bridge is present in the molecule, it is indicated by the number of digits in brackets. Structure **2** thus describes [2.2](2,5)furanophane. Additional de-

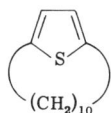

1

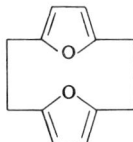

2

tails can be found in the work cited as well as in another communication on the subject.[5]

III. SYNTHESIS AND CHEMISTRY OF HETEROPHANES

A. Mononuclear π-Excessive Heterophanes

The syntheses of this class of heterophanes (the term *mononuclear* refers to the fact that the cyclophane contains only one bridged aromatic ring) have relied heavily on the classical methods of preparing furans, pyrroles, and thiophenes. Marchesini *et al.* described the synthesis of the [9](2,4)heterophanes **3a–3c** (Scheme 1).[6]

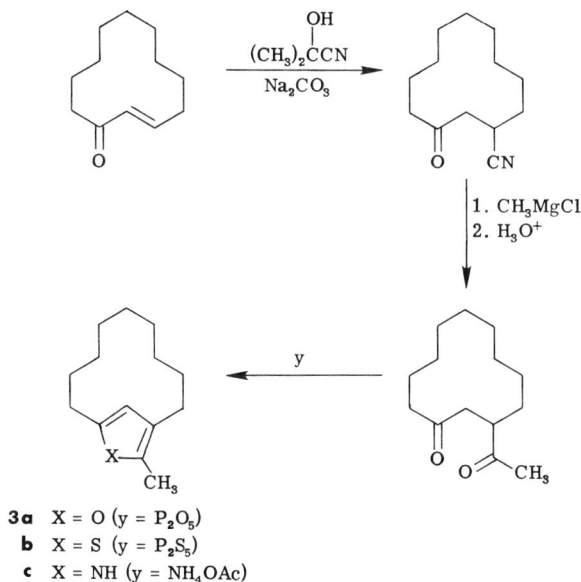

3a X = O (y = P_2O_5)
  b X = S (y = P_2S_5)
  c X = NH (y = NH_4OAc)

SCHEME 1

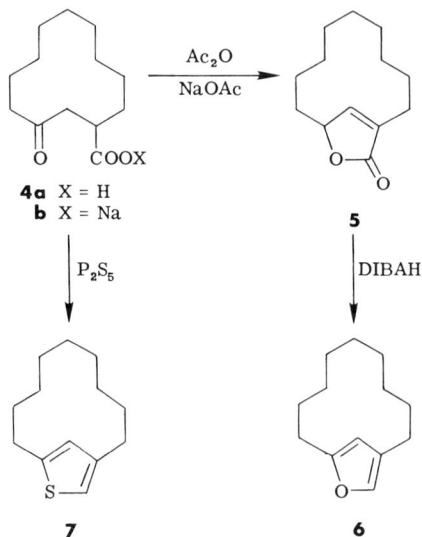

SCHEME 2

Similarly, the preparation of the unsubstituted thiophenophane and furanophane by a slightly different route has been reported (Scheme 2).[7] The keto acid **4a** is heated with a mixture of acetic anhydride and sodium acetate to give several products, one of which is the desired unsaturated lactone **5**. The crude mixture is treated with diisobutylaluminum hydride (DIBAH), giving the furanophane **6** in about 15% yield. When the sodium salt of the keto acid **4b** is heated with phosphorus pentasulfide, a 10% yield of the thiophenophane **7** is obtained.

The same workers treated 3-chloro-2-cyclododecenone with hydrazine and obtained [9](3,5)pyrazolophane (**8,** Scheme 3). Reaction of the same ketone with hydroxylamine hydrochloride affords a mixture of oximes, which when treated with sodium ethoxide in ethanol gives three products in comparable yields (**9a, 10a,** and **11**), including the desired [9](3,5)isoxazolophane (**11**).

Nozaki *et al.*[8] reported the synthesis of compounds **3a** and **8** by a different route involving the treatment of 2-cyclododecenone with hydrazine to give the pyrazoline **9b.** Dehydrogenation of **9b** with sulfur generates pyrazolophane **8.** The furanophane **3a** is obtained by reaction of 2-cyclododecenone with lithium actylide, acid-catalyzed isomerization, and hydration–cyclization (Scheme 4).

The same workers also reported syntheses of some [6](2,4)thiophenoand pyrrolophanes (Scheme 5).[9] Compound **10c,** obtained in a manner

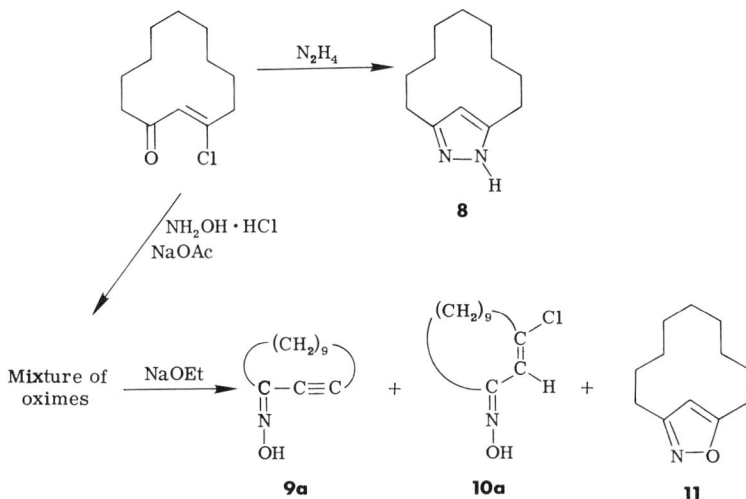

SCHEME 3

analogous to the formation of **10b** (Scheme 4), does not undergo dehydration to the furan upon treatment with mercuric sulfate and acid, as does its larger ring counterpart. Indeed, dehydration could not be effected. The corresponding thiophene **(12)**, however, is obtained as a labile yet distillable liquid, and the N-aryl pyrroles **13a** are obtained by treatment of the diketone precursor with aromatic amines.

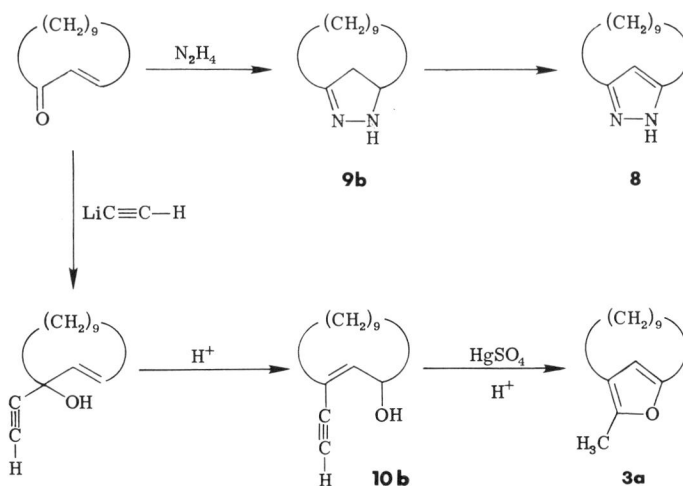

SCHEME 4

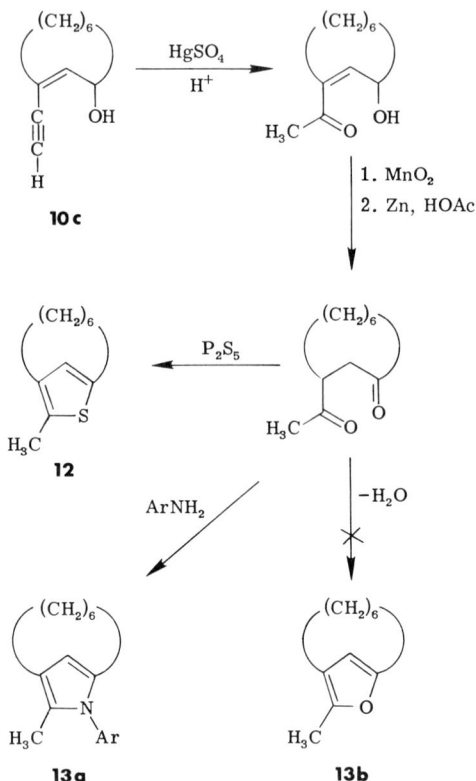

SCHEME 5

Compound **15** has the shortest bridge reported for any mononuclear phane.[9a] It is prepared by pyrolysis of **14.** The structure was confirmed by reduction to a pyrolidine, which was methylated to give the known compound homotropane **(16).** In an analogous manner compound **18** was prepared from **17.**

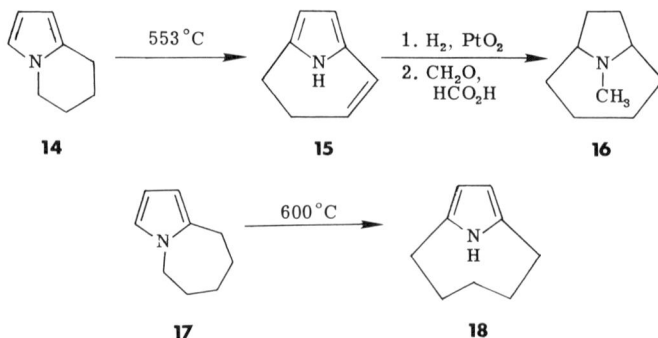

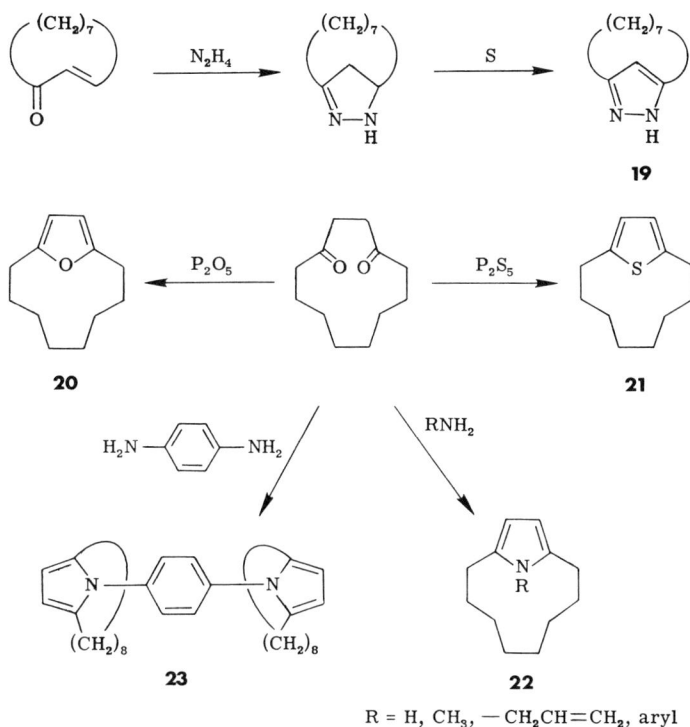

SCHEME 6

Other heterophanes prepared by Nozaki *et al.* include [7](3,5)pyrazo-lophane **(19)**,[10] [8](2,5)heterophanes **20–22**,[11,12] as well as some bis-he-terophanes such as **23** (Scheme 6).

Parham and co-workers have synthesized 3,5-bridged pyrazoles con-taining 6, 8, and 10 members and have discussed the mechanism of their formation.[13,14] The enol acetate **24** is treated with phenyl(trichloromethyl) mercury to give the cyclopropane **25**. Reaction of **25** with hydrazine affords a 45% yield of [10](3,5)pyrazolophane **(26)** along with about 5% of its isomer **27** (Scheme 7).

Pyrazoles can also be obtained from ethers **28** (Scheme 8) by ring open-ing to the diene **29** followed by cyclization to **26**.[15] Compound **29** was transformed by potassium *tert*-butoxide to **30**, which can also be cyclized to compound **26**. The [6](3,5)pyrazolophane **(32)**, along with some of its isomer **33**, is obtained from **31** in a similar manner.

Hunig and Hoch reported the preparation of pyrazoles **34** having 13-, 14-, and 15-membered bridges across the 3,5 positions[16] of the heterocy-clic ring. These compounds were prepared from the corresponding 1,3-diketone and hydrazine.

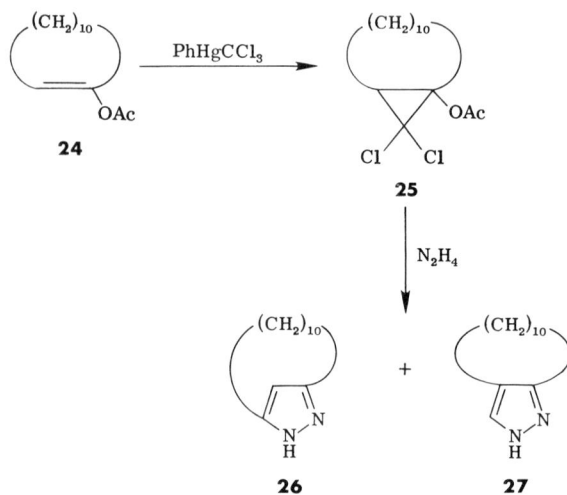

SCHEME 7

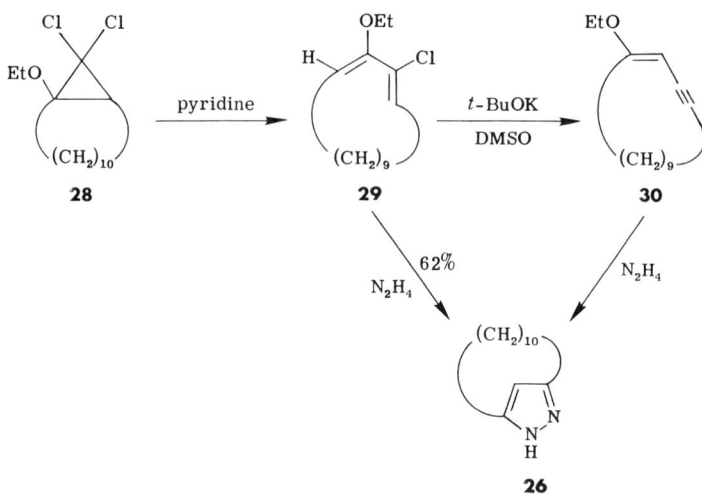

SCHEME 8

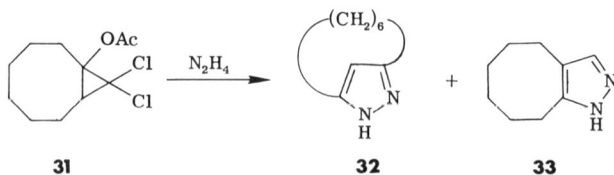

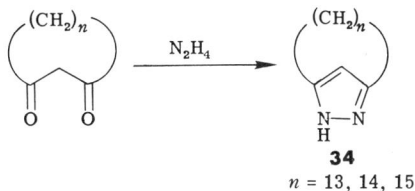

34

$n = 13, 14, 15$

Russian workers have expended considerable effort in the thiophenophane area. In some instances Friedel–Crafts reactions have been employed to effect the cyclization.[17–20] Reactions of this type usually

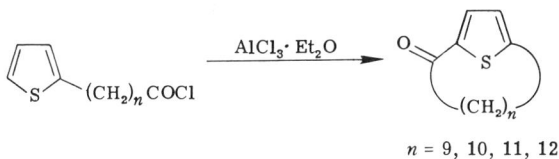

$n = 9, 10, 11, 12$

give a mixture of products including dimers and trimers. The same workers have employed the acyloin condensation.

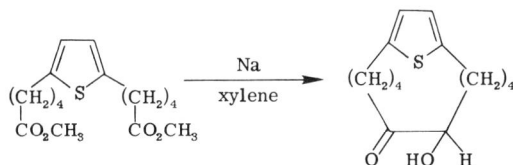

Intramolecular alkylation reactions have also been used in the thiophene series.[20–24]

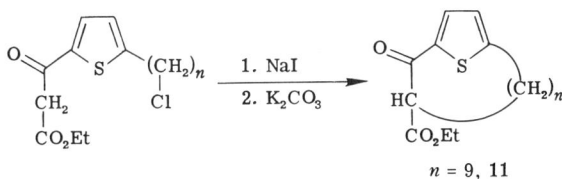

$n = 9, 11$

Hayward and Meth-Cohn reported the synthesis of macrocycles containing the benzimidazole moiety.[25,26] The synthesis of the imidazolone **37** along with other derivatives is shown in Scheme 9. Thus, compound **35** is formed by aromatic nucleophilic substitution. Its reaction with hydrochloric acid produces the nitrone **36**. Reaction of compound **36** with *p*-toluenesulfonyl chloride and sodium hydroxide effects a rearrangement to structure **37**. Compound **37** ring-opens with *n*-butyllithium to **38**, which, after hydrolysis, yields the diamine **39**. One of the several derivatives prepared was **40** ($n = 5$), obtained from the reaction of compound **39** with

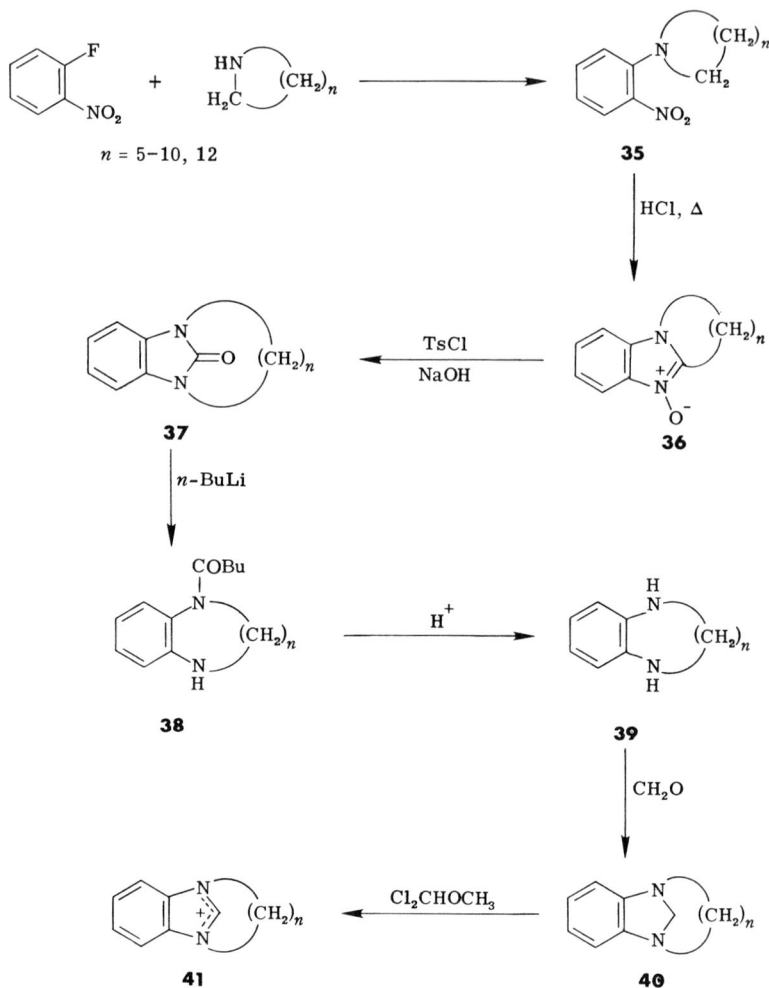

SCHEME 9

formaldehyde. The dihydrobenzimidozole reacts with dichloromethyl methyl ether to give the benzimidazolium ion **41**.

Benzimidazolone macrocycles such as **37** have also been produced by alkylation of benzimidazolone **42** itself.[27]

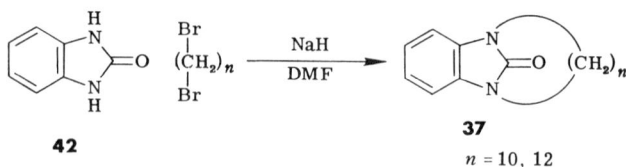

B. Multinuclear π-Excessive Heterophanes

The 1,6 Hofmann elimination has found widespread use in the syntheses of [2.2]phanes. Winberg and co-workers used this method to prepare the first reported [2.2]heterophane (Scheme 10).[28] Pyrolysis of the quaternary ammonium hydroxide **43** generates the intermediate 2,5-dimethylene-2,5-dihydrofuran **(44),** which is stable at −78°C. When heated to reflux in absolute ethanol in the presence of free-radical inhibitors, **44** affords [2.2](2,5)furanophane **(2)** in 73% yield. The substituted derivatives **45–48** are obtained from the appropriately substituted precursors.

In an analogous reaction these researchers obtained [2.2](2,5)-thiophenophane **(49)** in 19% yield by azeotropic removal of water from the reaction. Compound **50** was prepared by Fletcher and Sutherland by means of a "cross-breeding" reaction.[29] This involved the pyrolysis of equimolar amounts of the appropriate quaternary ammonium hydroxides in refluxing xylene.

The cross-breeding reaction method was reported earlier by Cram's group in the synthesis of [2.2](2,5)furanoparacyclophane **(51).**[30] The similar phane **52** was synthesized later by Whitesides *et al.*[31] This compound was needed for variable-temperature studies.

Researchers studying the effect of replacing a furan ring with a thiophene ring in these systems synthesized [2.2]paracyclo(2,5)-thiophenophane **(53)** in 1.6% yield by the Hofmann elimination process. Multilayered heterophanes containing furan or thiophene rings were also

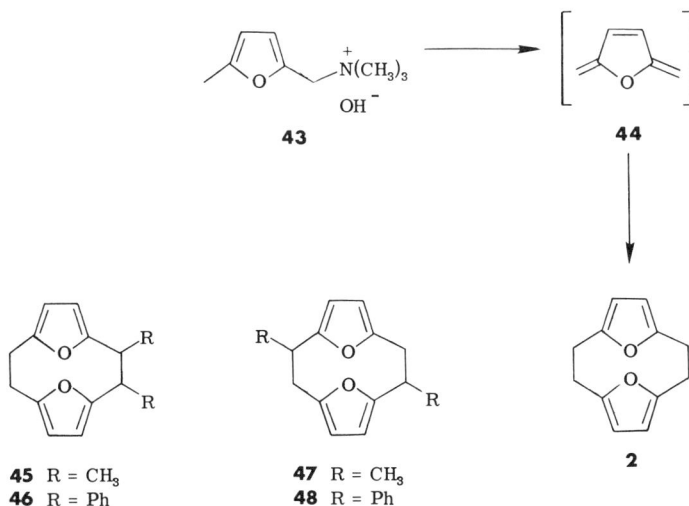

43   **44**

45 R = CH₃
46 R = Ph

47 R = CH₃
48 R = Ph

2

SCHEME 10

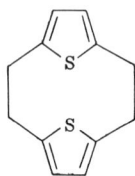 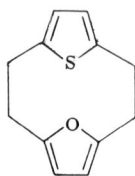 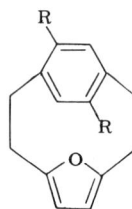

49   50   51 R = H
                                                     52 R = D

reported.[32] In a continuation of this research, the same group synthesized 4,7-dimethyl[2.2](2,5)furanoparacyclophane **(54)** and the thiophene ana-log **55**. The equivalent multilayered systems were also prepared.[33]

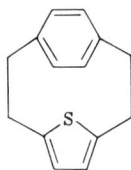 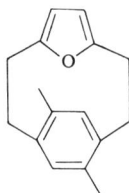 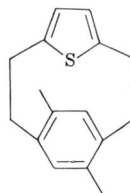

53   54   55

During studies on the reactions of singlet oxygen with strained aromatic systems, Wasserman and Keehn prepared [2.2](2,5)furano(1,4)naphtha-lenophane **(56)**[34] in 11% yield. The furanophane **2** was one of the by-products. The cross-breeding reactions yield all of the possible dimers in various yields.

When a mixture of the two quaternary ammonium hydroxides **57** and **58** is pyrolyzed in refluxing xylene, *anti*-[2.2](1,4)naphtha-leno(2,5)thiophenophane **(59a)** and its syn isomer **(59b)** are obtained in 4.1 and 0.3% yields, respectively,[35] along with compound **60**.

Heterophanes containing the anthracene nucleus can be obtained by the Hofmann elimination. Thus **61, 62,** and **64** give yields of 0.8, 2.8, and 5.5%, respectively.[35] [2.2](9,10)Anthraceno(2,5)furanophane **(63)** is ob-tained in 40% yield.[36]

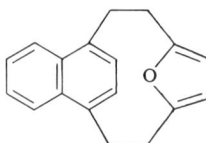 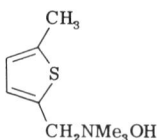 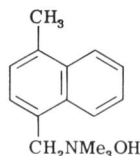

56   57   58

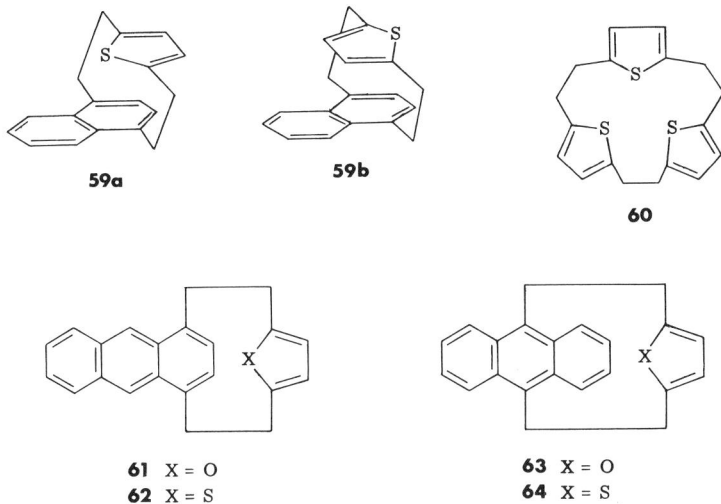

59a

59b

60

61 X = O
62 X = S

63 X = O
64 X = S

Unique, multilayered heterophanes such as **65** have also been prepared using the Hofmann elimination–cross-breeding reaction (Scheme 11).[37] Quadruple-layered heterophanes **66** have been prepared in an analogous manner.[38]

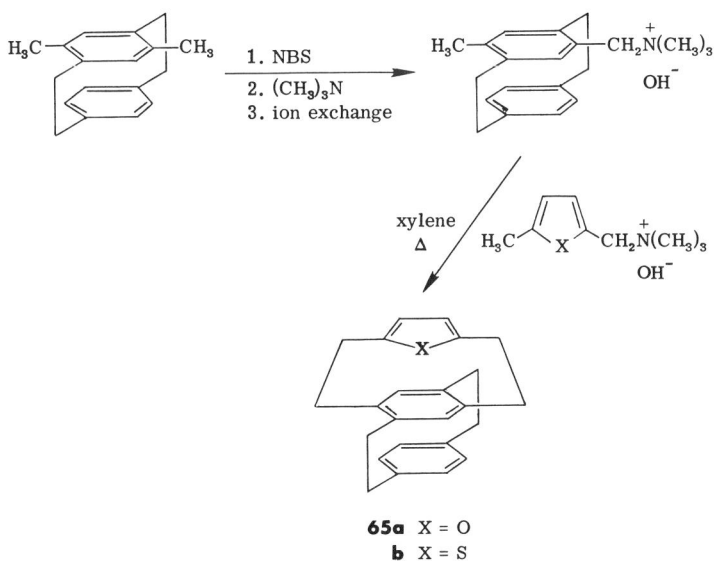

65a X = O
b X = S

SCHEME 11

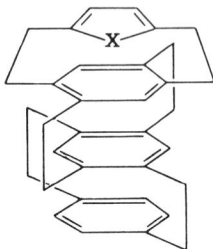

66a X = O
   b X = S

During the preparation of macrocyclic polyethers containing furanyl units, Cram obtained the strained heterophane **67** in 29% yield (Scheme 12).[39]

The successful syntheses of [2.2](2,5)oxazolophanes and [2.2](2,5)-thiazolophanes have been reported.[39a] [6.2]Furanophanes and [4.4]furanophanes are obtained via 1,8 Hofmann eliminations[40] which were also utilized for the [4.2] furanoparacyclophane indicated.

Wasserman and Keehn found that reaction of naphthalenofuranophane **56** with bromine in methanol gives, via **68,** the diketone **69.** Compound **69,** the geometry of which is unknown at this time, is also the product of singlet oxygen oxidation of **56.**[41]

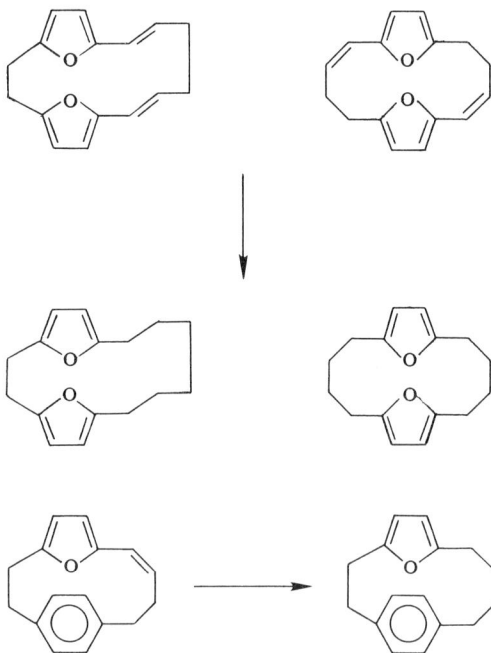

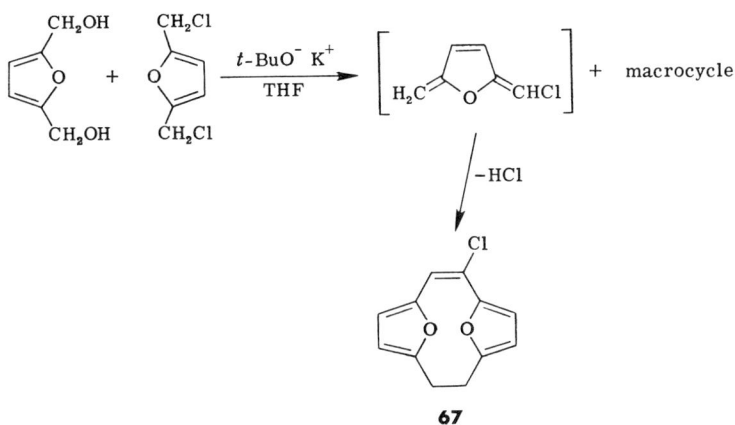

SCHEME 12

The Paal–Knorr method of synthesizing five-membered heterocyclic rings from 1,4-diketones has been extended to include cyclic systems.[42-45] Although Cram originally reported the hydrolysis of furanoparacyclophane **51** to give the diketoparacyclophane **70**,[30] it was Keehn and

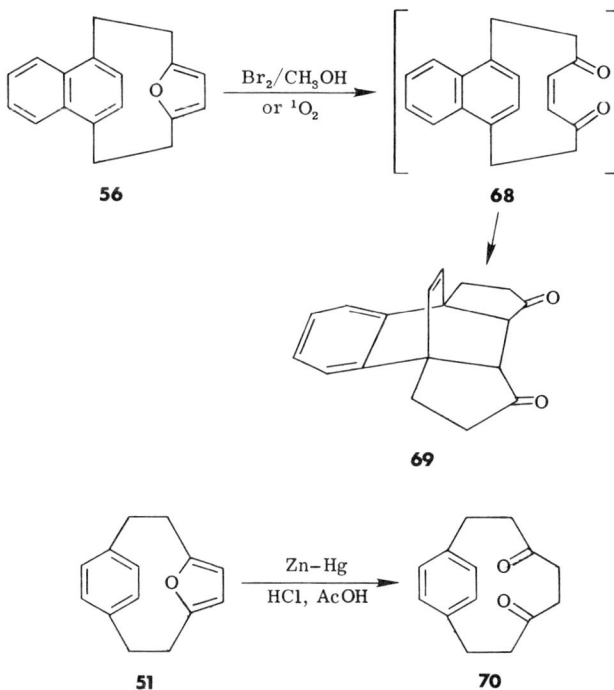

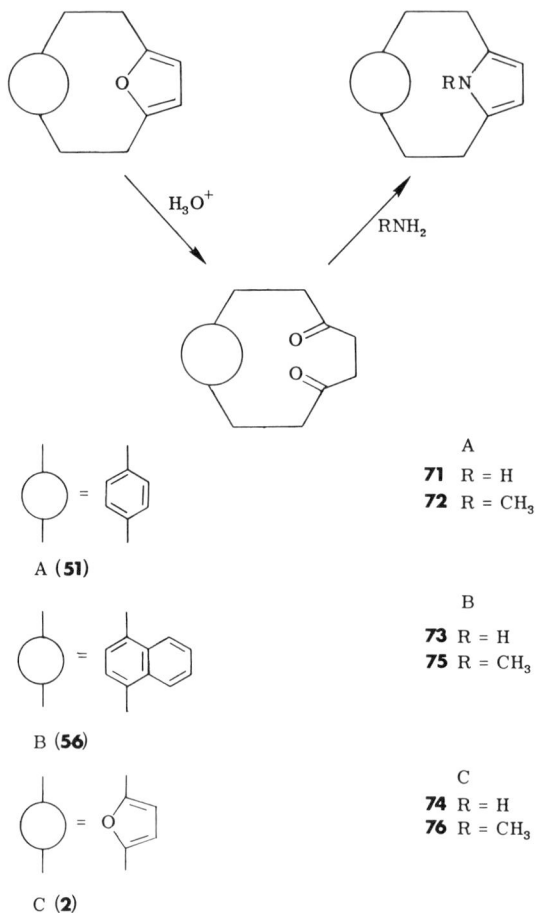

SCHEME 13

Haley who refined the method to deal with the more acid-sensitive furanophane **2.** They employed this method in synthesizing various conformationally interesting heterophanes (**71–75,** Scheme 13).[46] The deuteriated furanopyrrolophane **77,** the analog of **74,** was prepared for confor-

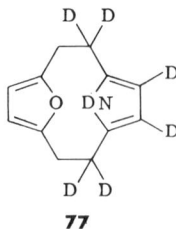

77

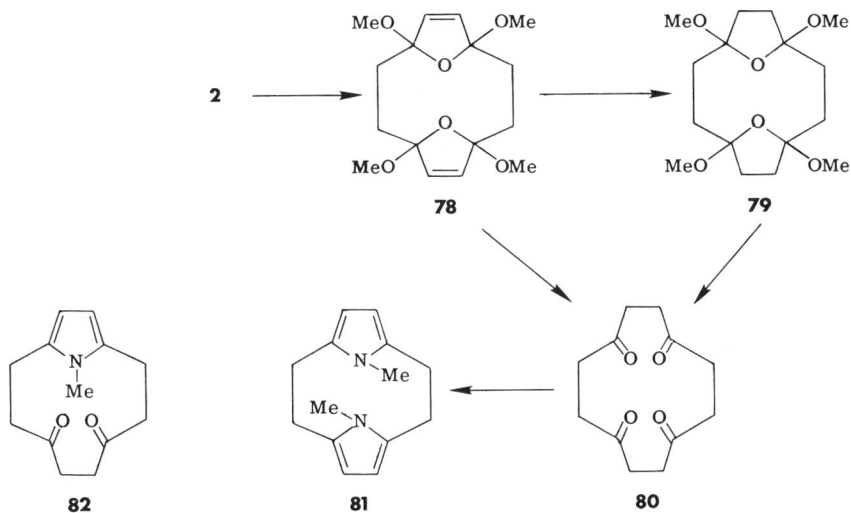

SCHEME 14

mation studies.[47] Wong and Paudler successfully used the cyclization method in preparing 76.[48]

The versatility of furanophane 2 as a synthetic precursor of other heterophanes has been further demonstrated by its conversion to the tetraketone 80, which can readily undergo Paal–Knorr cyclizations (Scheme 14).

The conversion of furanophane 2 to the tetramethoxy compound 78 followed by transformation to the acetal 79 had been previously described.[28] Wasserman and Bailey effected the acid hydrolysis of 79 to give the tetraketone 80 in a reported yield of 95%. They obtained N,N'-dimethyl[2.2](2,5)pyrrolophane (81) in 42% yield by condensation of 80 with an excess of N-methylamine.[49] The intermediate monocondensation product, diketo-N-methylpyrrolophane (82), can be isolated when a lesser amount of N-methylamine is allowed to react with the tetraketone.

Paudler and Stephan[50] utilized the tetramethoxy derivative 78 in the synthesis of the anti-aromatic pyrrolophane 85. Later studies showed that the tetramethoxy compound 78 can be converted directly to the tetraketone 80 by high-pressure hydrogenolysis and that the condensation of hydrazine with the tetraketone 80 gives the pyrrolophane 83 in greatly improved yields.[51] Conversion of the pyrrolophane 83 to the anti-aromatic annulene 85 is effected in two dehydrogenation steps via the isolable compound 84 (Scheme 15).[50,51]

Other pyrrolophanes synthesized by this method include N-(p-tolyl)[2.2](2,5)pyrrolophane (86), N-phenyl[2.2](2,5)pyrrolophane (87), N-

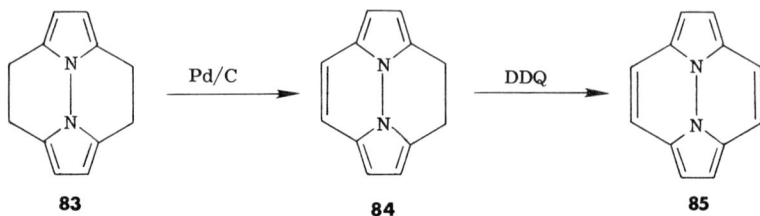

SCHEME 15

ethyl-N'-methyl[2.2](2,5)pyrrolophane **(88)**, N-ethyl[2.2](2,5)pyrrolo-
phane **(89)**,[48,52] N-(p-bromophenyl)[2.2](2,5)pyrrolophane **(90)**, N-(p-

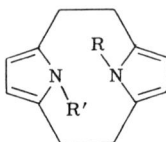

86 R = p-PhCH$_3$; R' = H   **89** R = C$_2$H$_5$; R' = H
87 R = Ph; R' = H   **90** R = p-BrPh; R' = H
88 R = C$_2$H$_5$; R' = CH$_3$ **91** R = p-BrPh; R' = CH$_3$

bromophenyl)-N'-methyl[2.2](2,5)pyrrolophane **(91)**, and the tris-
bridged pyrrolophane **92**.[48,53] The parent [2.2](2,5)pyrrolophane **95** could

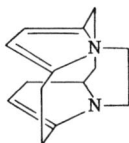

92

not be obtained by condensation of an excess of ammonia with tetraketone **80**. Instead, treatment of the N-benzylpyrrolophane **93** with sodium
in liquid ammonia generates the anion **94**, which gives the parent pyrrolophane **95** on addition of water (Scheme 16).[54]

  Because of the anomalous reaction of ammonia with tetraketone **80**, the
unsymmetric pyrrolophanes, which have one unsubstituted ring (i.e., an
N—H pyrrole ring), must be synthesized by prior condensation of the
substituted gaseous amine, as reported by Keehn and Haley in their synthesis of N-methyl[2.2](2,5)pyrrolophane **(96)**.[54] An additional cyclization
example is the synthesis of the other tris-bridged pyrrolophane **97** by
condensation of the tetraketone **80** with o-phenylenediamine.[52]

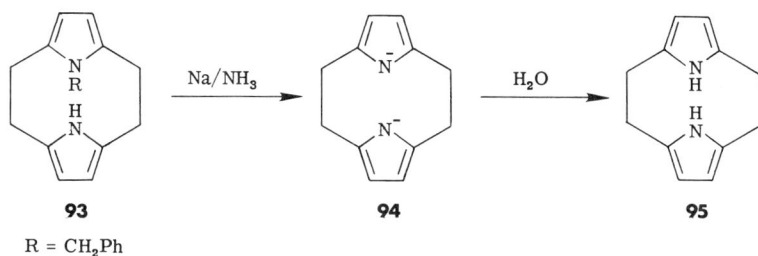

93

R = CH₂Ph

SCHEME 16

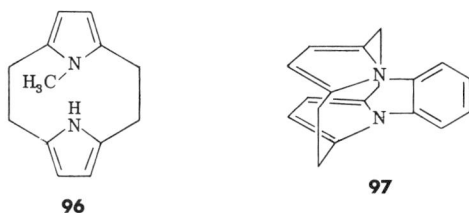

96

97

Htay and Meth-Cohn, in their report of the synthesis of macrocyclic benzimidazolones **(37)** (see previous section), indicated that the structure of the product depended on the value of n. Thus, when n is 5, 6, 7, or 8, only dimers **98** and trimers **99** are obtained (Scheme 17).[27]

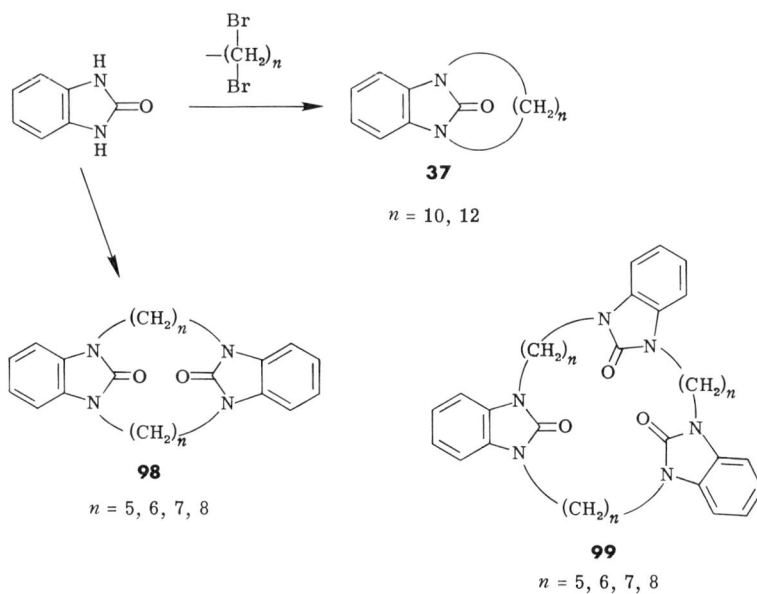

37

$n = 10, 12$

98

$n = 5, 6, 7, 8$

99

$n = 5, 6, 7, 8$

SCHEME 17

Bergman and Backvall reported the interesting derivative **100b** from their study of the Favorskii rearrangement of 3-substituted indoles.[55] Treatment of **100a** with sodium hydroxide gives a complex mixture from which **100b** is isolated.

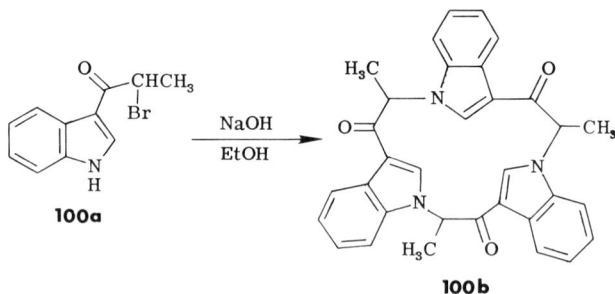

100a

100b

Some naturally occurring pigments with heterophane structures have been reported.[56,57] Compound **101** is an example. In others that have been reported R varies according to length and substitution of alkyl groups and

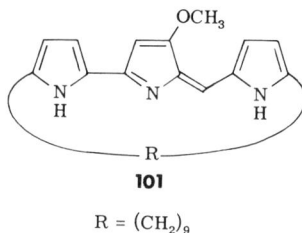

101

$R = (CH_2)_9$

oxygen-containing functions. These pigments are related to the naturally occurring compound metacycloprodigiosin **(102)**, which was isolated and synthesized by Wasserman and co-workers.[58,59]

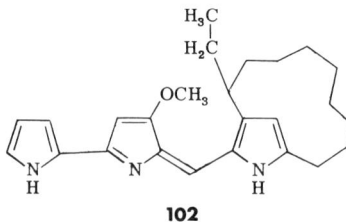

102

The novel heterophane **103** was synthesized by Gol'dfarb et al.[24] The recently synthesized[60] bis(indocarbocyanine) is the first known phane-type dye.

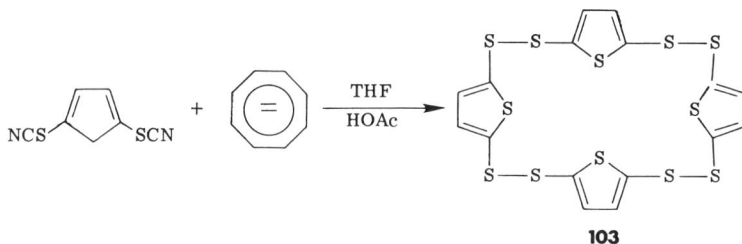

103

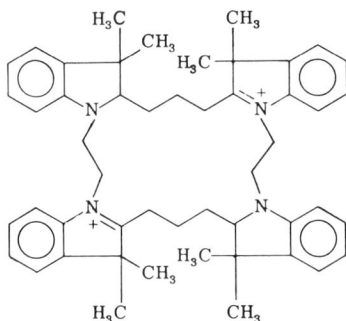

The transformations performed on these heterophanes are limited essentially to those compounds containing a furan or pyrrole ring. The chemistry of furanoparacyclophane **51** has been examined by Cram[30], Paudler,[61,62] and others.[64] The heterophane **51** reacts with bromine in methanol at low temperature to give intermediate **104**. If **104** is poured into water, the product is the *cis*-enedione **105**. Acid hydrolysis of compound **104** afforded the *trans*-enedione **106**. The cis isomer could be trans-

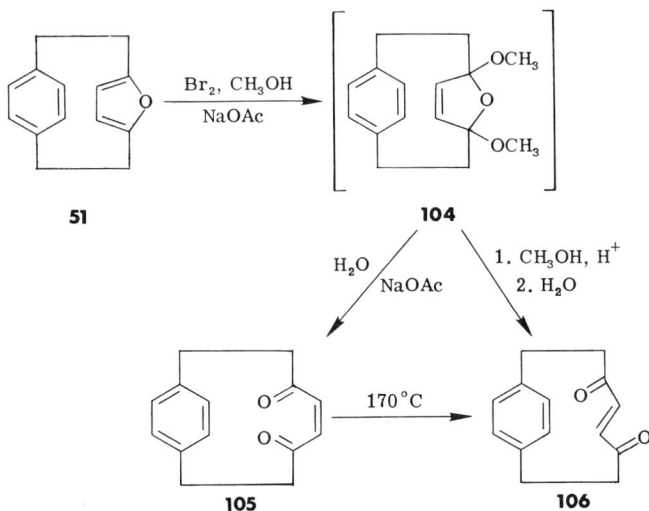

formed into the trans isomer by heating at 170°C.[63] Models indicate that
the trans isomer is indeed less hindered than the cis compound.

Wasserman examined the reaction of **51** with singlet oxygen.[64] After
hydrogenation of the reaction mixture, three products were isolated: **70,
107,** and **108** (Scheme 18). The structure of compound **108** was determined
by X-ray diffraction. The following steps were proposed to account for its
formation: decomposition of the initially formed endoperoxide **109** to **110,**
followed by rearrangement to compound **111,** isomerization to the ene-
dione **112,** intramolecular Diels–Alder reaction, and hydrogenation to
give compound **108** (Scheme 19).

When oxidation is carried out in methylene chloride, a product is obtained the structure of which is tentatively assigned as **113.**[64a]

The singlet oxygen oxidation of [2.2](2,5)furanophane was also examined by Wasserman, and the product formation was again shown to be
solvent dependent (Scheme 20).[64a,64b] When the reaction is performed in
methanol, compound **116** is obtained. This compound presumably arises
from the internal Diels–Alder reaction of **115.** In turn, compound **115**
results from the putative intermediate **114,** which must be reduced in
some unspecified fashion. The stereochemistry of **116,** endo or exo, was
not determined. Compound **116** was transformed into **117** by heating with
sulfuric acid. Compound **117** was then hydrogenated to the known hex-
ahydro-*as*-indacene **118.**

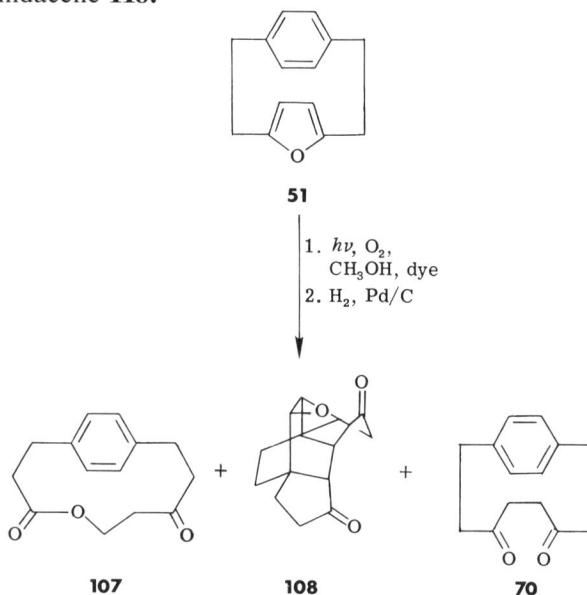

51

1. hv, O_2,
   CH_3OH, dye
2. H_2, Pd/C

107   +   **108**   +   **70**

SCHEME 18

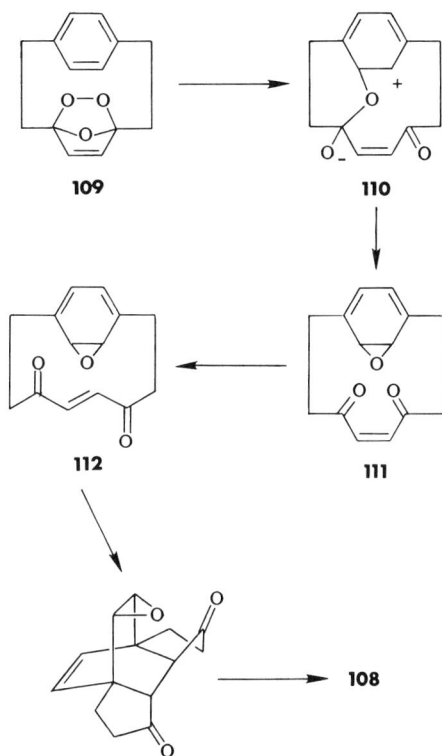

SCHEME 19

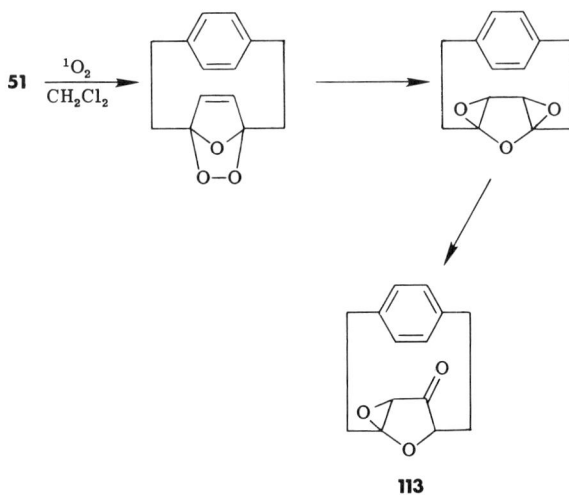

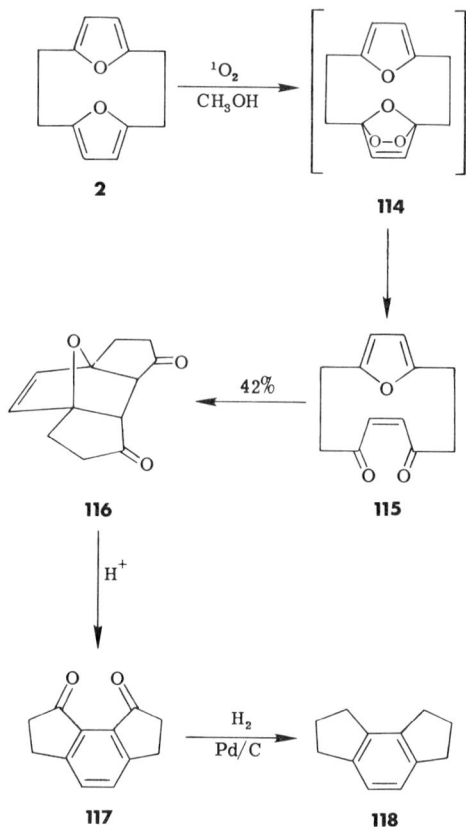

SCHEME 20

Other workers have isolated a by-product of this reaction, which is believed to have structure **119**.[64c] When the oxidation is carried out in methylene chloride, compound **121** is obtained, presumably from compound **120** via a rearrangement for which there are precedents. The structure of **121** was substantiated by an X-ray diffraction analysis.

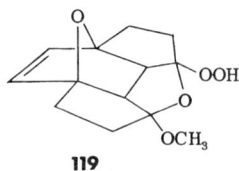

119

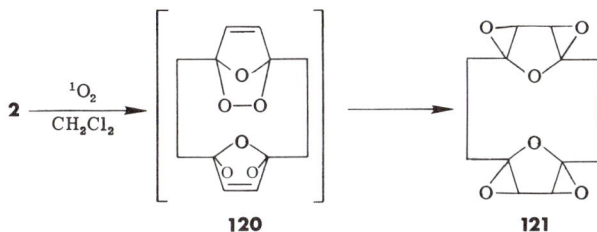

120   **121**

Compounds **116** and **117** were also obtained when oxidation of furanophane **2** was attempted with SeO$_2$ in ethanol.[62] They are presumed to arise from **115,** which is formed by oxidation of the initial hydrolysis product.

An unusual reaction, in light of Wasserman's investigation of the singlet oxygen reaction of furanoparacyclophane **51,** was that observed by Cram when **51** was left standing in sunlight. The phane dissolved in cyclohexane yielded compound **122,** with no trace of the furan portion of **51.**[63]

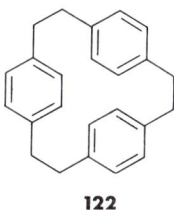

122

[2.2](2,5)Furanophane reacts in a Diels–Alder fashion with dimethyl acetylenedicarboxylate, yielding a 1 : 1 adduct.[63] None of the other common dienophiles reacts. The structure of the adduct was assigned on spectroscopic evidence (Scheme 21). The reaction gives, initially, the intermediate **123,** which has two sites of saturation that may react further. It is apparently the activated double bond of compound **123** that becomes attached to the remaining furan ring to give product **124.**

More recently, Kapicak and Battiste reported that compound **2** also reacts with benzyne, giving both a 1 : 1 **(125)** and 2 : 1 adduct **(126).**[65] None of the intramolecular product **(127)** is found. Attempts to transform **126** into one of the isomeric naphthalenophanes failed (Scheme 22).

Previously, Battiste and co-workers showed that compound **2** reacts with tetrachlorocyclopropene to yield a product the structure of which was determined to be **128** by X-ray diffraction (Scheme 23).[65a]

Wasserman and Kitzing examined the reaction of [2.2](2,5)furano-(1,4)naphthalenophane **(56)** with dimethyl acetylenedicarboxylate and obtained a 1 : 1 adduct.[66] By means of nmr and mass spectroscopy structure

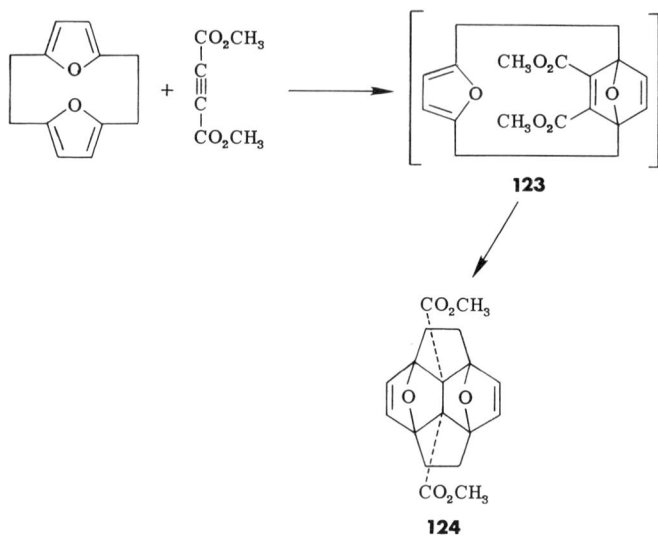

SCHEME 21

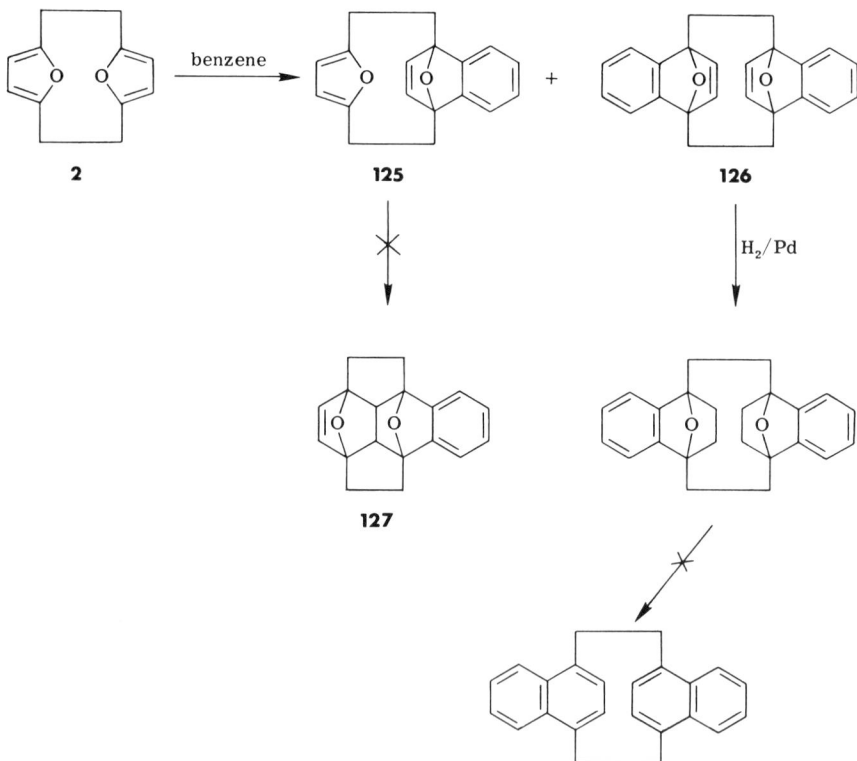

SCHEME 22

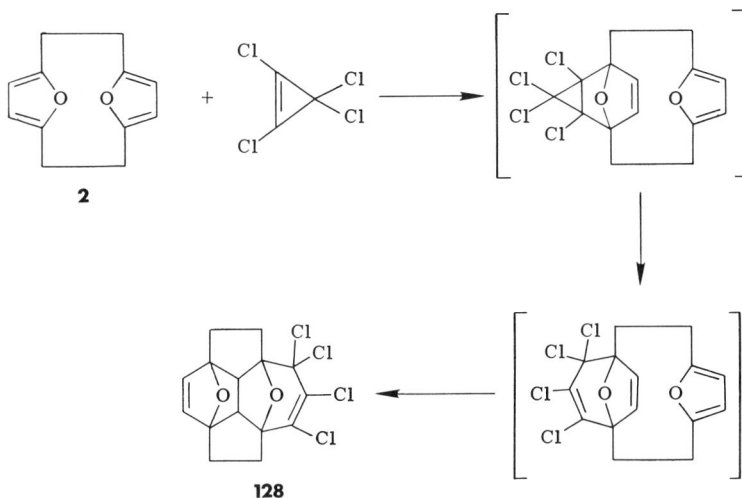

SCHEME 23

130 was assigned to this product (Scheme 24). Thus, the activated double bond of the initially formed oxabicyclo ring of **129** undergoes a second Diels–Alder reaction with the nonbridged six-membered ring of the naphthalene.

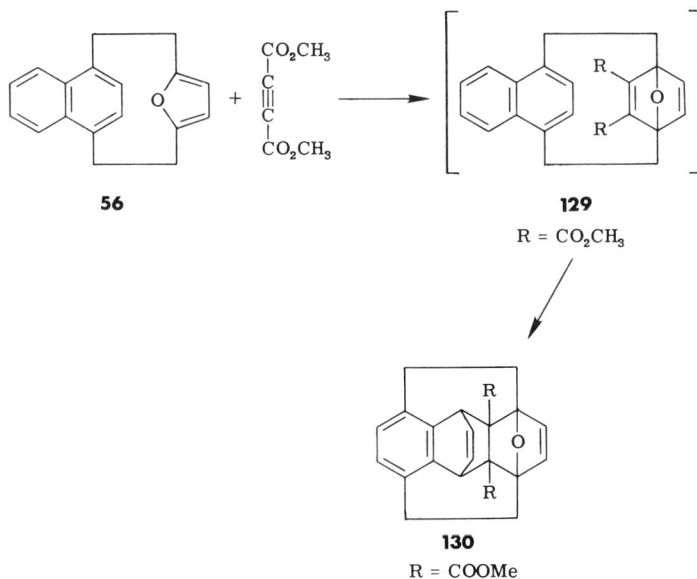

SCHEME 24

In contrast to these results, the internal cycloaddition that takes place after the formation of the initial adduct in the reaction of anthraceno-furanophane with dimethyl acetylenedicarboxylate occurs with the more electron-rich double bond. Thus, the 2 : 1 adduct **131** is formed.[67]

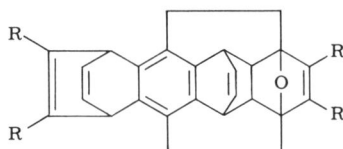

131

R = COOMe

The first electrophilic substitutions of multinuclear π-excessive heterophanes have been reported.[61,62] High yields of monoacetylated products **132, 133,** and **134** are obtained with acetic anhydride/boron trifluoride

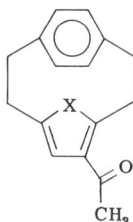

132 X = O
133 X = NH

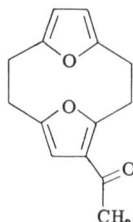

134

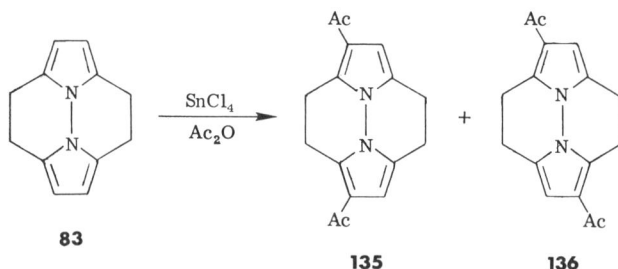

83

135   **136**

etherate and the appropriate heterophane under mild conditions. Acetylation of annulene **83** with stannic chloride[52] yields a mixture of **135** and **136.** Similarly, the following conversion has been accomplished:

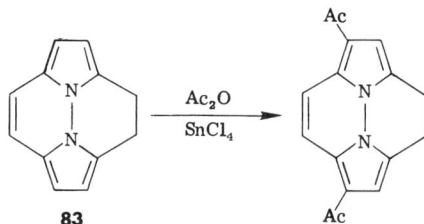

83

Several heterophanes have been shown to undergo a reversible two-electron reduction at a mercury electrode in dimethylformamide solution.[68] These compounds contain a conjugated perphery of $4n$ π electrons and thus are formally anti-aromatic hydrocarbons and form "aromatic" dianions of pronounced stability.

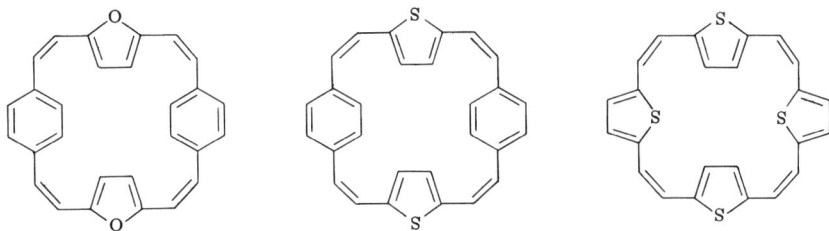

C. Mononuclear π-Deficient Heterophanes

The majority of the π-deficient mononuclear heterophanes have been prepared using the classical condensation reactions. Included among the heterophanes formed by these methods are pyrylophanium salts, pyridinophanes, and pyridazinophanes.

Balaban et al.[69] described the preparation of 15-methyl[10](2,6)pyrylophanium perchlorate (**137**) by diacylation of an olefin and its subsequent conversion to the corresponding pyridinophane (**138**, Scheme 25). The perchlorate was formed in low yield (1–2%). Compound **138**, 15-methyl-[10](2,6)pyridinophane, is a structural isomer of a naturally occurring heterophane, muscopyridine (**139**), isolated from the perfume gland of the musk deer and was first synthesized by Buchi[70] and more recently by Kumada[71] and Hiyama.[72]

Balaban also described the synthesis of another pyrylophanium perchlorate[73] (**140**, Scheme 26) by acetylation of cyclododecene. It, too, was easily converted to a pyridinophane **141**. Compound **140** reacts with

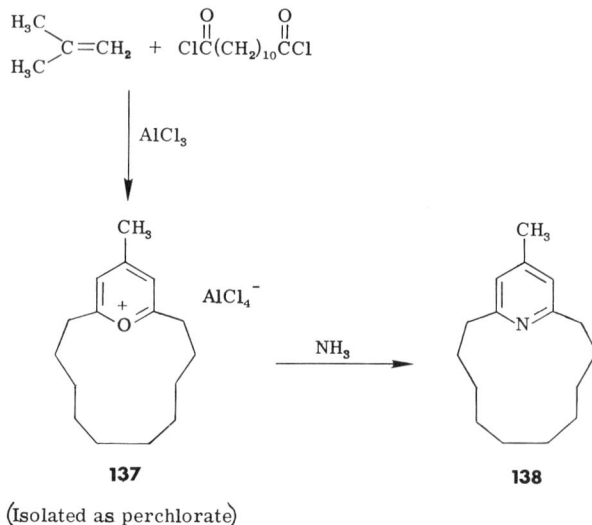

137

(Isolated as perchlorate)

138

SCHEME 25

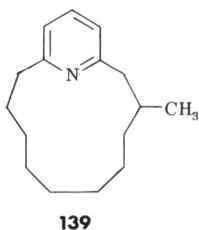

139

hydrazine or amines to give the pyridinium salts **142** and **143,** respectively. Reaction with phenylhydrazine gives the unexpected compound **144.**

Japanese workers have succeeded in preparing [7](2,6)pyridinophane **(145)** and [7](2,6)pyrylophanium perchlorate **(146)** by the reactions shown in Scheme 27.[74–76]

In a similar fashion Italian workers prepared the heterophanes **147** and **148** (Scheme 28) with the bridge at the 2,4 positions.[77] In addition, diketone **149** was converted to a mixture of 13-methyl[9](4,6)pyridazinophane **(150a)** and its tetrahydro derivative **150b.** Although the dihydro species was not observed, the authors speculated that it is formed and undergoes oxidation and/or disproportionation to give the observed products.

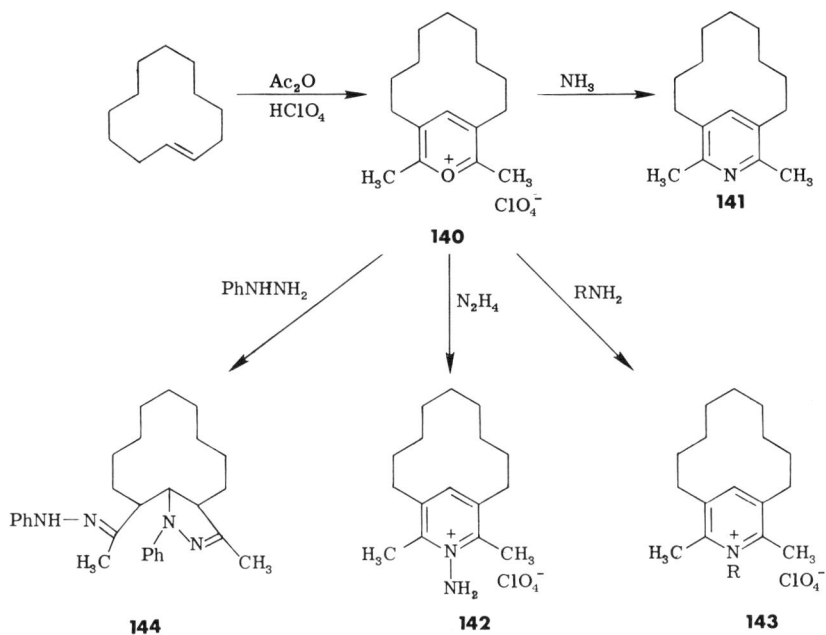

SCHEME 26

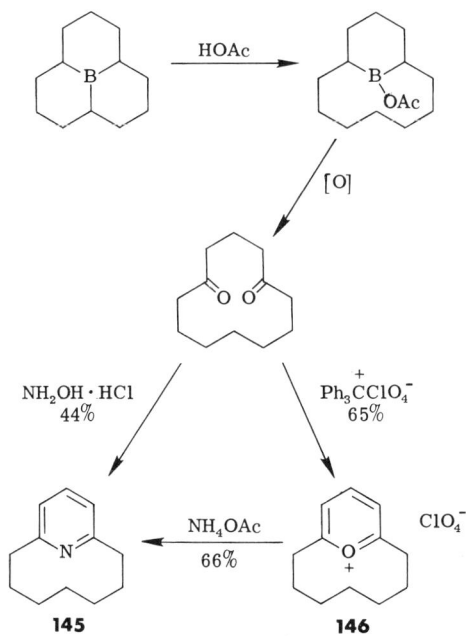

SCHEME 27

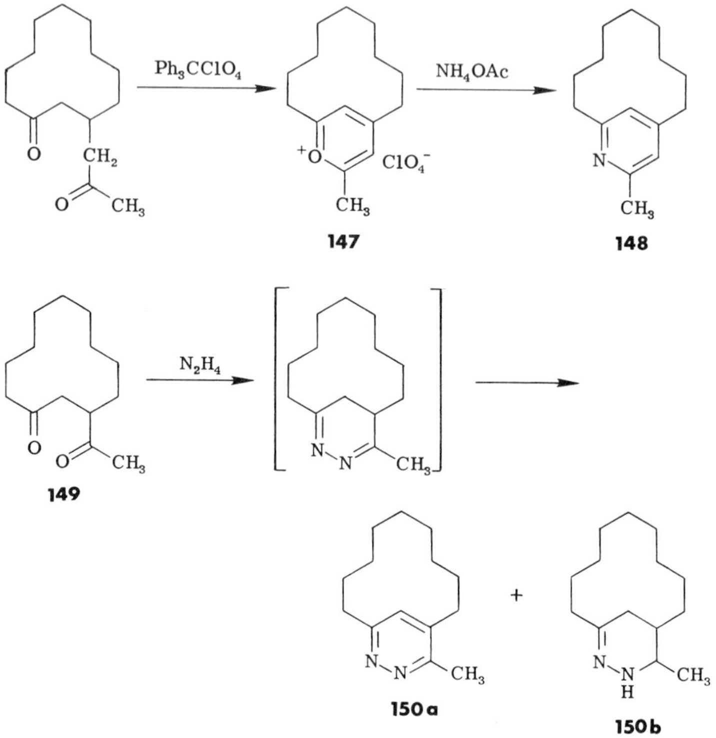

Scheme 28

Scheme 29

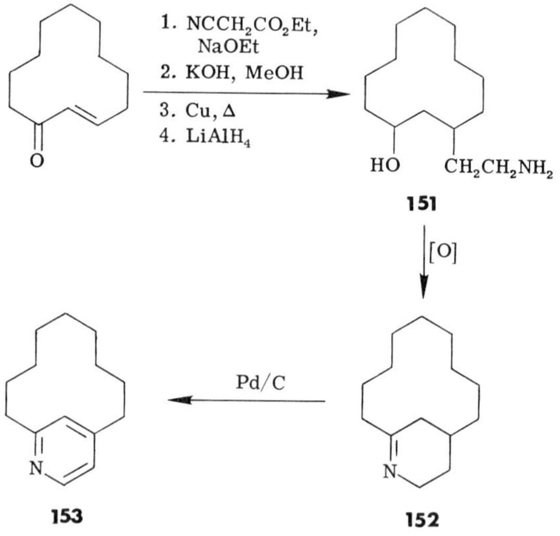

Before these syntheses, the same workers prepared [9](2,4)pyridinophane (153) by the route depicted in Scheme 29.[78] Jones oxidation of 151 causes spontaneous cyclization to 152, and dehydrogenation gives the desired compound 153.

More recently, Nozaki *et al.* synthesized [8](3,6)pyridazinophane (155), as shown.[79] The intermediate 154 spontaneously dehydrogenates on work-up to give the observed product.

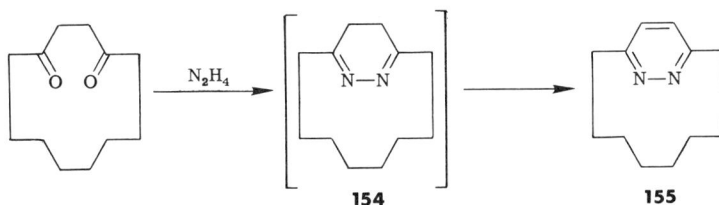

In 1968 Gerlach and Huber described the synthesis of (2,5)pyridinophane using the well-known condensation of β-aminovinyl ketones.[80] This reaction sequence is shown in Scheme 30. The synthesis involved acylation of acetylene using a diacid chloride followed by reaction with ammonia and cyclization of the bis(β-aminovinyl) diketone in an intramolecular reaction to yield the pyridinophanone 156. Wolff–Kishner reduction of the ketone gave the desired product 157.

Also isolated from the reaction mixture in the cases in which $n = 8$ or 9 are compounds that result from reactions between two molecules of the bis(β-aminovinyl) diketone. The structure of the dimer for $n = 9$ can be either 158 or 159. The correct structure (158) was assigned by conversion

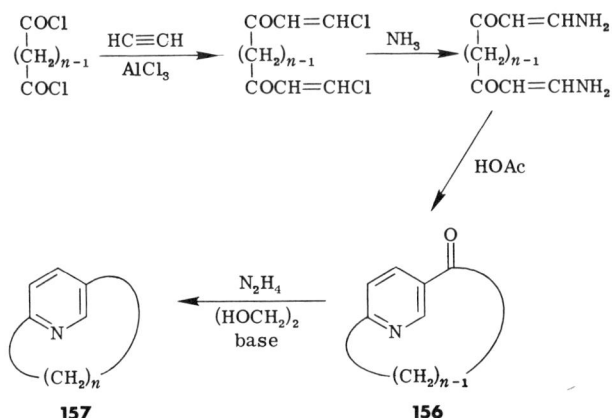

SCHEME 30

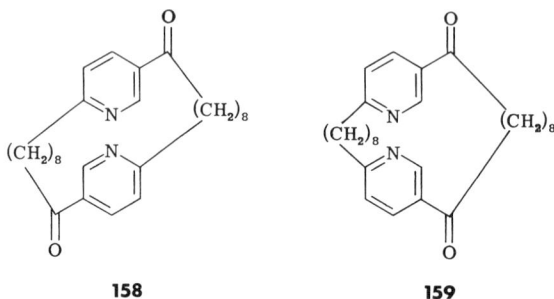

158   **159**

of both the dimer and the monomer **(160)** to the same hydroxy ester **(161)** by a Baeyer–Villiger oxidation (Scheme 31).

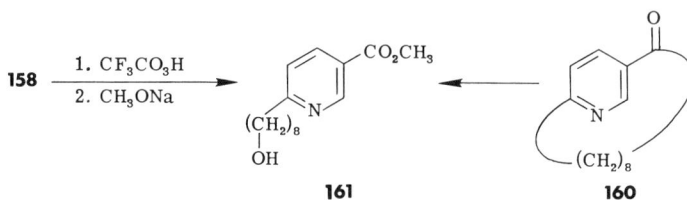

SCHEME 31

Parham and co-workers have published several reports on the syntheses of benzopyridinophanes.[81,82] Their method is delineated in Scheme 32. The indole **162,** prepared by a Fischer synthesis, reacts with phenyl(trichloromethyl)mercury to yield the dichlorocarbene adduct, which spontaneously loses hydrogen chloride to afford the heterophane **163.** The same

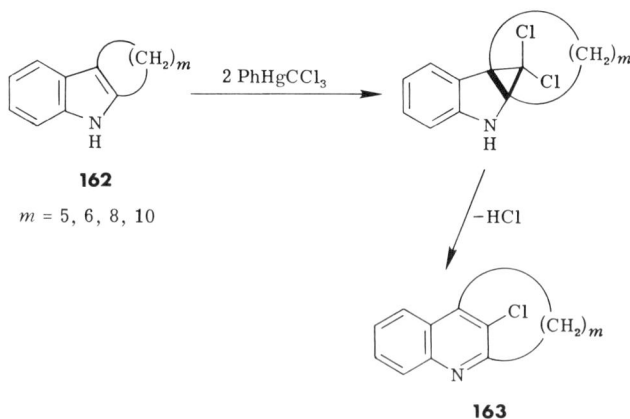

162

m = 5, 6, 8, 10

SCHEME 32

workers prepared the naphthopyridinophane **164** for antimalarial testing.[83]

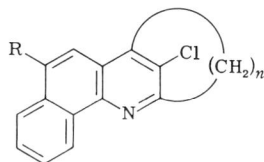

164

$n = 6, 10$

Wakefield and co-workers[84] prepared the two heterophanes **165a** and **165b** by a nucleophilic substitution reaction.

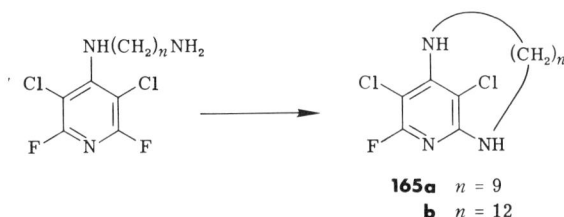

165a $n = 9$
b $n = 12$

Isele and Scheib described the synthesis of a heterophane having a pyridone nucleus **(166)**.[85] The method, outlined in Scheme 33, involves Michael addition of ammonia, coupling of the terminal acetylenes in the resulting compound, and finally catalytic hydrogenation.

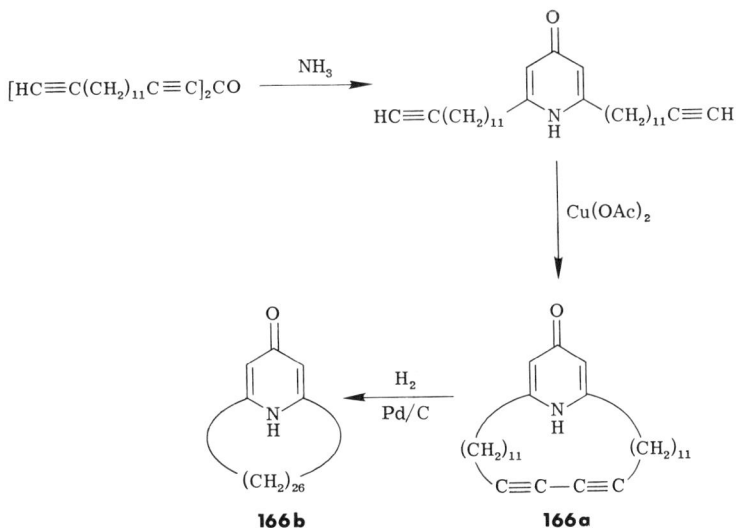

166b

166a

SCHEME 33

More recently, Japanese workers used a coupling reaction to prepare (2,6)pyridinophanes **167**, including [6](2,6)pyridinophane, which has the shortest bridge known in such a heterophane.[86] The coupling agent was dichloro[1,3-bis(diphenylphosphino)propane]nickel(II), Ni(dppp)Cl$_2$.

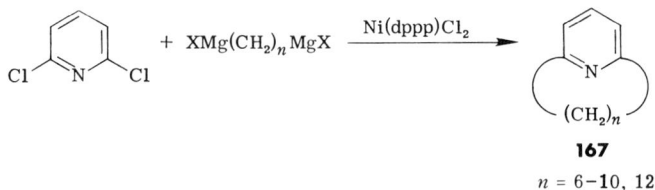

167

$n = 6-10, 12$

Chemical transformations on the methylene bridges of various pyridinophane derivatives are under investigation by Stutz et al.[86a]

The oxepinophane shown below was synthesized by a Diels–Alder condensation of an activated furan with octyne, producing an intermediate that, when irradiated, gives the oxaquadricyclane. Thermal isomerization affords the oxepinophane **168** in 62% yield.[87]

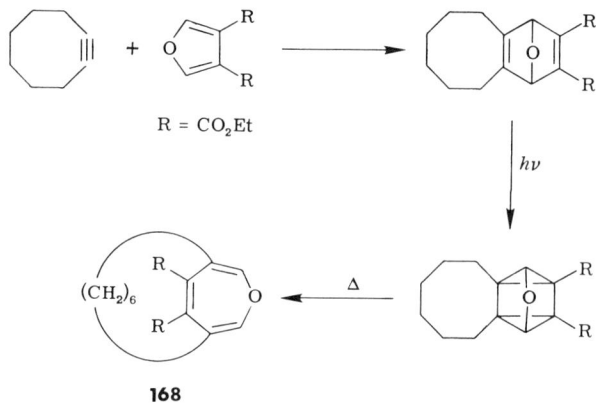

R = CO$_2$Et

$h\nu$

Δ

168

D. Multinuclear π-Deficient Heterophanes

Extrusion reactions may fall under the more general class of symmetry-controlled processes known as cheleotropic reactions. However, the concerted processes require a precise location of double bonds with respect to the species being extruded. The theory governing these processes has been treated in great detail elsewhere.[88–90]

In the synthesis of [2.2]phanes, the mechanism appears to be stepwise. Evidence of a stepwise process includes (a) the extreme conditions required and (b) the isolation of crossover products in alicyclic sulfones.

Herein lies the advantage of applying the method to strained cyclic systems. Because of the close proximity of the reactive intermediates produced by the extrusion reaction, recombination takes place rapidly and without crossover products to give the desired material in good yields. A review[91] has been published and includes cyclophane as well as some heterophane syntheses.

Rasmussen and Martel used this method to prepare [2.2](2,6)pyridinophane (169) in good yield.[92] Scheme 34 illustrates the synthetic sequence used in this method. The known dithiapyridinophane 171 is synthesized by the method of Vögtle and Schunder.[93] Thus,

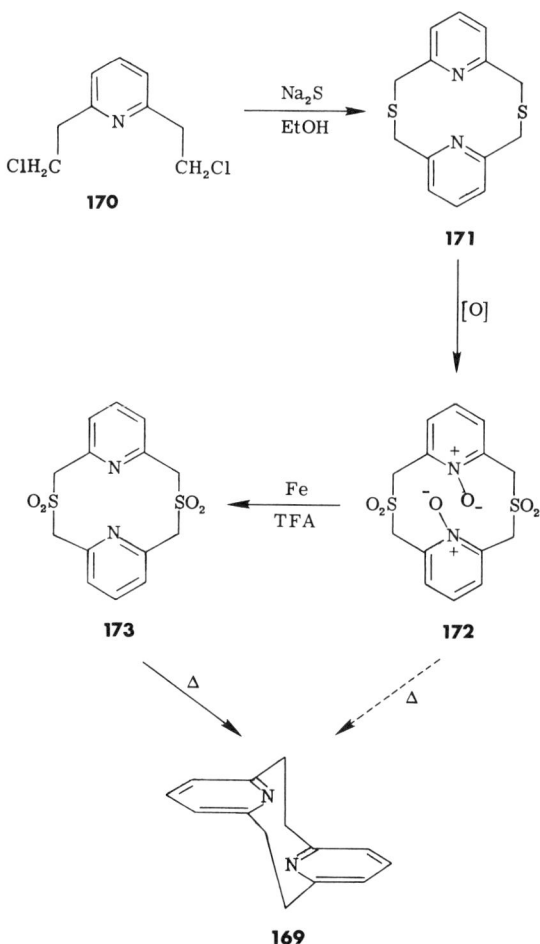

SCHEME 34

2,6-bis(chloromethyl)pyridine **(170),** when treated with sodium sulfide in ethanol, yields 19% purified bis(sulfide) **171.** This compound gives an almost quantitative yield of the bis(sulfone) bis-*N*-oxide **172** upon oxidation with excess *m*-chloroperbenzoic acid. The *N*-oxide functions can be selectively reduced with iron in trifluoroacetic acid (TFA), and the bis-(sulfone) **173** undergoes pyrolytic extrusion of sulfur dioxide (0.01 mm/ 680°C) to give the pyridinophane **169** in 46% yield.[92]

Although paracyclopyridinophane **174** can be obtained on a preparative scale by the Hofmann elimination,[94,95] the researchers had initially tried

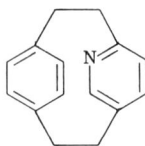

174

the pyrolysis of the appropriate bis(sulfone).[96] This method is successful only on a small scale, and the heterophane was identified by gas chromotography–mass spectrometry. The initial reaction also yields small quantities of **175** and **176.**

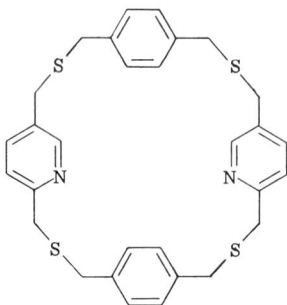   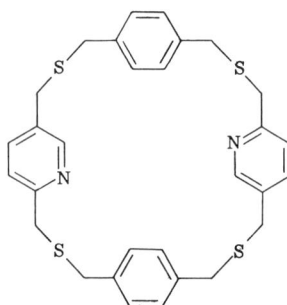

175   **176**

It was later shown that dithiaheterophanes can undergo photochemical extrusion of sulfur in triethyl phosphite.[97] This method was used to prepare the paracyclopyridinophane **174** in high yields.[97] This method has also been applied to pyrazinophane synthesis.[98a]

Scheme 35 illustrates the sequence of reactions used in the synthesis of [2.2]metacyclo(3,5)pyridinophane **(181).**[97] The bis(chloromethyl)pyridine **177** and bis(sulfide) **178** are mixed in ethanol to give 66% yield of the dithiaheterophane **179.** The bis(sulfide) **178** is employed in this case to prevent the formation of symmetric dimers. Photolysis of the dithiaheterophane **179** in triethyl phosphite (under nitrogen) at 25°C generates the metacyclopyridinophane **181** in 23% yield. The intermediate species **180**

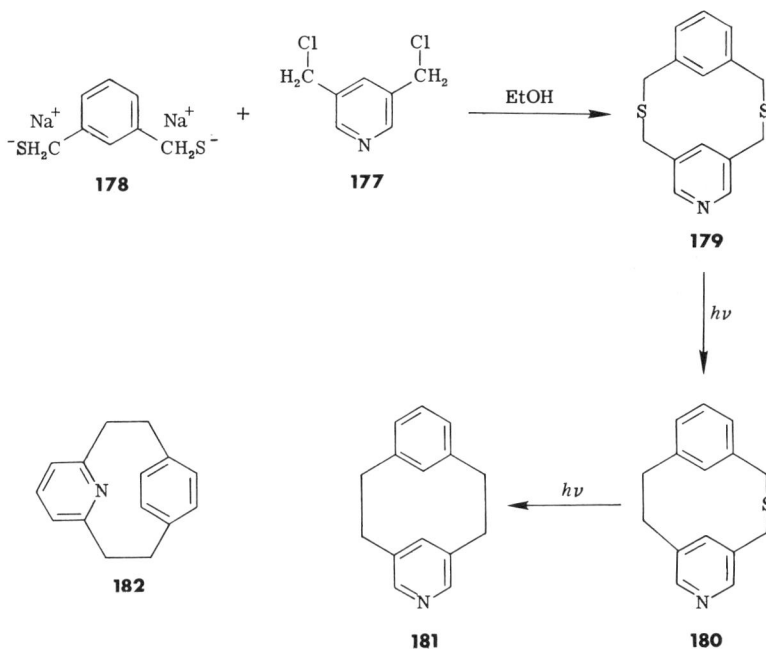

SCHEME 35

can be isolated from the reaction mixture, illustrating the stepwise nature of the reaction.

Other workers irradiated the appropriate dithiaheterophane in trimethyl phosphite and obtained [2.2]paracyclo(2,6)pyridinophane **(182)** in good yield.[99] It has also been shown (see Scheme 34) that bis(sulfone) N-oxides such as compound **172** may be pyrolyzed to give the appropriate heterophane without previous reduction of the N-oxide. Although the loss of the oxide during pyrolysis was not attempted on compound **172,** it was unsuccessful in the synthesis of heterophane **182.**[100]

The naphthalenopyridinophane **183** can be prepared by pyrolysis of the bis(sulfone) precursor in 65% yield.[101] Typically, extreme conditions (0.2 torr/500°C) are required.

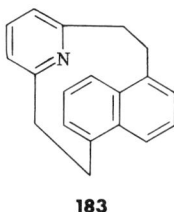

183

Relatively few heterophanes have been prepared by the Stevens rearrangement. They are related to the extrusion method in that the starting materials are the same. Thus, in the synthesis of the strained heterophane **186,** the dithiapyridinophane **171** is the starting compound. Methylation with Meerwein's reagent (trimethyloxonium tetrafluoroborate) followed by treatment with potassium *tert*-butoxide affords the rearrangement product **184** in high yield (Scheme 36).[102] Compound **184** is obtained as a mixture of separable isomers that form **185** on methylation. When mixture **185** is treated with 2,6-bis(*tert*-butyl) phenoxide ion, elimination of dimethyl sulfide occurs and [2.2](2,6)pyridinophane-1,9-diene **(186)** is obtained in 20% yield.

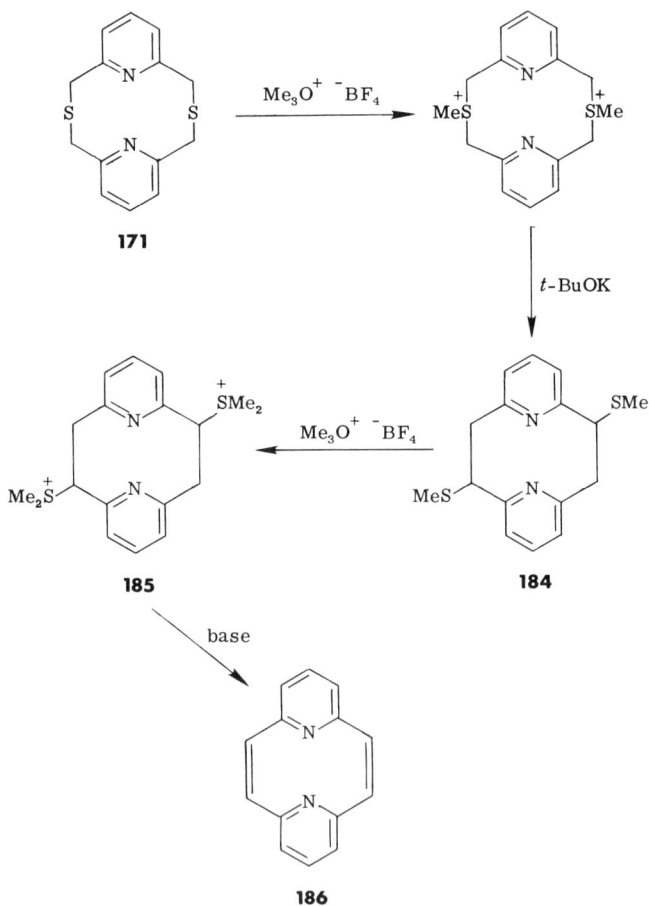

SCHEME 36

The Stevens rearrangement was used in the syntheses of [2.2]paracyclo(2,6)pyridinophane-1,9-diene **(187)**[100] and of the substituted heterophanes **189–194**. The synthetic versatility of the rearranged species **188** was shown by its conversion to the saturated heterophanes via Raney nickel desulfurization and to the unsaturated analogs via oxidation to the bis(sulfone), followed by pyrolysis (Scheme 37).[103]

In recent years it has been realized that organosodium or organopotassium compounds couple very efficiently with alkyl or benzylic halides. Although radicals are formed, no unwanted dimers are formed as they are in the Wurtz reaction. This is thought to be due to the formation of a solvent cage (Scheme 38).[104,105]

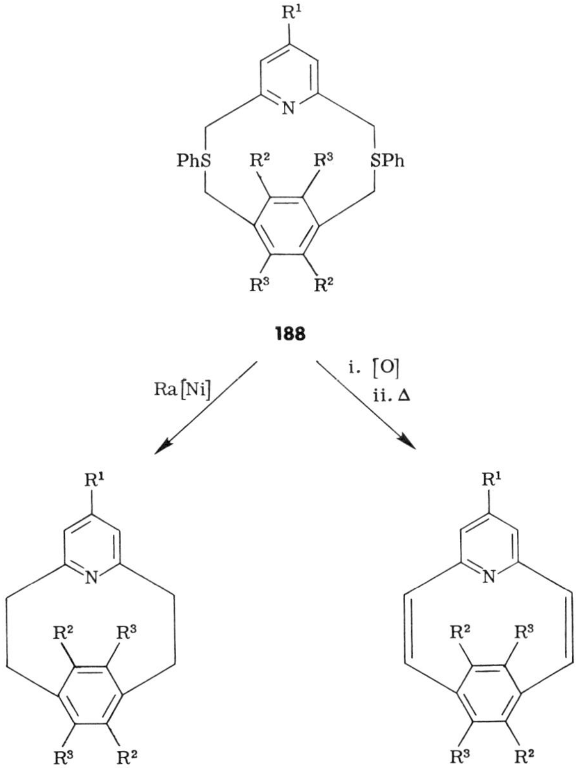

189 $R^1 = R^3 = H; R^2 = CH_3$
190 $R^1 = H; R^2 = R^3 = CH_3$
191 $R^1 = CH_3; R^2 = R^3 = H$

187 $R^1 = R^2 = R^3 = H$
192 $R^1 = R^3 = H; R^2 = CH_3$
193 $R^1 = H; R^2 = R^3 = CH_3$
194 $R^1 = CH_3; R^2 = R^3 = H$

SCHEME 37

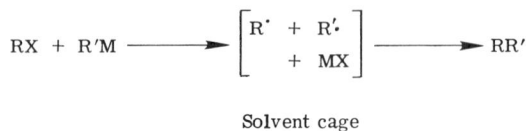

$$RX + R'M \longrightarrow \left[\begin{array}{c} R^{\cdot} + R'^{\cdot} \\ + MX \end{array} \right] \longrightarrow RR'$$

Solvent cage

SCHEME 38

Although the Wurtz reaction has been used extensively in the cy-clophane series,[106] it has found rather limited use in heterophane synthe-ses. Jenny and Holzrichter treated 3,5-bis(chloromethyl)pyridine **(177)** with sodium in tetrahydrofuran at low temperature and obtained a 2% yield of [2.2](3,5)pyridinophane **(195)**.[107]

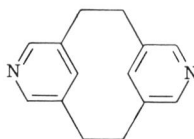

195

Boekelheide and Pepperdine reported the synthesis of the substituted metapyridinophane **197** by a similar reaction on the dibromo compound **196** (Scheme 39).[108]

The same workers reported that they had been able to separate a cyclic trimer and tetramer from the reaction mixture. These compounds were assigned structures **198,** [2.2.2](3,5)pyridinophane, and **199,** [2.2.2.2](3,5)pyridinophane, respectively.[109]

Compound **195** can be converted to 2,7-diazapyrene **(200)** by heating with palladium on carbon.[110] By employing a similar procedure, Jenny and Holzrichter prepared [2.2.2.2](2,6)pyridinophane and

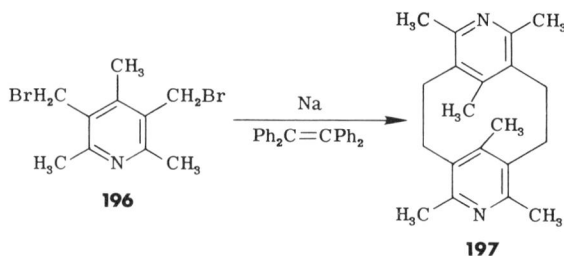

196   **197**

SCHEME 39

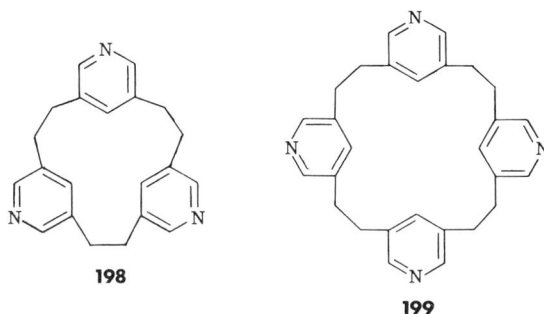

198

199

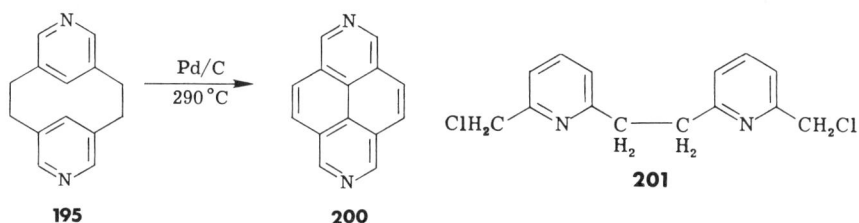

195   **200**   **201**

[2.2.2.2.2.2](2,6)pyridinophane by the dimerization and trimerization, respectively, of compound **201**.[111] Jenny and co-workers also prepared the
trimer and pentamer in the (2,6)pyridinophane series.[112]

The diazahexahydropyrene **202** was obtained in 30% yield from compound **197**. Oxidation to **203** failed, however (Scheme 40).[113] The synthesis of the azapyrene **206** via pyridinophane **205** was successful.[114] The
cyclization steps in the synthesis are presented in Scheme 41.

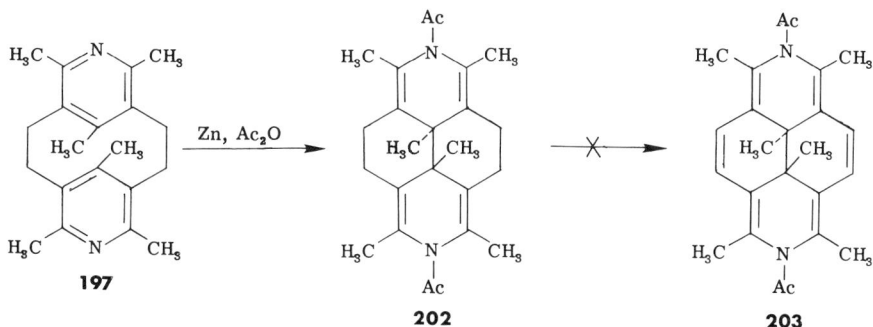

197   **202**   **203**

SCHEME 40

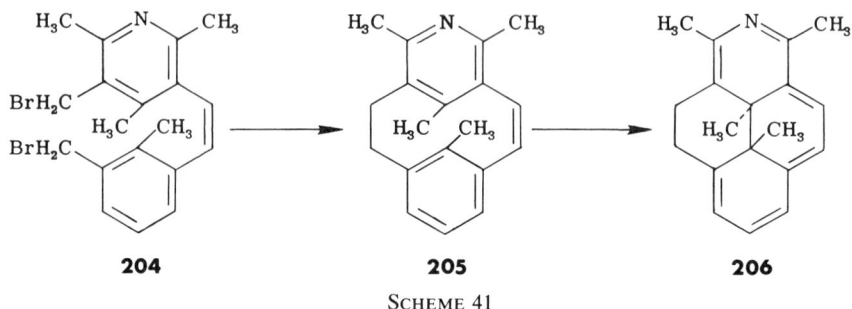

204   **205**   **206**

SCHEME 41

The first application of organometallic coupling reactions to the synthesis of [2.2]heterophanes was reported by Baker *et al.* Ring closure of **207** produced [2.2](2,6)pyridinophane (**169,** Scheme 42).[115] It was later shown that a bis coupling could be effected to prepare pyridinophane **169** from 2,6-bis(bromomethyl)pyridine (**208**).[102] The ring closure method of Baker was successfully employed in the synthesis of metacyclopyridinophane **209**.[114]

Scheme 43 illustrates the investigative work of Boekelheide and Pepperdine on the valence tautomerization of some metacyclopyridinophanes.[108] Thus, the (3,5)pyridinophane **205** undergoes a reversible isomerization to give compound **206**. The treatment of **205** with ruthenium on alumina in acid solution affords the salt **210,** which forms the free base

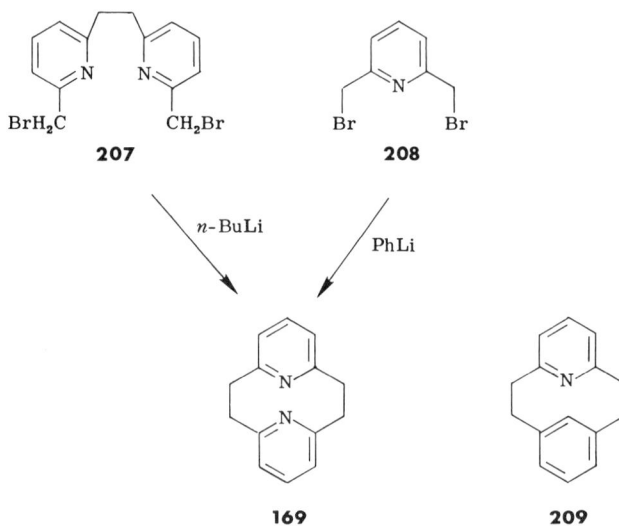

207   **208**

n-BuLi   PhLi

169   **209**

SCHEME 42

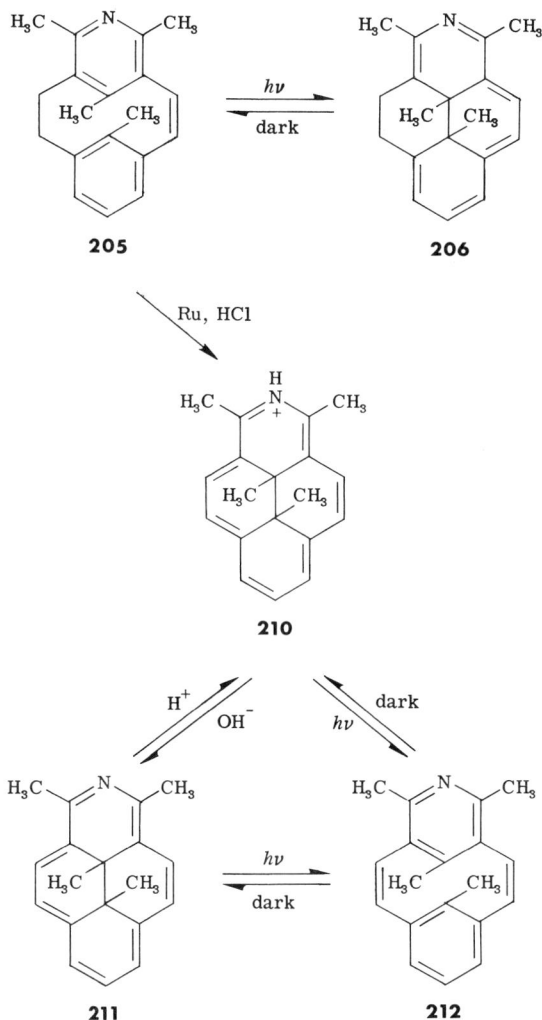

SCHEME 43

211 on treatment with hydroxide ion. When irradiated in an nmr tube with a 100-W Mazda lamp, compound **211** gives the valence tautomer **212** in approximately 65% yield. The reverse process is slow in this instance, and the valence isomer **212** can consequently be analyzed by its nmr spectrum. The isomerization is fast in the other cases, and the tautomers indicated are postulated on the basis of the disappearance of color from the starting solutions on irradiation.

Boekelheide *et al.* treated the 2,6-disubstituted pyridinophane **182** with *m*-chloroperbenzoic acid and obtained the *N*-oxide **213** in good yield.[42] These workers also prepared the fluoroborate salts **214** and **215** by treatment of the appropriate pyridinophanes with boron trifluoride etherate in ether solution. The complex **216** is prepared by the addition of antimony pentafluoride to pyridinophane **192** in deuteriochloroform in an nmr tube.[103]

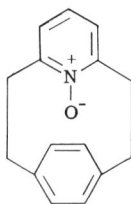

213

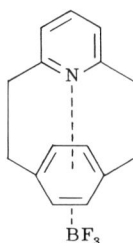

214

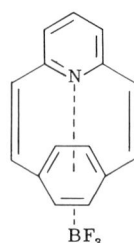

215

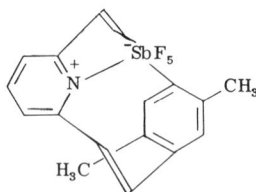

216

The preparation of the mono- and di-*N*-oxides of (2,6)pyridinophane **(169),** illustrated by structures **217** and **218,** respectively, have been reported.[102]

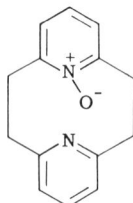

217

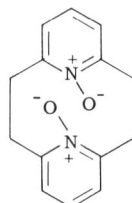

218

It has been shown[116] that it is possible to couple two aryl methyl carbons via the corresponding organocopper compounds. These are accessible by metal-exchange reactions on organolithium compounds (Scheme

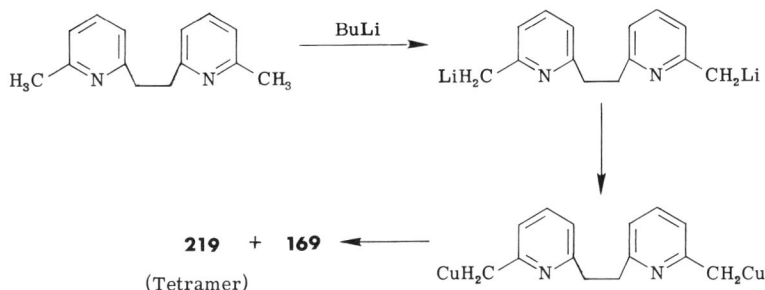

SCHEME 44

44). The reported yield of the desired heterophane is only 1%; tetramer **219** is the major product and is obtained in 4% yield.

The Hofmann elimination reaction was used by Bruhin and Jenny in their synthesis of [2.2](2,5)pyridinophanes.[94–96] Chromatography enabled

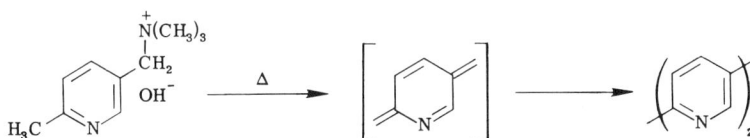

them to separate the reaction mixture into the four possible isomers **220–223.**

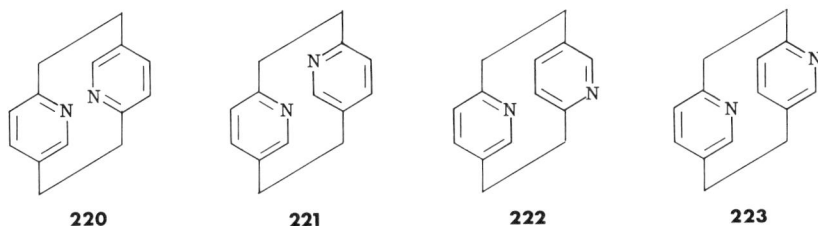

| **220** | **221** | **222** | **223** |

Vögtle *et al.* reported the synthesis of a number of cyclophanes (**226–229**)[117] and yet another series of heterophanes **171.**[118]

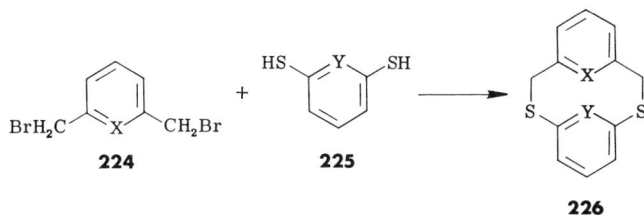

227   **228**   **229**

X = C—H, C—F, N

171

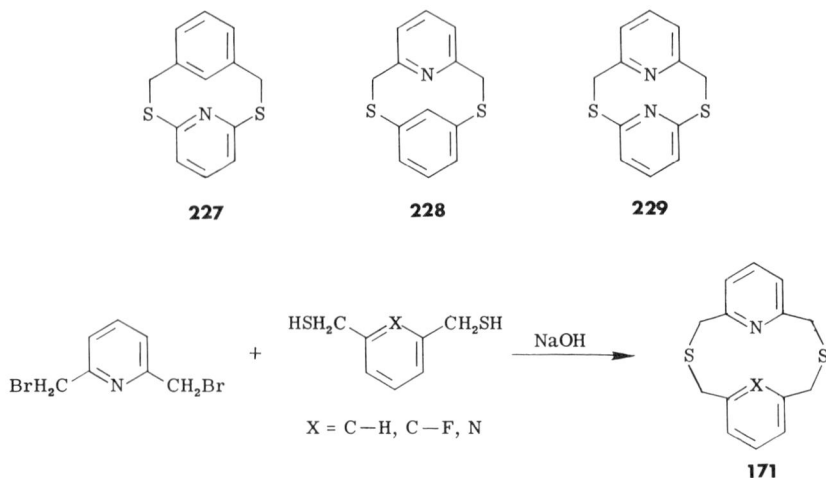

Undheim *et al.* synthesized compound **231** and found, by X-ray diffraction analysis, that the molecule is nonplanar.[119]

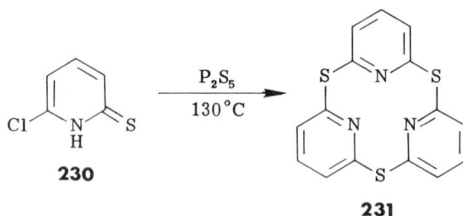

230   **231**

Russian workers have published numerous articles on the syntheses of polynuclear heterophanes bearing π-deficient heterocycles. These accounts are concerned mainly with molecules having one of two structural features: either a bridged 1,3,5-triazine moiety **(232)**[120-123] or a bridged diiminophthalimide residue **(233)**.[124-132] The bridging groups varied; a few are illustrated here.

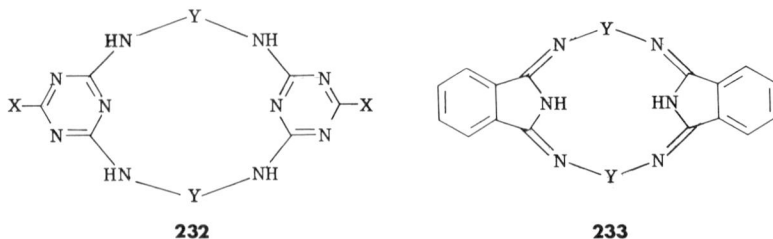

232   **233**

X = Cl, NH$_2$, PhNH, diazo compounds

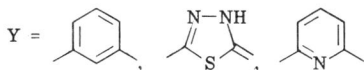

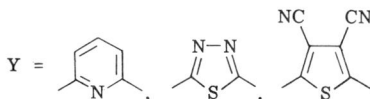

Compounds of type **232** are formed by reaction of the appropriate diamine with a dichloro-1,3,5-triazine. Those of type **233** and **234** are prepared by employing reactions similar to those shown here.

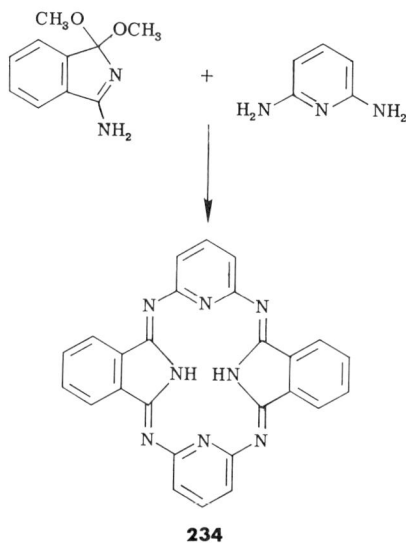

234

The *template effect* has been noted in the formation of some of these macrocycles.[133,134] The reaction illustrated here may be accomplished in 18 to 20% yields when performed in ethanol containing hydrochloric acid. When 0.03–0.1 M lithium perchlorate is used as well, the yield increases

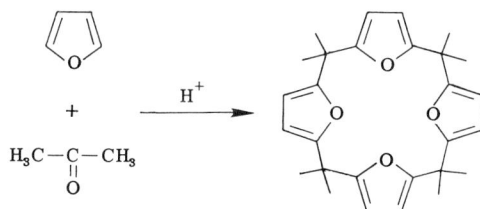

to 40 to 45%.[134] Iron(III) has been used in a similar template reaction to give the Shiff base **235** as the iron coordination compound.[135–137]

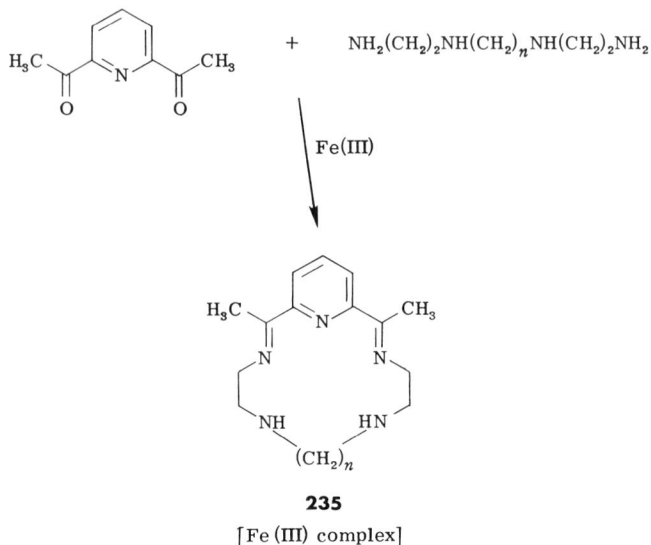

235

[Fe (III) complex]

Newkome and co-workers, utilizing the reactions shown in Scheme 45, prepared the benzopyridinophane **236**.[138]

The cobalt complex **238** was prepared by reaction of **237** with cobalt(II) chloride in anhydrous ethanol.

The quaternized and reduced pyridinophanes **239–242** were synthe-

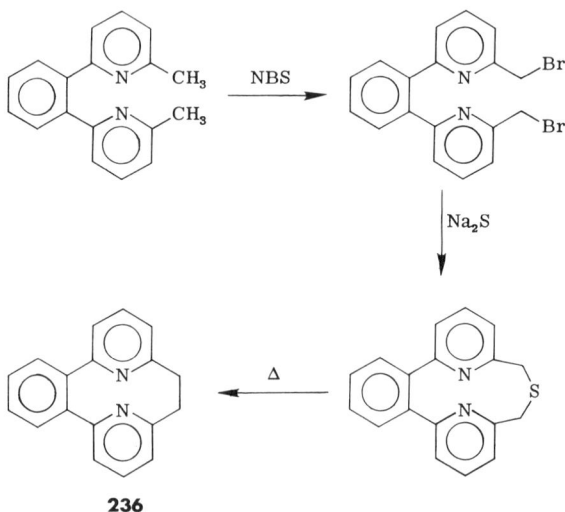

236

SCHEME 45

237   **238**

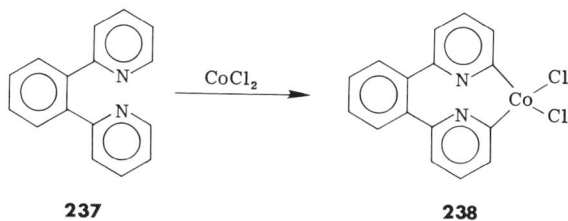

sized[139] via a modification of a method described by Dittmer[139a] *et al.* for the preparation of related systems.

The multibridged pyridinophanes **243** and **244** were synthesized by Boekelheide[140] in an adaptation of the method used in the cyclophane series.[141]

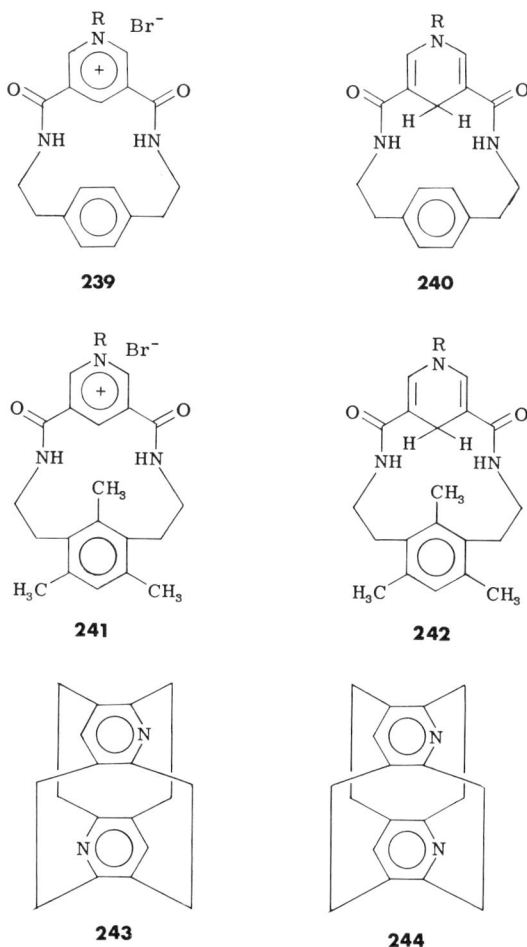

239   **240**

241   **242**

243   **244**

E. Multinuclear π-Excessive/π-Deficient Heterophanes

Two π-excessive/π-deficient heterophanes **(245** and **246)** have been reported by the authors.[48,62,142,143] Furanopyridinophane **245** was synthe-

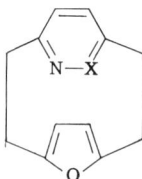

245 X = CH
246 X = N

sized by the Hoffman elimination method, and the furanopyridazinophane **246** was synthesized by the Paal–Knorr cyclization. The interesting annulenophane **247** is a coproduct of the reaction.[62,143] Oxidation of **246** with

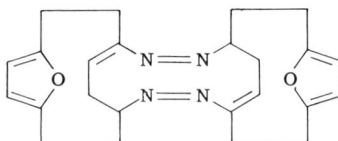

247

m-chloroperbenzoic acid yields the *N*-oxide **248** in high yield.[143] Diketopyridazinophane **249** is a by-product in this oxidation.[143]

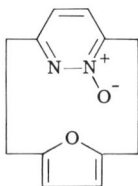 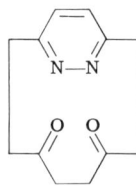

248   **249**

F. Annulenes

The literature abounds with reports of the use of the Wittig reaction in the syntheses of phanes and annulenes, as well as of nonconjugated large-

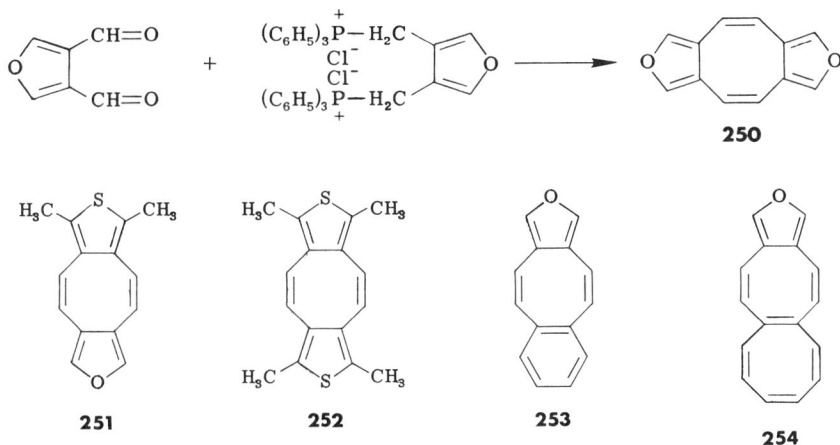

SCHEME 46

ring systems.[144-146] A few [2.2]heterophanes have been synthesized by this method. However, they are confined to the ortho-substituted type. Scheme 46 illustrates this method. It should also be noted that the orthophanes do not provide some of the prerequisites for the interest shown in the phane class of compounds. That is, the rings are not rigidly held face to face. Thus, there is little if any strain in these systems and no transannular interactions. Interest in these systems is generally confined to the large-ring systems that constitute the cavity of the phane. Questions arise regarding the aromaticity of these rings when they are fully conjugated. Herein lies the usefulness of the Wittig reaction, which forms a double bond between the reacting centers and therefore immediately provides a fully conjugated eight-membered ring when used to prepare [2.2]ortho-substituted heterophanes.

The synthesis illustrated in Scheme 46 is typical. The dicarboxaldehyde and the bisphosphonium salt are mixed in hot, absolute dimethylformamide, and a solution of lithium ethoxide in absolute ethanol is added dropwise to afford a 44% yield of the orthoheterophane **250**. Heterophanes **251–254** were similarly prepared.[144,147] Compounds **255** and **256**, synthesized by the Hofmann elimination,[28] have received little attention because no conjugation is present in the cavity.

Many other annulenes containing heterocyclic rings such as furan,[148-151] dihydropyridine,[152] and others[153] have been synthesized along with various annulenones.[154-159] This activity has resulted from an interest in the diatropic or paratropic character of these species. No attempt is made in this chapter to review this subject thoroughly. The syntheses of these compounds, which generally involve a bis-Wittig reaction, have been reviewed.[144] An interesting discussion of annulenes can also be found in the book by Garratt.[160]

IV. CONFORMATIONAL ASPECTS

The conformational properties of the cyclophanes are perhaps the most widely studied aspects of this class of compounds. Indeed, these studies have provided the impetus for much of the synthetic work in the area.

A. Nuclear Magnetic Resonance Spectroscopy

The technique most frequently employed in these studies has been ¹H-nmr, especially variable-temperature nmr, which allows an evaluation of energy barriers between conformations to be made.

In 1970 Paudler and Stephan reported the synthesis of compound **83,** which can be converted to a dihydro **(84)** or a completely unsaturated derivative **(85),** a [12]annulene, as indicated in Scheme 15. The X-ray structure determination of this annulene has been reported[50,51] and indicates complete delocalization around the periphery. The ¹H-nmr spectrum of **85** shows a shielding of the protons relative to **83** and **84,** indicating the existence of a paramagnetic ring current. This suggests that **85** is an anti-aromatic compound. The nmr spectrum of **84** shows a deshielding of the pyrrole protons relative to **83,** supporting the idea that there is a diamagnetic ring current encompassing the unsaturated periphery of the molecule. Thus, **84** may be considered a diaza[14]annulene.

Flitsch and Peeters[161] have prepared the annulene **259** by the route shown in Scheme 47. Compound **259** exhibits a strong *diamagnetic* ring current. All the ring protons absorb in the range δ 8.9–9.2. In comparison, the olefinic protons of compound **258** absorb at δ 6.93 and those of **257** absorb at δ 6.85–6.98.

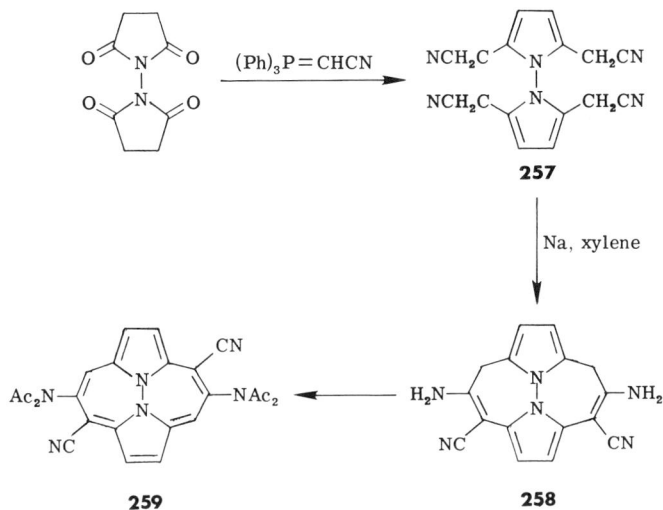

Scheme 47

Flitsh *et al.*[162] described the synthesis of **260** and **261** and similar compounds,[163] as shown in Scheme 48. Compound **260** shows what appears to be strong diatropic character because the olefinic protons have the following chemical shifts: δ 8.25 (m,1H), 7.58 (s,1H), 6.75–6.45 (m,4H). In contrast, compound **261** is paratropic, its olefinic hydrogens absorbing at considerably higher field than **260**: δ 6.18 (s,2H), 5.52 (s,2H), 5.33 (s,2H).

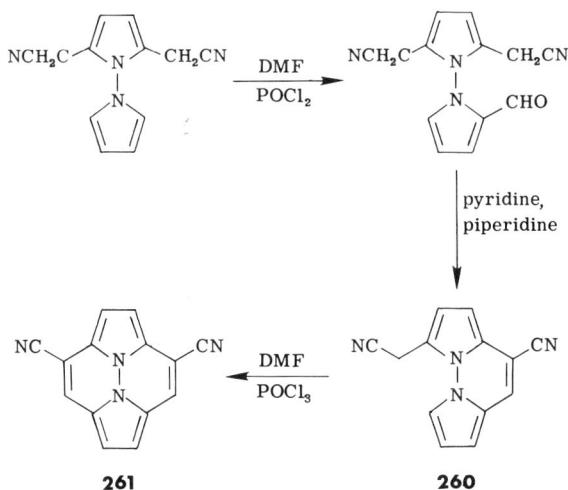

Scheme 48

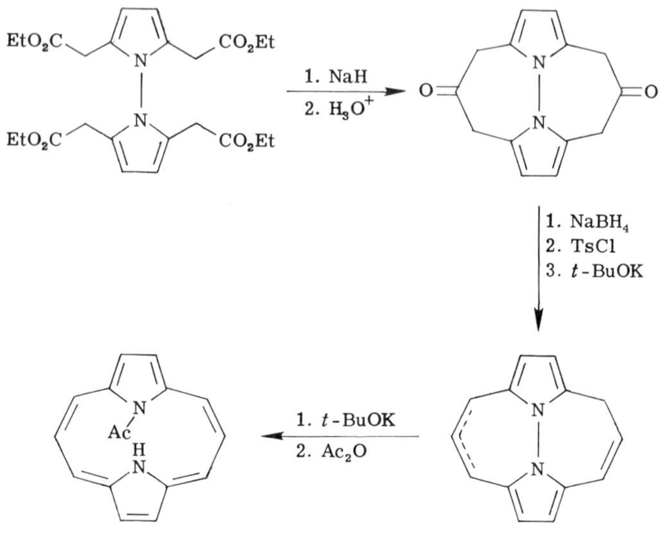

SCHEME 49

The terms *diatropic* and *paratropic* refer to the observation of a diamagnetic or paramagnetic ring current, respectively, in an annulene.

The interesting annulene **263** is obtained as described in Scheme 49.[163a] The diolefin **262** is isomerized when treated with potassium *tert*-butoxide to give an unstable annulene, which, when treated with acetic anhydride, gives **263**. The ring protons absorb at δ 7.7–8.4 (m, 10H), the N—H at δ −3.1, and the acetyl group at δ −1.34 (s,3H).

The [6](2,4)heterophanes **12** and **13** have been examined in this manner.[9] In all four instances the methylene hydrogens resonate at δ <1.10.

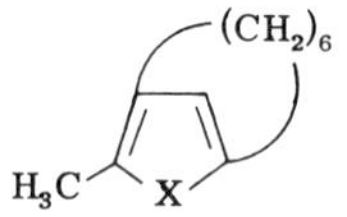

12 X = S
13 X = NPh, N(*o*-tolyl),
         N(*p*-tolyl)

The shielding of the methylene protons relative to paraffinic methylenes occurs in all of the mononuclear heterophanes. This can be explained if one assumes that the methylene bridge lies above the aromatic nucleus and is subject to the shielding effect of the diamagnetic ring current. The anisotropic influence of the heteroatom undoubtedly plays a role as well. However, the relative contributions of these two effects are difficult to assess. No variable-temperature study was conducted in this case. These heterophanes also show a bathochromic shift in their uv absorption spectra in comparison with corresponding open-chain models. This is presumably due to the nonplanarity of the ring (see Section IV,B).

Heterophane **19**, [7](3,5)pyrazolophane,[10] exhibits absorptions for one proton at δ 1.96 and for four other protons at δ 0.5–1.0. The extremely

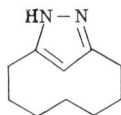

19

shielded absorption for one proton, tentatively assigned as one of the protons at C-4, indicates that the molecule is fixed in one conformation. No coalescence was noted up to a temperature of 205°C.

Parham and Dooley[13,14] provided ¹H-nmr data on the two pyrazolophanes **26** and **32**. Compound **26** shows a broad band for 16 methylene hydrogens at δ 0.7–2.0, indicating some transannular shielding by the

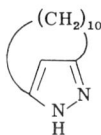   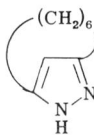

26   **32**

aromatic ring. Compound **32** was not available in its pure form but was obtained as 27% of an inseparable mixture along with its isomer **33**.

The ¹H-nmr spectrum of the mixture shows a broad absorption for eight hydrogens at δ 1.0–1.4. Thus, it contrasts with compound **19**, studied by Nozaki, who reported a high-field absorption for one proton at δ −1.96. One might have expected similar behavior for compound **32**, but this was apparently not observed. Similarly, a lack of methylene shielding has been noted in [6](2,6)pyridinophane.

Compounds **20** and **21** were examined by means of variable-temperature ¹H-nmr[11,12]. A shift of some of the methylene protons to higher field as the temperature is lowered is observed, indicating a flexibility at room temperature viewed as flipping of the methylene chain to alternate faces of the aromatic ring. Compound **20** exhibits an nmr spectrum at −97°C that appears to be similar to that of compound **21** at room temperature. Both compounds show a high-field absorption for two protons. Thus, compound **21** has the larger barrier to flipping of the bridge. At −97°C the heterophane **21** shows broad absorption at δ −1.5 for one proton, suggesting that it is being frozen into one conformer. Compound **22** could not be examined at low temperature because of its insolubility.

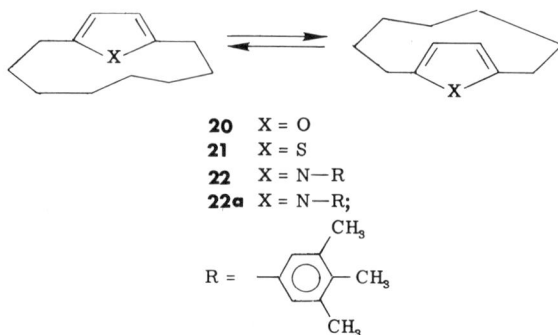

20	X = O
21	X = S
22	X = N—R
22a	X = N—R;

Compound **22** (R = H) is inert to hydrogen deuterium exchange for the proton attached to nitrogen even when strong bases such as sodium hydride are used. It thus appears that the methylene bridge effectively surrounds the nitrogen, preventing the approach of base.

Alkyl and aryl derivatives of **22** were also examined for conformational preferences.[12] One case is shown (**22a**). In this molecule the aryl protons appear as a singlet at δ 6.83. At −45°C the signal separates into two singlets at δ 7.41 and 6.27, respectively. It is suggested that free rotation of the aryl ring is inhibited at low temperature. Furthermore, the polymethylene chain forces the aryl ring out of the plane with the pyrrole ring such that one of the aryl hydrogens is located in the shielding region of the pyrrole, while the other is in the deshielding region. The barrier to rotation was calculated to be 6.9 kcal/mol.

Tamao *et al.*[86] reported the synthesis of [6](2,6)pyridinophane **(167)** and discussed its [1]H-nmr spectrum. The most striking feature of the spectrum

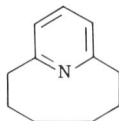

167

is the lack of strong transannular shielding. The methylene protons absorb in the region at δ 1.3–2.83. This is in sharp contrast to [7](2,6)pyridinophane.[74–76] In addition, the C-1 and C-6 methylene protons give sharp triplets, indicating a rapid conformational mobility of the bridge. At −92.5°C two of the four protons on C-3 and C-4 appear as a multiplet at δ

0.71. The coalescence temperature of −40°C relates to an energy barrier of 11.0 kcal/mol, a value that is higher than that for the seven-membered bridged compound.

Nozaki *et al.*[74-76] examined the conformational behavior of [7](2,6)pyridinophane **(145)**. At room temperature this molecule exhibits a quintet for two protons at δ 0.16, indicating conformational flexibility. At −111°C the absorption pattern is split into two bands; one is at δ −1.40, and the other is obscured by interference from other protons. The coalescence temperature is −75.5°C, corresponding to an energy barrier ($\Delta G_c^{\ddagger}$) of 9.0 kcal/mol. For comparison, Vögtle found the energy barrier in the dithiaheterophane **264** to be <10.2 kcal/mol. Some derivatives of **145** containing substituents

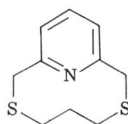

264

on the bridge have also been examined. It should be noted that the *N*-oxide of **145** could not be prepared. This was presumably due to steric hindrance. In contrast, the *N*-oxides of [10](2,6)pyridinophane and [8](3,6)pyridazinophane were prepared in fair yield.[70]

Bradamante *et al.*[77] examined the ¹H-nmr spectra of compounds **147, 148,** and **150a.** These workers found that the methylene proton absorptions are not as highly shielded as those in the [9]heterophane **6** (δ 0.36).

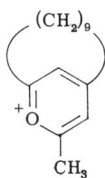   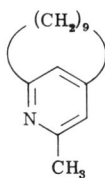   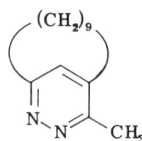

147   **148**   **150a**

The methylenes in **147, 148,** and **150a** absorb in the range 0.7–1.5 and are thus still shielded relative to paraffinic methylenes. It consequently appears that the geometry of the six-membered ring allows the bridge to

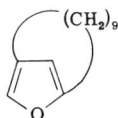

6

assume conformations that result in less effective shielding by the aromatic ring. Compounds **147** and **148** have a temperature-insensitive ¹H-nmr spectrum down to −90°C, and compound **150a** is insensitive to −30°C, a temperature at which it crystallizes.

Parham and co-workers[81,82] examined molecules such as **163**. They found that the methylene protons have a very complex nmr pattern, with

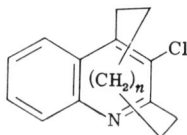

163

n = 2, 4, 6

a chemical shift range from δ −0.41 to 3.95 in the case of m = 4. When the chlorine is removed, the chemical shift of the methylenes shows greater equivalence (δ 0.23–1.92), indicating that the chlorine atom provides considerable constraint to molecule rotation. In addition, these compounds show a bathochromic uv effect and loss of fine structure as n decreases.

Balaban[73] described the ¹H-nmr spectrum of compounds **140** and **141**.

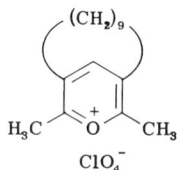

140

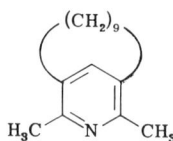

141

These compounds show no strong shielding effect for the methylene protons (δ 1.07–1.67). The benzylic protons appear as a triplet, again suggesting rapid interconversion of conformers.

The benzylic protons of compound **155** show a complex multiplet of δ 3.0, indicating no flipping of the polymethylene bridge.[79] Two absorptions for two protons each at δ −0.3 and 0.5 were assigned to the protons at C-4 and C-5. The latter proton must lie above the ring.

The ¹H-nmr spectrum shows no change up to 200°C. However, upon cooling, the absorption at δ −0.3 disappears and reappears at −98°C at δ −1.75. This was interpreted as a pseudorotation of the polymethylene chain interconverting the two enantiomeric structures **155a** and **155b**.

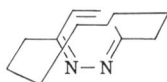

155a

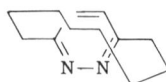

155b

Fletcher and Sutherland[29] studied the conformational aspects of heterophanes **2**, **49**, and **50**. The methylene protons of **2** in hexafluorobenzene appear as sharp A_2B_2 patterns, which coalesce at 63°C. The ¹H-nmr

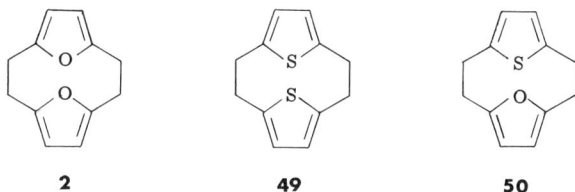

2   **49**   **50**

spectra of compounds **49** and **50** are unchanged at 200°C. The difference in behavior is apparently due to the larger steric requirement of the sulfur atom as well as the presence of the longer C—S bonds.

The dichotomy between oxygen and sulfur, insofar as their steric requirements are concerned, has been noted in several other heterophanes. For instance, in compounds **53, 65,** and **66** (X = S) the ¹H-nmr spectra are

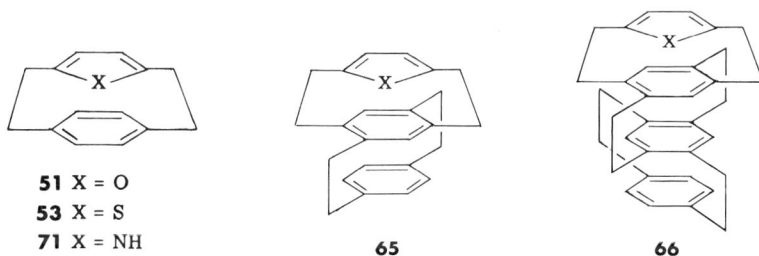

51 X = O
53 X = S
71 X = NH

65   **66**

independent of temperature, whereas if X = O the compounds show coalescence temperatures of −39, −58, and −62°C, respectively.[37,38] Similarly, the coalescence temperature of [3.3](2,5)thiophenophane is 105°C, whereas [3.3](2,5)furanothiophenophane remains freely rotating even at −90°C.[164]

In the case of compound **71**[165] the coalescence temperature is 105°C. On this basis one can suggest that the relative steric requirements of these heteromoieties increase in the order O ≤ N—H ≤ S. It should also be mentioned that compound **74** has a temperature-independent ¹H-nmr spectrum.[47,165]

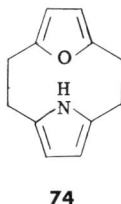

74

In phanes in which aromatic nuclei can have transannular effects on protons, the characteristic result is an upfield shift, as illustrated here. It is noteworthy that in rigid systems the "internal" protons are shielded more than the "external" ones, although both types are shielded. One notable exception is the furanophane **2** discussed earlier. Thus, the absence of significant shielding in ring protons indicates that these protons

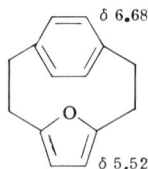

51

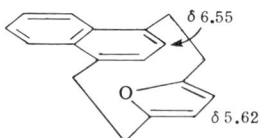

56

are "external." This is illustrated by compound **56,** which exhibits no coalescence, even at low temperatures. The absence of significant shielding in the furanoid protons indicates the anti stereochemistry[12,13] for this compound. These considerations were used to deduce the stereochemistry of the phanes discussed in this section. The results are listed in Table I.

On the basis of this analysis, the two naphthalenothiophenophanes **59a** and **59b** can be ascribed anti and syn stereochemistry, respectively.[35] Similarly, the anthracenophanes **61** and **62** are assigned anti stereochemistry. The [9,10]-bridged anthracenophanes **63** and **64** are also shown to be rigid rotamers.

Although variable-temperature studies were not carried out on the pyridinophanes **220–223,** the four possible isomers have been isolated.[94,95]

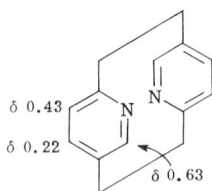

220

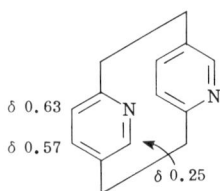

222

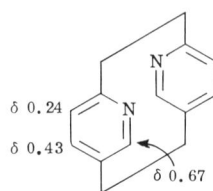

221

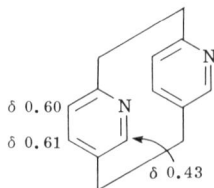

223

TABLE I

Stereochemistry and T_c Values for Selected Cyclophanes

Compound	Stereochemistry	Coalescence temperature	References
2	Anti	63°C	167
245	Syn	Greater than 110°C	142
49 X = Y = S	Anti	Greater than 200°C	29
50 X = S; Y = O	Anti	Greater than 200°C	29
51 R = H; X = O	Free rotation	−40°C	32, 165
53 R = H; X = S	Rigid	No change to 150°C	32, 165
54 R = CH₃; X = O	Free rotation	−29°C	33
55 R = CH₃; X = S	Rigid	No change to 150°C	33
56 X = O	Anti	Not found	34, 35
59a X = S	Anti	Not found	34, 35
59b X = S	Syn	Not found	34, 35

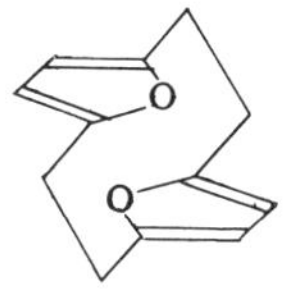

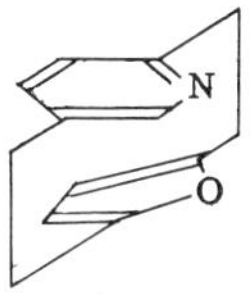

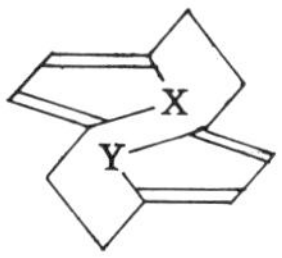

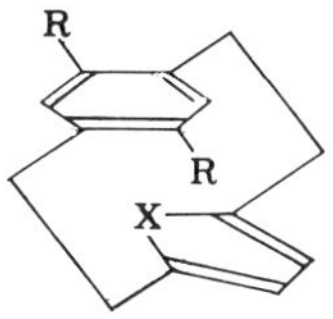

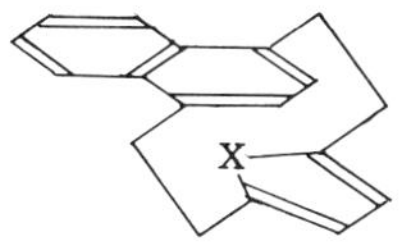

(continued)

TABLE I (*Continued*)

Compound	Stereochemistry	Coalescence temperature	References
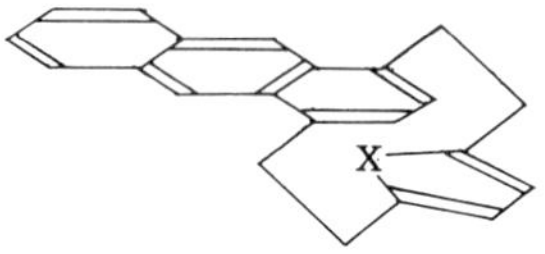			
61 X = O	Anti	Not found	35
62 X = S	Anti	Not found	35
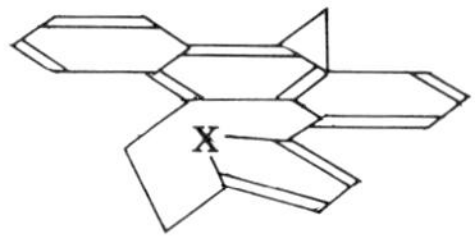			
63 X = O	Rigid	Not found	35
64 X = S	Rigid	Not found	35
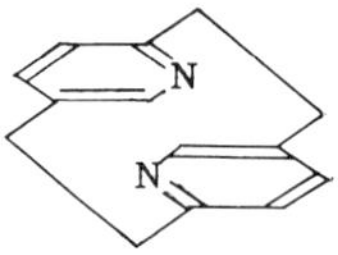			
220	Four isomers	Rigid at room temperature	—
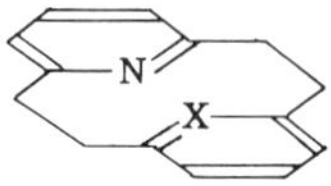			
169 X = N	Free rotation	13.5°C	167
209 X = CH	Anti	No change up to 200°C	167
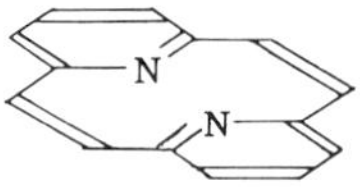			
186	Anti	No change between −80°C and 100°C	102
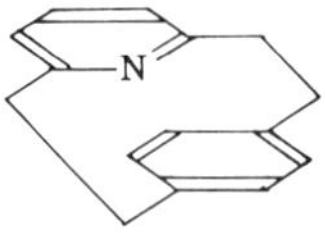			
182	Free rotation	−50°C	99
		−43.5°C	100
		−29°C	103

TABLE I (*Continued*)

Compound	Stereochemistry	Coalescence temperature	References
187	Free rotation	Not found	100
189 $R^1 = R^3 = H$; $R^2 = CH_3$	Free rotation	$-25°C$	103
190 $R^1 = H$; $R^2 = R^3 = CH_3$	Free rotation	$-25°C$	103
71 R = H	Anti	105°C	165
72 R = CH$_3$	Anti	Greater than 190°C	165
73	Anti	Greater than 190°C	165
X = O	Syn and anti	80°C	
X = S	Syn and anti	>150°C	—

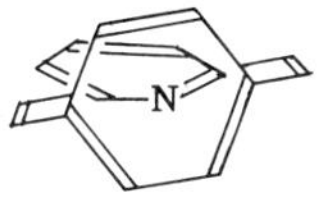

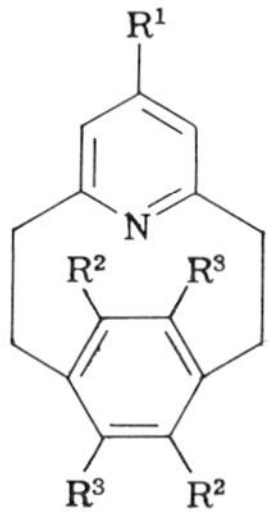

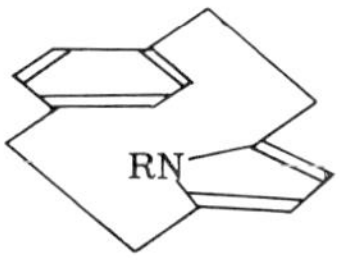

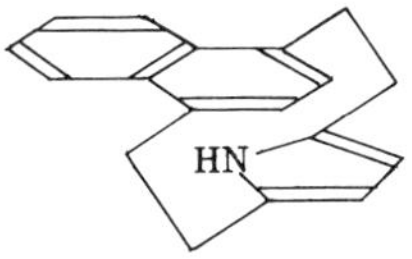

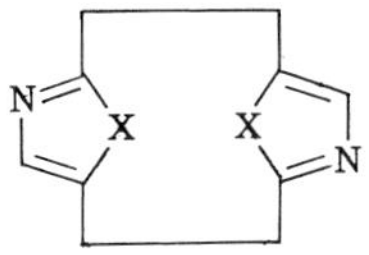

(*continued*)

TABLE I (*Continued*)

Compound	Stereochemistry	Coalescence temperature	References
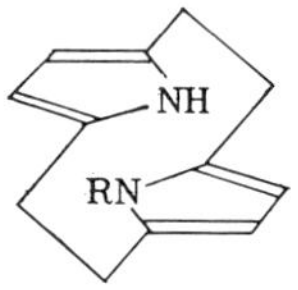			
95 R = H	Anti	Greater than 190°C	54
96 R = CH$_3$	Anti	Greater than 190°C	54
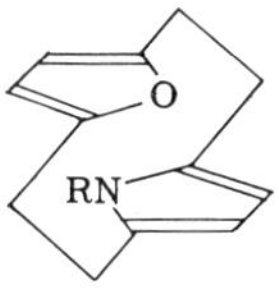			
74 R = H	Anti	No change from −35°C to 190°C	47, 165
76 R = CH$_3$	Anti	No change from −60°C to 80°C	48
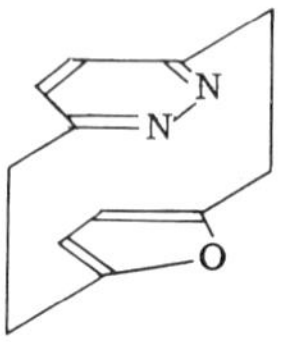			
246	Syn	Greater than 110°C	143

The nmr spectra of these isomers provide direct evidence of transannular interactions in these systems. The numbers indicate the upfield shifts, relative to 2,5-lutidine, experienced by these compounds. These shifts are due to the phane structure. In addition to this the location of the nitrogen in the opposing ring affects the chemical shift of the protons in the ring under consideration. Thus, the pseudo-ortho proton in **220** is deshielded by 0.20 ppm relative to the equivalent proton in **222**, which is too far from the opposing nitrogen to experience any field effect. The pseudo-ortho proton in **221** is deshielded by 0.18 ppm relative to the equivalent proton in **223**. This gives an average deshielding effect of 0.19 ppm for protons located pseudo-ortho to a nitrogen atom. Similar considerations yield a value of 0.38 ppm for protons experiencing pseudo-geminal deshielding. These values were corroborated by the nmr analysis of the paracyclopyridinophane **174,** which shows a deshielding of 0.17 ppm for the pseudo-ortho protons and 0.40 ppm for the pseudo-geminal protons.[96]

The coalescence temperature of paracyclopyrrolophane **71** is reported to be 105°C, in contrast to that of the *N*-methyl analog **72** (greater than 109°C) and that of the naphthalenopyrrolophane **73** (greater than 190°C), which has anti stereochemistry.[165]

The absorptions at "normal" positions in the ¹H-nmr spectra for the pyrrole protons of [2.2](2,5)pyrrolophanes[166] such as **81** suggest that these molecules have the indicated anti conformation, in which there is a minimum of transannular shielding.

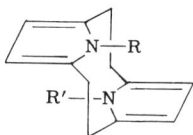

81 R = R′ = CH₃
53 R = R′ = H
96 R = CH₃; R′ = H

The ¹H-nmr spectrum of compound **96**[54] is temperature independent. The spectrum of compound **95** could not be examined because of its lack of thermal stability.[54]

The anisotropic field effect of ether-type oxygens has been shown to be operative in the anomalous degree of shielding experienced by the α-methylenic proton in furanopyridinophane.[142] Thus, it appears that the furan ring is tilted so that the oxygen lies closer to the nitrogen side of the pyridine ring. This was confirmed by X-ray crystallography.[52]

Furanopyridazinophane **246** was shown to have an nmr spectrum that remained unchanged in the temperature range −60–110°C.[143]

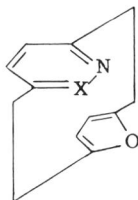

245 X = CH
246 X = N

In the meta-substituted heterophane series, pyridinophane **169** exhibits coalescence at 13.5°C, in contrast to metacyclophane itself, the nmr spectrum of which remains unchanged up to 200°C.[167] Pyridinophane-1,9-

diene (186) shows no spectral changes in the temperature range -80–$100°$ C.[102] The unsaturated bridges in the molecule do not allow the ring flipping that occurs in the saturated analog 169. Although this rationale seems logical, it does not hold for the paracyclopyridinophanes 182 and 187. The coalescence temperature of the saturated compound 182 was measured and found to be $-50°C$[99] (or $-43.5°C$,[100]), whereas the unsaturated system 187 undergoes ring flipping even at $-110°C$.[100] A word of caution in explaining this result is necessary. An X-ray analysis of 187 showed that the rings may not be flipping in solution if the solution conformation is the same as that in the solid state, in which a perpendicular orientation of the two rings is observed (these data are presented in the next section in tabular form). It should also be noted that the benzenoid analogs of 182 and 187 also exhibit a more facile ring flipping in the unsaturated system, and these rings are *not* perpendicular to each other.[42] The presence of an unsaturated bridge in the furanophane 67 also facilitates ring flipping relative to furanophane 2 (coalescence temperature 30°C for 25; 63°C for 2).[39] Further research on these systems prompted a reinvestigation of the paracyclopyridinophane 182, which was then reported to show coalescence in its nmr spectrum at $-29°C$ in deuterioacetone.[103] It was also observed that substituents on the aromatic rings exert little influence on the ring-flipping process in these systems (Table I).

The metacyclopyridinophane 209 does not show coalescence in its nmr spectrum up to 200°C. This system should be compared with pyridinophane 169, which has been discussed (coalescence temperature 13.5°C).[167]

N-Oxidation of [2.2](2,5)furano(3,6)pyridazinophane (246) afforded compound 248, the [1]H-nmr spectrum of which revealed that one α-methyl-

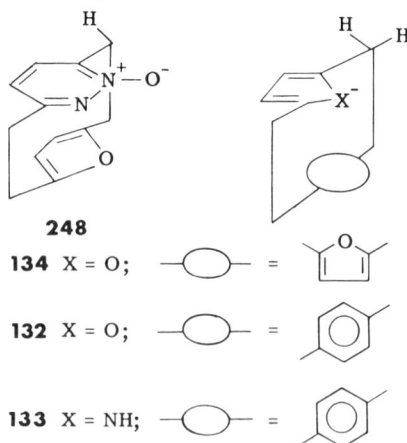

248

134 X = O; ⟨○⟩ =

132 X = O; ⟨○⟩ =

133 X = NH; ⟨○⟩ =

enic hydrogen experienced a deshielding field effect that agreed in magnitude with the calculated value.[62,143] Similar effects are also noted in the acetylated compounds **132–134**.[61,62] It is interesting that the degree of α-methylene proton deshielding in the furanoparacyclophane derivative **132** is less than that in **133** and **134,** due to rotation of the substituted ring, which causes averaging of the field effect over both methylenic hydrogens. The addition of optically active shift reagent to nmr samples caused resolution of the rigid isomers of **133** and **134.** The freely rotating acetylfuranoparacyclophane **132** could not be resolved due to rapid interconversion of enantiomers.[61,62]

A slow rate of hydrogen–deuterium exchange in pyrrolophanes **89, 93,** and **96** (about 2 days in Ch_3OD) has been noted.[52] A possible reason for the slow exchange is outlined here. The nitrogen is forced into a tetrahedral bonding arrangement by the incoming deuterium. This forces the hydrogen toward the opposite pyrrole ring, causing considerable crowding.[52]

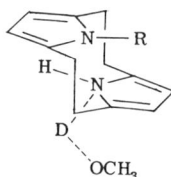

Vögtle and Effler[117] examined cyclophanes of type **276.** They found that **226** is conformationally rigid up to a temperature of 180°C, and the others show various barriers to flipping. Thus, $^vG_c^{\ddagger}$ values of 20.5, ≥20.5, and <13.6 kcal/mol were found for **227, 228,** and **229,** respectively.

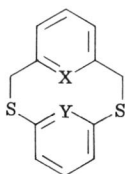

226 X = Y = CH
227 X = CH; Y = N
228 X = N; Y = CH
229 X = Y = N

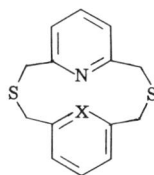

171a X = N
  b X = CF
  c X = CH

Vögtle and Schunder also examined compounds **171a–171c.**[118] Compounds **171a** and **171c** are conformationally mobile at room temperature, whereas **171b** has a coalescence temperature of about 190°C.

B. Ultraviolet Analysis

The uv spectra of unsaturated systems can be explained with reference to Scheme 50. The left side of the scheme illustrates the formation of the π orbitals by combination of the p atomic orbitals of the constituent atoms. The transition observed in the uv region is the $\pi-\pi^*$ transition, as shown. The other diagrams illustrate the molecular orbitals formed when two isolated π systems are conjugated. The consequent decrease in the energy of the $\pi-\pi^*$ transition is shown. When the conjugation is increased further, as occurs when there are transannular interactions, the energy of this transition decreases further, resulting in a bathochromic shift of the absorption band in the uv spectrum. A complication arises when this method is used to investigate transannular interactions in phanes. This is due to the strain in these systems, which causes a distortion of the planarity of the aromatic rings. This distortion increases the energy of the bonding molecular orbitals, with a consequent decrease in the energy of the antibonding orbitals, resulting in a decrease in the transition energy. Thus, a bathochromic shift may be the result of transannular interactions or a distortion of the planarity of the aromatic rings. Bathochromic shifts in mononuclear species are due solely to ring distortion.

Cram used this to circumvent the difficulty in his analyses of transannular interactions in the paracyclophanes.[168] Thus, examination of singly bridged phanes such as **265,** where m was varied from 9 to 12, showed bathochromic shifts only in the bands at 265 and 273 nm. These shifts could therefore be attributed to distortion of the rings from planarity. When a series of paracyclophanes **(266)** was examined, the shifts in the

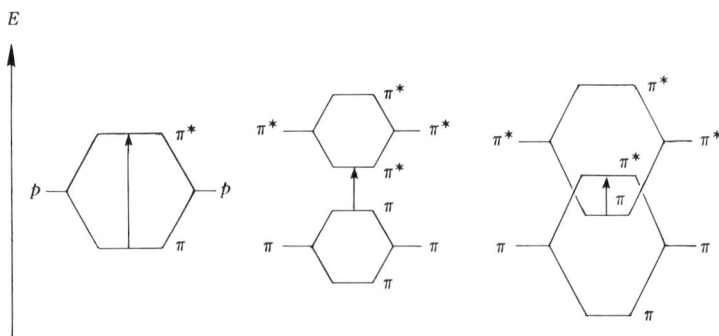

SCHEME 50

265   **266**

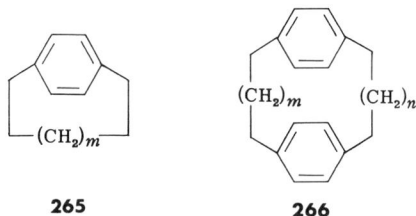

bands at 223 nm could be ascribed to the effect of transannular interactions.

A second method used by Cram was to examine the uv absorption spectra of the 1:1 tetracyanoethylene complexes of phanes such as **266.** Because it was already known[169] that the long-wavelength absorption of such complexes in alkylbenzenes is correlated to the electron-releasing capacity of the π-base of the complex, bathochromic shifts in these bands were evidence of transannular interactions in the phanes.

Table II summarizes the details in the following discussion. The long-wavelength absorption of furanophane **2** (222 nm)[28] exhibits a bathochromic shift of 2 nm relative to its constituent monomer, 2,5-dimethylfuran. Although this does not indicate a strong $\pi-\pi$ interaction between the rings, the nmr spectrum of this compound, discussed earlier, indicates a transannular interaction that is possibly due to the lone pair of electrons on the furan oxygen.

The uv absorption spectrum of (2,5)thiophenophane **(49)** shows two bands at 245 nm ($\varepsilon = 7700$) and 275 nm ($\varepsilon = 5720$), compared with the spectrum of 2,5-dimethylthiophene, which has one band at 238 nm ($\varepsilon = 7250$).[28,171] Not only is there an overall red shift in the spectrum, but a new band has appeared. The evidence for transannular interactions in this system is therefore irrefutable.

The mixed system, furanopyridinophane **245,** has been discussed in the literature along with evidence for the planarity of the furan ring.[51] The absorption band of the furan ring shows a bathochromic shift of 4 nm relative to 2,5-dimethylfuran.

Furanoparacyclophane **51** seems to demonstrate transannular interactions in that the furanoid absorption was reported to be at 226 nm, which was shifted by 6 nm to the red relative to dimethylfuran.[30,63] However, the solvent used in this case was absolute ethanol, and the benzene in this solvent casts some doubt on the results. Other workers measured the uv spectrum of this compound, and their results, obtained from a cyclohexane solution, diagrammatically indicated a red shift of all the bands relative to the analogous bands in the constituent monomers.[32] The sulfur

TABLE II
Ultraviolet Spectra of Selected [2.2]Heterophanes

Compound	λ_{max}, nm ($\log_{10} \varepsilon$)	Solvent	References
	220 (3.90)	EtOH	170
	216 (3.80)	Hexane	170
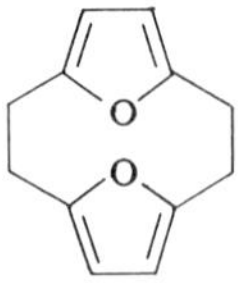 **2**	222 (4.17)	EtOH	28
	237 (3.86)	Hexane	171
	238 (4.8)	Isooctane	171
	231 (3.87)	Isooctane	171
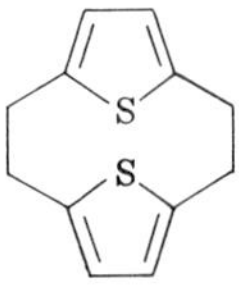 **49**	245 (3.89), 275 (3.75)	EtOH	28
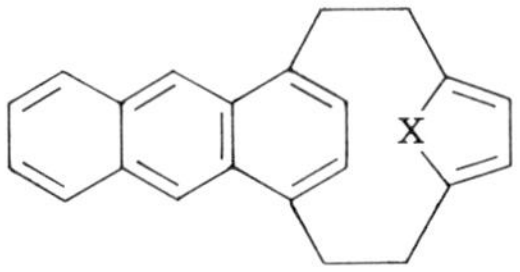 **61** X = O	Diagrammatic red shifts	—	35
62 X = S	Diagrammatic red shifts	—	35
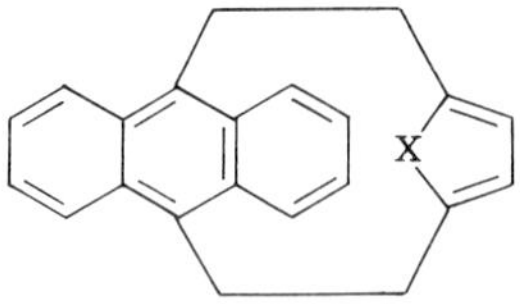 **64** X = S	Diagrammatic red shifts	—	35

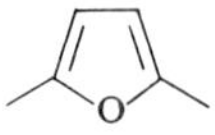

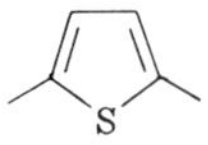

TABLE II (*Continued*)

Compound	λ_{max}, nm ($\log_{10} \varepsilon$)	Solvent	References
71 R = H	—	Absolute EtOH	46
72 R = Me	—	Absolute EtOH	46
74	—	Absolute EtOH	46
73	243 (sh)	—	172
76	220, 270	—	48
245	224	EtOH	142, 170

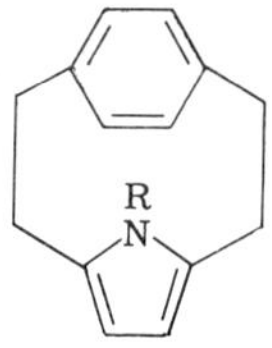

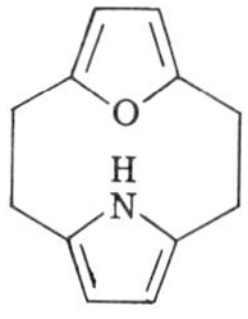

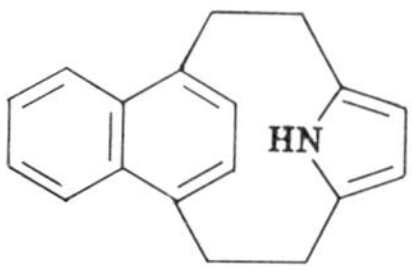

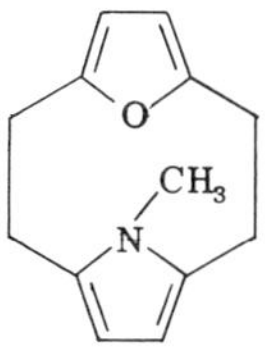

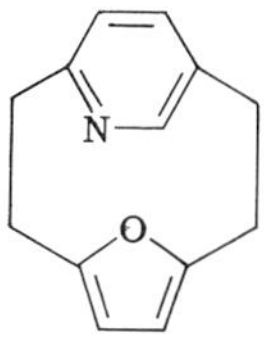

(*continued*)

TABLE II (*Continued*)

Compound	λ_{max}, nm ($\log_{10} \varepsilon$)	Solvent	References
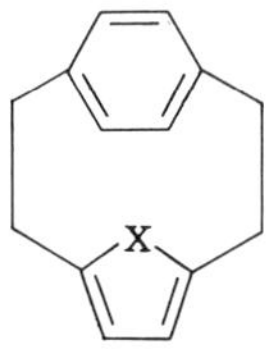 **51** X = O	226.5 (3.9)	Absolute EtOH	30
53 X = S	Diagrammatic[a]	Cyclohexane	32
	Diagrammatic[a]	Cyclohexane	33
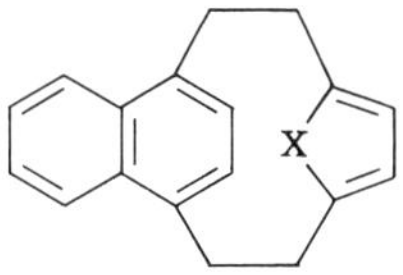 **56** X = O	Diagrammatic red shifts	—	35
59a X = S (anti)	Diagrammatic red shifts	—	35
59b X = S (syn)	Diagrammatic red shifts	—	35
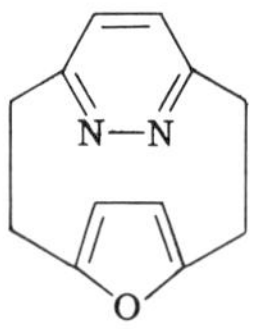 **246**	224 (3.80)	EtOH	62
	222 (3.81)	Hexane	143

[a] No values given; the spectra were reproductions.

analog of this compound (**53**) showed similar results. The long-wavelength absorptions of the tetracyanoethylene complexes of the substituted heteroparacyclophanes were analyzed along with the uncomplexed molecules, and these analyses unequivocally demonstrated the increased basicity of these molecules due to transannular interactions.[33]

The analysis of the uv spectrum of furanonaphthalenophane **16** is similar to that of furanopyridinophane[142,143] in that the planarity of the furan ring in these systems has been shown to be undistorted in the solid state, yet both compounds show bathochromic shifts in the furanoid absorption relative to dimethylfuran.[35] Although no values for these shifts in compounds **56, 59, 61, 62,** and **64** were quoted, the diagrams leave little doubt that the shifts are due to transannular interactions. This conclusion has been corroborated by X-ray and photoelectron spectral data, which are discussed in Sections IV,C and IV,E, respectively.

The uv spectra of paracyclopyrrolophanes **71** and **72** have been reported[46] and, although bathochromic shifts were indicated, the solvent was absolute ethanol, and the pyrrole absorptions were somewhat obscured. This was also the case for the furanopyrrolophane **74**. The spectrum of naphthalenopyrrolophane **73,** however, showed a distinct shoulder at 243 nm, indicative of transannular interactions in this system. This band is not present in either of the constituent monomers.[171]

The uv spectrum of N-methyl[2.2](2,5)furano(2,5)pyrrolophane **(76)** was investigated and found to exhibit transannular interactions. Although it was unchanged in the 220-nm range relative to its constituent monomers, a new band at 270 nm indicated the existence of transannular interactions in this system.[48]

A bathochromic shift in the uv spectrum of **246** in comparison with that of 3,6-dimethylpyridazine and 2,5-dimethylfuran was also observed.[62,143]

C. X-ray Analysis

Relatively few heterophanes have been analyzed by X-ray diffraction. This is due partly to the limited usefulness of these analyses. However, in some cases the solid-state structure can corroborate postulates concerning the solution conformation of the molecule in question.

As illustrated in Table III the syn configuration postulated for furanopyridinophane **245** was verified by this method.[52] Similarly, *anti*-furanonaphthalenophane was shown to be the correct structure in the solid state.[172] The other analyses are listed in the table. It should be noted that, in mixed heterophanes, the five-membered ring experiences little or no distortion from planarity. This trend implies that the bathochromic shifts discussed earlier in connection with the uv spectra of these systems are much more likely to be due to transannular interactions. This conclusion is corroborated in the section dealing with the photoelectron spectra of these molecules (Section IV,E).

TABLE III

X-ray Analyses

Compound	Ring–ring angle (deg)	Distortion from planarity[a] (Å)	Reference
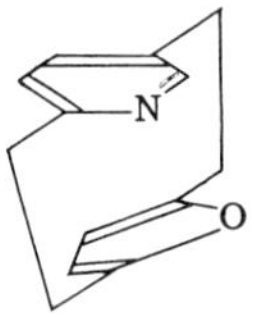 **245**	23	0.172 0.02	51
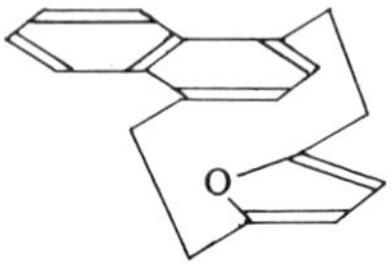 **56**	22	0.177 0.00	172
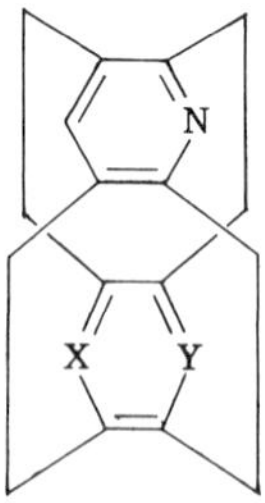 **243** X—N; Y = CH (anti) **244** X = CH; Y = N (syn)	2	0.12 0.12	173
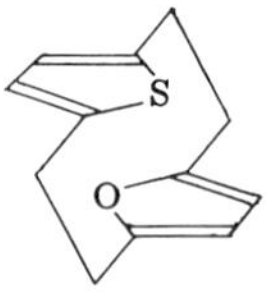 **250**	0	— —	174
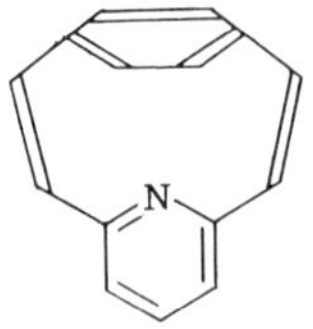 **187**	90	0.222 0.00	175

[a] Upper value refers to upper ring as drawn.

D. Infrared Spectroscopy

There is only one example in the literature of ir spectroscopy being used to demonstrate the existence of transannular interactions.[103] Thus, the C—D frequency shift for a broad range of amines whose pK_a values were known was measured in deuteriochloroform. The correlation of pK_a to C—D shift was very good and, when utilized to obtain the pK_a values of the pyridinophanes **182, 189,** and **190,** gave the following results. (*a*) Pyridinophane **182** is 1.20 pK_a units more basic than 2,6-dimethylpyridine; (*b*) substitution on the benzenoid ring by methyl groups caused a decrease in the basicity of these heterophanes. The conclusion was that the phane structure increases the basicity of these compounds by electron donation from the benzene ring to the pyridine ring. The increase in the number of methyl substituents, although increasing the electron density of the benzene ring, sterically inhibits the transfer of this electron density to the pyridine ring (Table IV).

TABLE IV
Infrared Spectral Studies

Compound	pK_a	ν (cm^{-1})
Pyridine	5.23	28
2-Methylpyridine	5.97	39
3-Methylpyridine	5.68	32
2,6-Dimethylpyridine	6.75	47
2,4,6-Trimethylpyridine	7.43	53

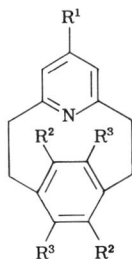

182 R^1 = R^2 = R^3 = H	7.95	58
189 R^1 = R^3 = H; R^2 = CH$_3$	7.48	53
190 R^1 = H; R^2 = R^3 = CH$_3$	7.38	52

E. Photoelectron Spectroscopy

Boekelheide and co-workers first utilized photoelectron spectroscopy to examine [2.2]heterophanes[103] and observed nothing unusual about the spectra of the pyridinophanes **182, 189,** and **190.** However, increasing attention has been devoted to this method in the literature, and it is now clear that the significant depression of the ionization potentials of the cyclophanes is due to transannular interaction.[176] A detailed theoretical discussion shows that this depression cannot be due to the increase in substitution of the aromatic rings, because the orbitals in question are orthogonal. Thus, a greater lowering of the ionization potentials is observed on going from paracyclophane to superphane than on going from p-xylene to hexamethylbenzene. This is due solely to transannular interactions in these systems. In the heterophane series, a similar trend has been reported. Although the report contains no detailed discussion of the results, the similarity between the geometry of these systems and that of the cyclophanes leads one to conclude that the lowering of the ionization potentials in the heterophanes is also due to transannular interactions.[170] The only assumption made here is the validity of Koopman's theorem. Consideration of adiabatic ionization potentials does not affect this conclusion. The actual ionization potentials obtained are readily available from the original article. The systems exhibiting a decrease in ionization potential due to transannular interactions are **2, 49, 51, 56, 59, 63, 71, 73, 74, 95,** and **174.**

REFERENCES

1. C. J. Brown and A. C. Farthing, *Nature (London)* **164,** 915 (1949).
2. B. H. Smith, "Bridged Aromatic Compounds." Academic Press, New York, 1964.
3. K. M. Smith, *Aromat. Heteroaromat. Chem.* **2,** 423 (1974).
4. A. D. Adler, V. Varodi, and P. George, *Enzyme* **17,** 43 (1974).
5. F. Vögtle and P. Newmann, *Tetrahedron Lett.* p. 5329 (1969).
6. S. Bradamante, A. Marchesini, and G. Pagani, *Chim. Ind. (Milan)* **55,** 962 (1971).
7. S. Bradamante, R. Fusco, A. Marchesini, and G. Pagani, *Tetrahedron Lett.* p. 11 (1970).
8. S. Fujita, T. Kawaguti, and H. Nozaki, *Bull. Chem. Soc. Jpn.* **43,** 2596 (1970).
9. S. Fujita, T. Kawaguti, and H. Nozaki, *Tetrahedron Lett.* p. 1119 (1971).
9a. J. M. Patterson, J. Brasch, and P. Dunchko, *J. Am. Chem. Soc.* **27,** 1652 (1962).
10. S. Fujita, Y. Hayashi, and H. Nozaki, *Tetrahedron Lett.* p. 1645 (1972).
11. H. Nozaki, T. Koyama, T. Mori, and R. Noyori, *Tetrahedron Lett.* p. 2181 (1968).
12. H. Nozaki, T. Koyama, and T. Mori, *Tetrahedron* **25,** 5357 (1969).

13. W. E. Parham and J. F. Dooley, *J. Am. Chem. Soc.* **89,** 985 (1967).

14. W. E. Parham and J. F. Dooley, *J. Org. Chem.* **33,** 1476 (1968).

15. W. E. Parham and R. J. Sperley, *J. Org. Chem.* **32,** 926 (1967).

16. S. Hunig and H. Hoch, *Justus Liebigs Ann. Chem.* **716,** 68 (1968).

17. S. Z. Taits *et al., Khim. Geterotsikl. Soedin.* p. 170 (1972); *Chem. Abstr.* **76,** 153469 (1972).

18. Ya. L. Gol'farb *et al., Zh. Obshch. Khim.* **29,** 3564 (1959); *Chem. Abstr.* **54,** 19639d (1960).

19. Ya. L. Gol'farb *et al., Izv. Akad. Nauk SSSR, Ser. Khim.* p. 1451 (1963); *Chem. Abstr.* **59,** 15243e (1963).

20. Ya. L. Gol'farb, S. Z. Taits, and L. I. Belen'ku, *Tetrahedron* **19,** 1851 (1963).

21. S. Z. Taits and Ya. L. Gol'farb, *Izv. Akad. Nauk SSSR, Ser. Khim.* p. 1289 (1963); *Chem. Abstr.* **59,** 13990a (1963).

22. Ya. L. Gol'farb, S. Z. Taits, and V. N. Bulgakova, *Izv. Akad. Nauk SSSR, Ser. Khim.* p. 1299 (1963); *Chem. Abstr.* **59,** 13990g (1963).

23. S. Z. Taits *et al., Izv. Akad. Nauk SSSR, Ser. Khim.* p. 2536 (1975); *Chem. Abstr.* **84,** 59421 (1976).

24. Z. V. Todres, F. M. Stoyanovich, Ya. L. Gol'dfarb, and D. N. Kursanov, *Khim. Geterotsikl. Soedin.* p. 632 (1973); *Chem. Abstr.* **79,** 66117 (1973).

25. R. J. Hayward and O. Meth-Cohn, *J. Chem. Soc., Perkin Trans. 1* p. 212 (1975).

26. R. J. Hayward and O. Meth-Cohn, *J. Chem. Soc., Perkin Trans. 1* p. 219 (1975).

27. M. M. Htay and O. Meth-Cohn, *Tetrahedron Lett.* p. 79 (1976).

28. H. E. Winberg, F. S. Fawcett, W. E. Mochel, and C. W. Theobald, *J. Am. Chem. Soc.* **82,** 1428 (1960).

29. J. R. Fletcher and I. O. Sutherland, *Chem. Commun.* p. 1504 (1969).

30. D. J. Cram and G. R. Knox, *J. Am. Chem. Soc.* **83,** 2204 (1961).

31. G. M. Whitesides, B. A. Pawson, and A. C. Cope, *J. Am. Chem. Soc.* **90,** 639 (1968).

32. N. Osaka, S. Mizogami, T. Otsubo, Y. Sakata, and S. Misumi, *Chem. Lett.* p. 515 (1974).

33. S. Mizogami, T. Otsubo, Y. Sakata, and S. Misumi, *Tetrahedron Lett.* **29,** 2791 (1971).

34. H. H. Wasserman and P. M. Keehn, *Tetrahedron Lett.* **38,** 3227 (1969).

35. S. Mizogami, N. Osaka, T. Otsubo, Y. Sakata, and S. Misumi, *Tetrahedron Lett.* **10,** 799 (1974).

36. H. Wynberg and R. Helder, *Tetrahedron Lett.* **45,** 4317 (1971).

37. S. Mizogami, T. Otsubo, Y. Sakata, and S. Misumi, *Tetrahedron Lett.* p. 2791 (1971).

38. N. Osaka, S. Mizogami, T. Otsubo, Y. Sakata, and S. Misumi, *Chem. Lett.* p. 515 (1974).

39. J. M. Timko and D. J. Cram, *J. Am. Chem. Soc.* **96,** 7159 (1974).

39a. S. H. Mashraqui and P. M. Keehn, *J. Am. Chem. Soc.* **104,** 4461 (1982).

40. S. H. Kusefoglu and D. T. Longone, *Tetrahedron Lett.* **27,** 2391 (1978).

41. H. H. Wasserman and P. M. Keehn, *Tetrahedron Lett.* p. 3227 (1969).

42. S. Bradamante, A. Marchesini, and G. Pagani, *Chim. Ind. (Milan)* **53,** 267 (1971).

43. S. Fujita, T. Kawaguti, and H. Nozaki, *Tetrahedron Lett.* **16,** 1119 (1971).

44. H. Nozaki, T. Koyama, T. Mori, and R. Noyori, *Tetrahedron Lett.* **18,** 2181 (1968).

45. H. Nozaki, T. Koyama, and T. Mori, *Tetrahedron Lett.* **25,** 5357 (1969).

46. J. F. Haley and P. M. Keehn, *Tetrahedron Lett.* **41,** 4017 (1973).

47. S. M. Rosenfeld and P. M. Keehn, *J. Chem. Soc., Chem. Commun.* p. 119 (1974).

48. C. Wong, Ph.D. Thesis, University of Alabama, University (1975).

49. H. H. Wasserman and D. T. Bailey, *Chem. Commun.* p. 107 (1970).

50. W. W. Paudler and E. A. Stephan, *J. Am. Chem. Soc.* **92,** 4468 (1970).

51. J. L. Atwood, D. C. Hrncir, C. Wong, and W. W. Paudler, *J. Am. Chem. Soc.* **96,** 6132 (1974).
52. R. L. Mahaffey, Ph.D. Thesis, University of Alabama, University (1977).
53. R. L. Mahaffey, J. L. Atwood, M. B. Humphrey, and W. W. Paudler, *J. Org. Chem.* **41,** 2693 (1976).
54. J. F. Haley and P. M. Keehn, *Tetrahedron Lett.* **21,** 1675 (1975).
55. J. Bergman and J. Backvall, *Tetrahedron Lett.* p. 2899 (1973).
56. N. N. Gerber, *Tetrahedron Lett.* p. 809 (1970).
57. N. N. Gerber, *J. Heterocycl. Chem.* **10,** 925 (1973).
58. H. H. Wasserman *et al., Tetrahedron, Suppl.* **8** (Part II), 647 (1966).
59. H. H. Wasserman, G. C. Rodgers, and D. D. Keith, *J. Am. Chem. Soc.* **91,** 1263 (1968).
60. I. L. Mushkalo, G. G. Dyadyusha, and L. S. Turova, *Tetrahedron Lett.* **21,** 2977 (1980).
61. M. D. Bezoari and W. W. Paudler, *J. Org. Chem.* (in press).
62. M. D. Bezoari, Ph.D. Thesis, University of Alabama, University (1981).
63. D. J. Cram, C. S. Montgomery, and G. R. Knox, *J. Am. Chem. Soc.* **88,** 515 (1966).
64. H. H. Wasserman and A. R. Doumaux, *J. Am. Chem. Soc.* **88,** 4517 (1966).
64a. H. H. Wasserman and R. Kitzing, *Tetrahedron Lett.* **60,** 5315 (1969).
64b. H. H. Wasserman and A. R. Doumaux, Jr., *J. Am. Chem. Soc.* **88,** 4611 (1966).
64c. T. J. Katz, V. Balogh, and J. Schulman, *J. Am. Chem. Soc.* **90,** 734 (1968).
65. L. A. Kapicak and M. A. Battiste, *Chem. Commun.* p. 930 (1973).
65a. M. A. Battiste, L. A. Kapicak, M. Mathew, and G. J. Palenik, *Chem. Commun.* p. 1536 (1971).
66. H. H. Wasserman and R. Kitzing, *Tetrahedron Lett.* p. 3343 (1969).
67. H. Wynberg and R. Helder, *Tetrahedron Lett.* **45,** 4317 (1971).
68. K. Ankner, B. Lamm, B. Thulin, and O. Wennerström, *J. Chem. Soc., Perkin Trans. 2* p. 1301 (1980).
69. A. T. Balaban, M. Gavat, and C. D. Nenitzescu, *Tetrahedron* **18,** 1079 (1962).
70. K. Biemann, G. Buchi, and B. H. Walker, *J. Am. Chem. Soc.* **79,** 5558 (1957).
71. K. Tamao, S. Kodama, T. Nakatsuka, Y. Kiso, and M. Kumada, *J. Am. Chem. Soc.* **97,** 4405 (1975).
72. H. Saimoto, T. Hiyama, and H. Nozaki, *Tetrahedron Lett.* **21,** 3897 (1980).
73. A. T. Balaban, *Tetrahedron Lett.* p. 4643 (1968).
74. H. Nozaki, S. Fujita, and T. Mori, *Bull. Chem. Soc. Jpn.* **42,** 1163 (1969).
75. S. Fujita and H. Nozaki, *Bull. Chem. Soc. Jpn.* **44,** 2827 (1971).
76. S. Fujita, K. Imamura, and H. Nozaki, *Bull. Chem. Soc. Jpn.* **45,** 1881 (1972).
77. S. Bradamante *et al., Chim. Ind.* (*Milan*) **55,** 962 (1973).
78. A. Marchesini *et al., Tetrahedron Lett.* p. 671 (1971).
79. T. Hiyama, S. Hirano, and H. Nozaki, *J. Am. Chem. Soc.* **96,** 5287 (1974).
80. H. Gerlach and E. Huber, *Helv. Chim. Acta* **51,** 2027 (1968).
81. W. E. Parham, R. W. Davenport, and J. B. Biasotti, *Tetrahedron Lett.* p. 557 (1969); *J. Org. Chem.* **35,** 3775 (1970).
82. W. E. Parham, K. B. Sloan, and J. B. Biasotti, *Tetrahedron* **27,** 5767 (1971); *J. Org. Chem.* **38,** 927 (1973).
83. W. E. Parham, D. C. Egberg, and S. S. Slagar, *J. Org. Chem.* **37,** 3248 (1972).
84. D. Moran, M. N. Patel, N. A. Tahir, and B. J. Wakefield, *J. Chem. Soc., Perkin Trans. 1* p. 2310 (1974).
85. G. L. Isele and K. Scheib, *Chem. Ber.* **108,** 2312 (1975).
86. K. Tamao *et al., J. Am. Chem. Soc.* **97,** 4405 (1975).

86a. H. Reinshagen, G. Schulz, and A. Stutz, *Monatsh. Chem.* **110**, 575 (1979).
87. W. Tochtermann and P. Rosner, *Tetrahedron Lett.* **21**, 4905 (1980).
88. R. B. Woodward and R. Hoffman, "The Conservation of Orbital Symmetry." Akademie-Verlag, Berlin, 1971.
89. G. B. Gill, *Q. Rev., Chem. Soc.* **22**, 338 (1968).
90. R. E. Lehr and A. P. Marchand, "Orbital Symmetry: A Problem-Solving Approach." Academic Press, New York, 1972.
91. F. Vögtle and L. Rossa, *Angew. Chem., Int. Ed. Engl.* **18**, 515 (1979).
92. H. J. J.-B. Martel and M. Rasmussen, *Tetrahedron Lett.* **41**, 3843 (1971).
93. F. Vögtle and L. Schunder, *Chem. Ber.* **102**, 2677 (1969).
94. J. Bruhin and W. Jenny, *Chimia* **25**, 238 (1971).
95. J. Bruhin and W. Jenny, *Chimia* **25**, 308 (1971).
96. J. Bruhin and W. Jenny, *Chimia* **26**, 420 (1972).
97. J. Bruhin, W. Kneubuhler, and W. Jenny, *Chimia* **5**, 277 (1973).
98. J. Bruhin and W. Jenny, *Tetrahedron Lett.* **15**, 1215 (1973).
98a. H. A. Staab and W. K. Appel, *Liebigs Ann. Chem.* p. 1065 (1981).
99. V. Boekelheide, I. D. Reingold, and M. Tuttle, *J. Chem. Soc., Chem. Commun.* p. 406 (1973).
100. V. Boekelheide, K. Galuszko, and K. S. Szeto, *J. Am. Chem. Soc.* **96**, 1578 (1974).
101. M. W. Haenel, *Tetrahedron Lett.* p. 4007 (1978).
102. V. Boekelheide and J. A. Lawson, *Chem. Commun.* p. 1558 (1970).
103. I. D. Reingold, W. Schmidt, and V. Boekelheide, *J. Am. Chem. Soc.* **101**, 2121 (1979).
104. M. March, "Advanced Organic Chemistry," 2nd ed. McGraw-Hill, New York, 1977.
105. F. Vögtle and P. Neumann, *Synthesis* p. 85 (1973).
106. D. J. Cram, C. K. Dalton, and G. R. Knox, *J. Am. Chem. Soc.* **85**, 1088 (1963).
107. W. Jenny and H. Holzrichter, *Chimia* **21**, 509 (1967).
108. V. Boekelheide and W. Pepperdine, *J. Am. Chem. Soc.* **92**, 3684 (1970).
109. W. Jenny and H. Holzrichter, *Chimia* **22**, 139 (1968).
110. W. Jenny and H. Holzrichter, *Chimia* **22**, 247 (1968).
111. W. Jenny and H. Holzrichter, *Chimia* **22**, 306 (1968).
112. W. Jenny and H. Holzrichter, *Chimia* **23**, 158 (1969).
113. V. Boekelheide and W. Pepperdine, *J. Am. Chem. Soc.* **92**, 3684 (1970).
114. J. Bruhin and W. Jenny, *Chimia* **25**, 238 (1971).
115. W. Baker, K. M. Buggle, J. F. W. McOmie, and D. A. M. Watkins, *J. Chem. Soc.* p. 3594 (1958).
116. Th. Kauffmann, G. Beissner, W. Sahm, and A. Woltermann, *Angew. Chem., Int. Ed. Engl.* **9**, 808 (1970).
117. F. Vögtle and A. H. Effler, *Chem. Ber.* **102**, 3071 (1969).
118. F. Vögtle and L. Schunder, *Chem. Ber.* **102**, 2677 (1969).
119. K. R. Reistad, P. Groth, R. Lie, and R. Undheim, *Chem. Commun.* p. 1059 (1972).
120. V. F. Borodkin and Ya. G. Vorob'ev, *Izv. Vyssh. Ucheb. Zaved., Khim. Khim. Tekhnol.* **15**, 1750 (1972); *Chem. Abstr.* **78**, 97620 (1973).
121. R. P. Smirnov et al., *Izv. Vyssh. Ucheb. Zaved., Khim. Khim. Tekhnol.* **16**, 1062 (1973); *Chem. Abstr.* **79**, 105221 (1973).
122. V. F. Borodkin et al., *Izv. Vyssh. Ucheb. Zaved., Khim. Khim. Tekhnol.* **16**, 1772 (1973); *Chem. Abstr.* **80**, 70791 (1974).
123. V. F. Borodkin and A. V. Makarycheva, *Izv. Vyssh. Ucheb. Zaved., Khim. Khim. Tekhnol.* **18**, 238 (1975); *Chem. Abstr.* **83**, 28204 (1975).
124. V. F. Borodkin et al., *Tr. Ivanov. Khim.-Tekhnol. Inst.* **14**, 141 (1972); *Chem. Abstr.* **79**, 115546 (1973).

125. P. V. Gubin *et al., Tr. Ivanov. Khim.-Tekhnol. Inst.* **14,** 21 (1972); *Chem. Abstr.* **82,** 43485 (1975).
126. M. I. Al'yanov *et al., Tr. Ivanov. Khim.-Tekhnol. Inst.* **12,** 139 (1970); *Chem. Abstr.* **77,** 34473 (1972).
127. R. P. Smirnov *et al., Tr. Vses. Mezhvuz. Nauchno-Tekh. Konf. Vopr. Sint. Primen. Org. Krasitelei, 1900* 17 (1970); *Chem. Abstr.* **76,** 14518 (1972).
128. L. M. Fedorov *et al., Izv. Vyssh. Ucheb. Zaved., Khim. Khim. Tekhnol.* **15,** 466 (1972); *Chem. Abstr.* **77,** 48426 (1972).
129. N. A. Kolesnikov *et al., Izv. Vyssh. Ucheb. Zaved., Khim. Khim. Tekhnol.* **16,** 1084 (1973); *Chem. Abstr.* **79,** 105222 (1973).
130. A. P. Snegireva and V. F. Borodkin, *Izv. Vyssh. Ucheb. Zaved., Khim. Khim. Tekhnol.* **17,** 1364 (1974); *Chem. Abstr.* **82,** 97999 (1975).
131. P. V. Gubin *et al.,* USSR Patent 352,895 (1972); *Chem. Abstr.* **78,** 97734 (1973).
132. V. F. Borodkin and R. D. Komarov, USSR Patent 411,087 (1974); *Chem. Abstr.* **80,** 108593 (1974).
133. G. R. Newkome and J. M. Robinson, *Chem. Commun.* p. 831 (1973).
134. M. Chastrelte and F. Chastrelte, *Chem. Commun.* p. 534 (1973).
135. J. D. Curry and D. H. Busch, *J. Am. Chem. Soc.* **86,** 592 (1964).
136. S. M. Nelson and D. H. Busch, *Inorg. Chem.* **8,** 1859 (1969).
137. M. G. B. Drew *et al., J. Chem. Soc., Dalton Trans.* p. 2507 (1975).
138. G. R. Newkome, J. M. Roper, and J. M. Robinson, *J. Org. Chem.* **45,** 4380 (1980).
139. H. J. VanRamesdonk, J. W. Verhoeven, U. K. Pandit, and T. J. deBoer, *Tetrahedron Lett.* **21,** 1549 (1980).
139a. D. C. Dittmer and B. B. Blidner, *J. Org. Chem.* **38,** 2873 (1973).
140. H. C. Kang and V. Boekelheide, *Angew. Chem., Int. Ed. Engl.* **20,** 571 (1981).
141. V. Boekelheide and G. Ewing, *Tetrahedron Lett.* p. 4245 (1978).
142. C. Wong and W. W. Paudler, *J. Org. Chem.* **39,** 2570 (1974).
143. M. D. Bezoari and W. W. Paudler, *J. Org. Chem.* **45,** 4584 (1980).
144. K. P. C. Vollhardt, *Synthesis* p. 765 (1975).
145. A. G. Anastassiou and H. S. Kasmai, *Adv. Heterocycl. Chem.* **23,** 55 (1978).
146. G. R. Newkome, J. D. Sauer, J. M. Roper, and D. C. Hager, *Chem. Rev.* **77,** 513 (1977).
147. W. Carruthers and M. G. Pellatt, *J. Chem. Soc., Perkin Trans. 1* p. 1136 (1973).
148. P. J. Beeby, R. T. Weavers, and F. Sondheimer, *Angew. Chem., Int. Ed. Engl.* **13,** 138 (1974).
149. R. T. Weavers and F. Sondheimer, *Angew. Chem., Int. Ed. Engl.* **13,** 139 (1970).
150. T. M. Cresp and M. V. Sargent, *J. Chem. Soc., Perkin Trans. 1* p. 1786 (1973).
151. A. B. Holmes and F. Sondheimer, *Chem. Commun.* p. 1434 (1973).
152. P. J. Beeby, J. M. Brown, P. J. Garratt, and F. Sondheimer, *Tetrahedron Lett.* p. 599 (1974).
153. J. M. Brown and F. Sondheimer, *Angew. Chem., Int. Ed. Engl.* **13,** 337 (1974).
154. T. M. Cresp and M. V. Sargent, *Chem. Commun.* p. 1457 (1971).
155. T. M. Cresp and M. V. Sargent, *J. Chem. Soc., Perkin Trans. 1* p. 2961 (1973).
156. H. Oagawa *et al., Tetrahedron* **30,** 1033 (1974).
157. H. Oagawa *et al., Tetrahedron Lett.* p. 3889 (1974).
158. T. M. Cresp and M. V. Sargent, *J. Chem. Soc., Perkin Trans. 1* p. 2145 (1974).
159. T. M. Cresp and M. V. Sargent, *Chem. Commun.* p. 101 (1974).
160. P. J. Garratt, "Aromaticity." McGraw-Hill, New York, 1971.
161. V. W. Flitsch and H. Peeters, *Chem. Ber.* **106,** 1731 (1973).
162. V. W. Flitsch and H. Lerner, *Tetrahedron Lett.* p. 1677 (1974).

163. W. Flitsch and W. Schulten, *Chem. Ber.* **114,** 620 (1981).

163a. V. W. Flitsch and H. Peeters, *Tetrahedron Lett.* p. 1461 (1975).

164. Y. Miyahara, T. Inazu, and T. Yoshino, *Chem. Lett.* p. 397 (1980).

165. S. Rosenfled and P. M. Keehn, *Tetrahedron Lett.* p. 4021 (1973).

166. H. H. Wasserman and D. T. Bailey, *Chem. Commun.* p. 107 (1970).

167. I. Gault, B. J. Price, and I. O. Sutherland, *Chem. Commun.* p. 540 (1967).

168. D. J. Cram, *Rec. Chem. Prog.* **20,** 71 (1959).

169. R. E. Merrifield and W. D. Phillips, *J. Am. Chem. Soc.* **80,** 2778 (1980).

170. B. Kovac, M. Allan, E. Heilbronner, J. P. Maier, R. Gleiter, M. W. Haenel, P. M. Keehn, and J. A. Reiss, *J. Electron Spectrosc. Relat. Phenom.* **19,** 167 (1980).

171. J. P. Philips, H. Feuer, and B. S. Thayagarjan, eds., "Organic Electronic Spectral Data," Vol. 10. Wiley, New York, 1968.

172. M. Corson, B. M. Foxman, and P. M. Keehn, *Tetrahedron* **34,** 1641 (1978).

173. A. W. Hanson, *Cryst. Struct. Commun.* **10,** 751 (1981).

174. B. Kamenar and C. K. Prout, *J. Chem. Soc.* p. 4845 (1965).

175. L. H. Weaver and B. W. Mathews, *J. Am. Chem. Soc.* **96,** 1581 (1974).

176. B. Kovac, M. Mohraz, E. Heilbronner, V. Boekelheide, and H. Hopf, *J. Am. Chem. Soc.* **102,** 4314 (1980).

CHAPTER **7**

Condensed Benzenoid Cyclophanes

JAMES A. REISS

Department of Organic Chemistry
La Trobe University
Bundoora, Victoria, Australia

I. INTRODUCTION

Several years before the first recorded synthesis[1] of [2.2]paracyclo-
phane **(1)**, a report of an unsuccessful attempt to prepare a condensed

benzenoid cyclophane, [2.2](2,7)naphthalenophane-1,11-diene (2), had been published.[2] It was considered that the diene 2 would prove to be a useful precursor of the fully condensed aromatic compound coronene (3), but it was not until about 30 years later that this objective was achieved.[3] In the intervening years a large variety of condensed aromatic cyclophanes have been reported, and their chemistry and physical properties determined.

At the time of the publication of B. H. Smith's book, *Bridged Aromatic Compounds,* in 1964,[4] fewer than 20 cyclophanes derived from condensed aromatic compounds had been recorded. Most of these were alkyl-bridged naphthalenediols, two exceptions being [6](2,8)naphthaleno[2]paracyclophane[5] (4) and [2.2](2,7)naphthalenophane.[6,7] (5). Compound 5 represented the only true [2.2]cyclophane derived from a condensed benzenoid aromatic compound at that time.

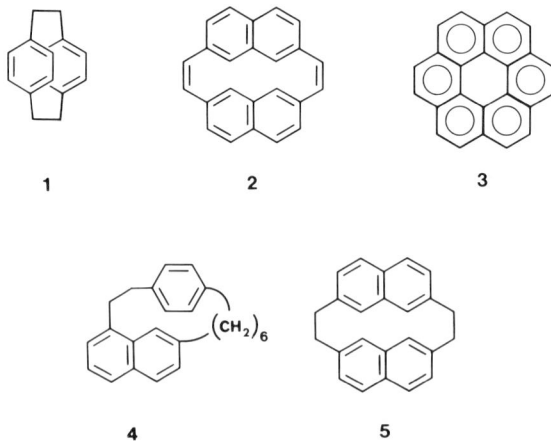

1   2   3

4   5

To some extent the chemistry of [2.2]cyclophanes from condensed aromatic molecules has paralleled that of the [2.2]meta-, [2.2]metapara-, and [2.2]paracyclophanes. There are examples of reactions and properties that are a direct extension and extrapolation of the simple bridged systems. Conversely, new areas of chemical interest became apparent with the larger systems, which were previously unrecognized or inaccessible in the smaller cyclophanes. Twelve years ago our group at La Trobe University saw the possibility of using cyclophanes and cyclophanedienes such as compound 2 to obtain polycyclic aromatic hydrocarbons similar to coronene (3). Although structure 3 was by no means an objective of major importance, the lower and higher members of the series of cyclic hydrocarbons to which coronene belongs were all but unknown. In the last decade a limited number of these nonplanar, twisted, and saddle-shaped molecules have been prepared, but the field has barely been touched. That aromatic rings are moderately flexible is now taken for granted, and the only apparent limitation to obtaining distorted aromatic compounds would appear to be the synthetic pathways available. Our experience here has shown that it is possible to make these "bent and battered" compounds, and we believe that target structures such as the higher circulenes and aromatic Möbius molecules are possibly within reach, perhaps from cyclophane precursors.

[2.2]Cyclophanes have been synthesized from condensed aromatic molecules for a number of reasons. These include (a) the determination of chemical and physical properties of bridged aromatic molecules having large proximate aromatic residues; (b) their use as precursors of topologically interesting compounds such as the circulenes, propellicenes, and paddlanes; (c) the determination of strain energies and dynamic properties of large bridged compounds; (d) their use as precursors of annulated annulenes; (e) the study of charge transfer between adjacent aromatic residues; and (f) optical activity studies using helicene-derived cyclophanes.

This chapter focuses in particular on the synthesis and reactions of [2.2]cyclophanes from condensed aromatic molecules. Wherever appropriate, related compounds, including, for example, [2.3]- and [3.3]cyclophanes, heterocyclic analogs, or bridged aromatic compounds incorporating the biphenyl moiety, are discussed. A number of other areas formally related to cyclophanes derived from condensed aromatic hydrocarbons are not considered in depth in this chapter. These include (a) naturally occurring cyclophane biphenyls of the myricanol variety,[8] (b) 1,1'-binaphthyls and their use in crown ethers,[9] and (c) a number of cyclophanes better viewed as annulene derivatives.[10] The organization of this chapter has a structural basis, with cyclophanes derived from common aromatic residues being grouped together.

II. GENERAL SYNTHETIC ROUTES
TO CYCLOPHANES

A number of general synthetic routes have found favor in the synthesis of cyclophanes from condensed aromatic substrates. The most widely applied routes are summarized in the equations below, in which simple aromatic derivatives are used as examples for illustrative purposes only. References to specific compounds can be found in the following sections.

1. Dithia intermediates. This route has been widely used and offers many alternatives.

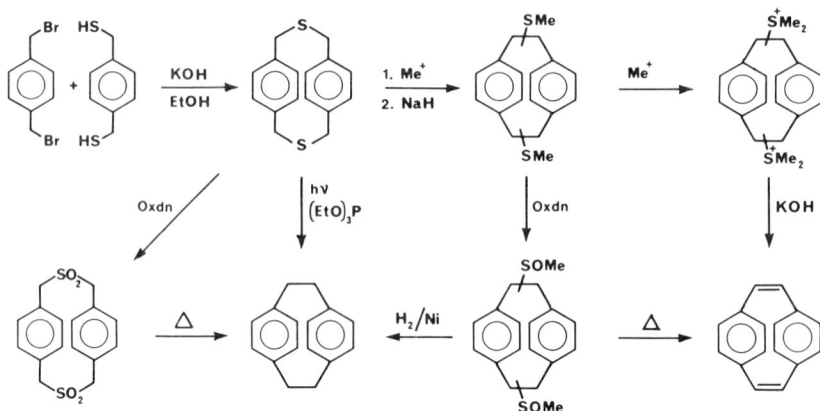

2. Cross-breeding Hofmann elimination. This method is limited in that mixtures of products may result.

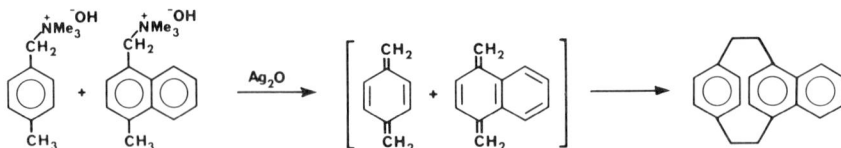

3. Wittig reaction. This reaction has been used for the synthesis of a number of macrocyclic cyclophanes and, although it is essentially a one-step reaction, the procedure suffers from low yields.

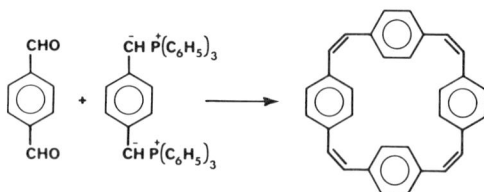

4. Succinoylation. This reaction sequence involves many steps and has had only limited application.

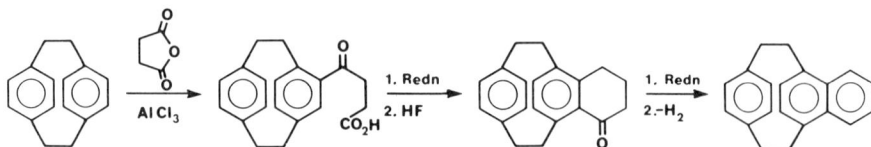

Other methods, including the acyloin reaction, the oxidative coupling of acetylenes, and the Paal–Knorr reaction, are discussed in relevant sections.

III. NAPHTHALENE-DERIVED CYCLOPHANES

A. (1,3)Naphthalene Cyclophanes

There are several reports dealing with [2.2]cyclophanes derived from 1,3-disubstituted naphthalenes. Wasserman and Keehn reported the formation of 4-methoxy[2.2](1,3)naphthalenoparacyclophane (6) as a solvolysis–rearrangement product of the reaction between [2.2](1,4)naphthalenoparacyclophane (7) and singlet oxygen.[11] The structure assignment of 6 was based partly on its temperature dependent ^{1}H-nmr spectrum. The spectra of 6 obtained in hexachloro-1,3-butadiene clearly showed coalescence of signals due to the H_c and H_d protons in the range of 25–100°C, suggesting a rapid interconversion of the two forms 6a and 6b, which is a characteristic of the metaparacyclophane structure.[12–15] The activation energy parameters for this exchange process were not given.

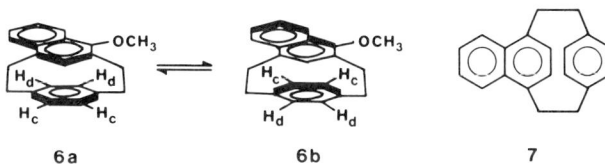

A general approach that has now become a standard synthetic route to [2.2]cyclophanedienes uses as a key intermediate a dithia[3.3]cyclophane. This intermediate can be ring-contracted by either a Stevens rearrangement or Wittig rearrangement, and the resulting product then subjected to a Hofmann elimination to produce the diene.[16–18] Many cyclophanedienes have been prepared by these methods, and two excellent reviews[19] summarizing the scope of the preparations have been published. An illustration of this approach is the synthesis of [2.2](1,3)naphthalenometacyclophane-1,11-diene (10, Scheme 1),[20] which indicates the high yields possible with this procedure. Cyclophane 10 could be easily converted to the dihydropyrene 11 under thermal or photochemical conditions.

Subsequent papers[21,22] reported syntheses directed toward the isolation of related cyclophanes such as 12 and 14 and dihydropyrenes 13 and 15

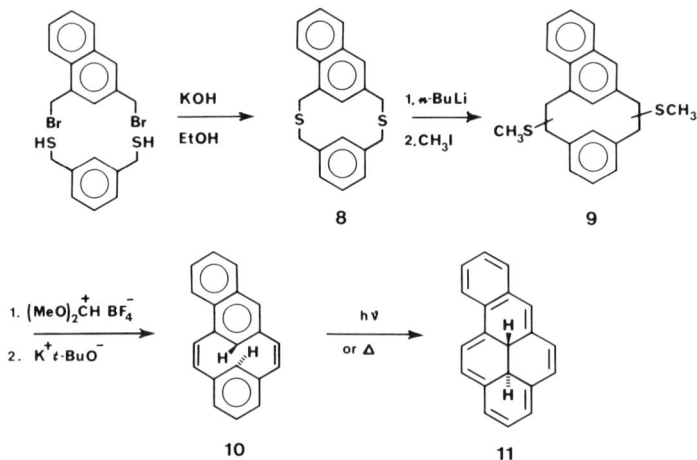

SCHEME 1

using the same general approach of Scheme 1. In each case the dimethyl-dihydropyrenes **13** and **15** were obtained directly (and in good yield) as the sole product, each arising from the precursor cyclophane by a facile electrocyclic ring closure. The ¹H-nmr data obtained for the dihydro-pyrenes gave good support to the hypothesis that equivalent Kekulé structures lead to stronger diatropism (aromaticity).

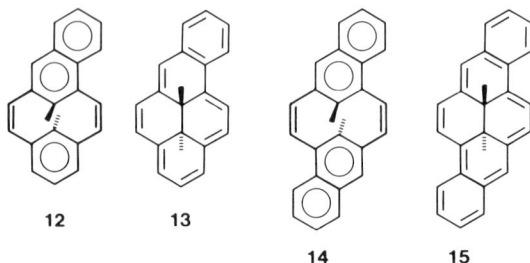

    12   13   14   15

There are few other reports concerned with cyclophanes from 1,3-disubstituted naphthalenes. Parham[23-25] reported the preparation and re-actions of a series of alkyl-bridged 1,3-naphthalenes, generally with a substituent in the 2 position, e.g., compound **16.** The reactivity and rear-rangement reactions of such compounds have been examined as a func-tion of ring size, and other papers in this series have examined heterocy-clic analogs of **16** such as the quinoline derivatives. Grice and Rees[26,27] also reported the synthesis of **17** and its rearrangement in acetic acid to **18.**

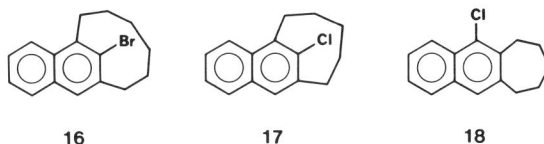

| 16 | 17 | 18 |

B. [2.2](1,4)Naphthalene Cyclophanes

The simplest [2.2]cyclophane of this series, [2.2](1,4)naphthalenopara-cyclophane **(22)**, was first prepared in low yield from [2.2]paracyclophane **(19)** by Cram *et al.*[28] who used the annulation procedure shown in Scheme 2. Wasserman and Keehn[11] reported the synthesis of **22** by the dimerization of *p*-xylylenes formed *in situ* from the pyrolysis of the quaternary ammonium salts **23** and **24** (Scheme 3). However, on the basis of the high yields of [2.2]cyclophanes obtained from dithia[3.3]cyclophanes by the photochemical extrusion of sulfur in triethyl phosphite.[29] It would appear that the latter method, namely **25** to **22,** is the optimum route to **22.**[30] This photochemical reaction for the ring contraction of cyclophanes is essentially an application of a method developed by Corey and Block.[31]

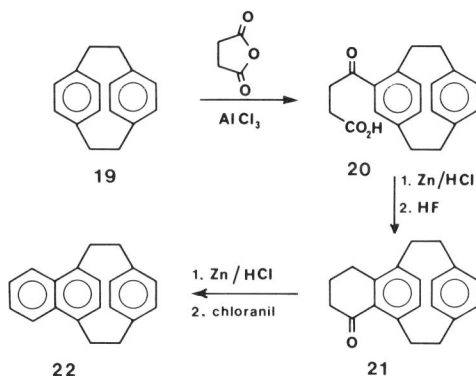

SCHEME 2

Electron spin resonance spectra of the radical-anions of **22a** and its octadeuterio derivative **22b** have been recorded.[30] This paper formed part of a larger study on radical-anions of [2.2]paracyclophanes.[32]

The naphthalenophane **22a** was subjected to photooxidation using a 150-W floodlamp, methylene blue sensitizer, and oxygen for 12 days. Two

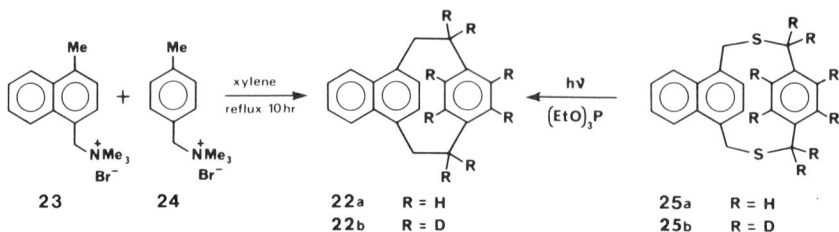

SCHEME 3

oxidation products (**6** and **26**) were obtained in a total of 15% yield, and the pathway shown in Scheme 4 was suggested as a likely route to these compounds.[11]

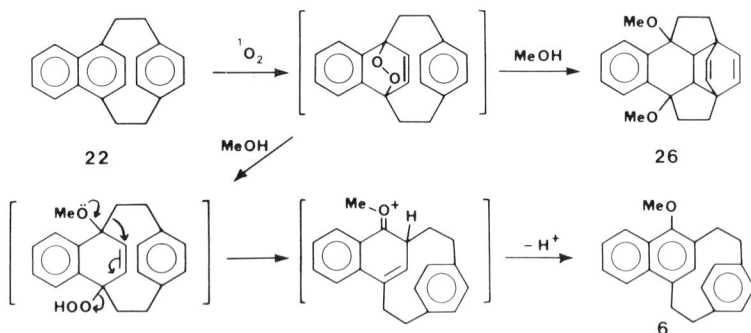

SCHEME 4

A considerable number of investigations have been made into the chemistry of the [2.2](1,4)naphthalenophanes. The anti isomer **29** was first reported in 1963 to be formed by a dimerization reaction of a benzo-1,4-xylylene intermediate (**28**) generated *in situ* from **27** with azeotropic distillation of water from a xylene solution.[28] The reaction conditions gave 1-hydroxymethyl-4-methylnaphthalene as the major product (95%), with only 3% of the desired compound (**29**, Scheme 5). The anti configuration of **29** was confirmed by an alternative nine-step synthesis[28] from [2.2]paracyclophane (**19**) using an annulation procedure similar to that shown in Scheme 2. This route gave an even lower yield (0.07%) of **29**. However, by its nature the multistep approach could have produced only the anti isomer, and [1]H-nmr spectra were consistent with the structure proposed.[28] With minor modifications the dimerization route shown in Scheme 5 could be used to produce good yields of both the anti and syn

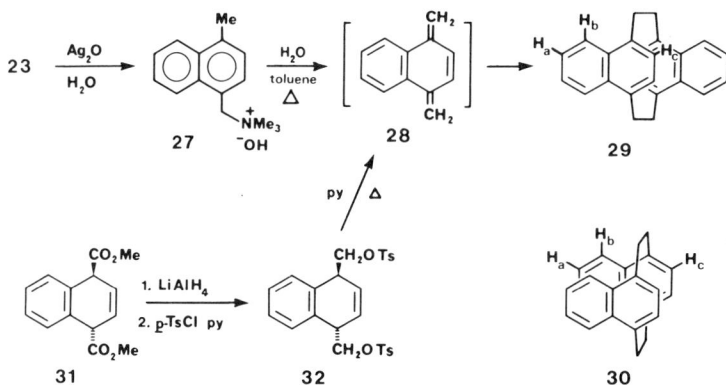

isomers **29** and **30**. Wasserman and Keehn[33] effected the decomposition of the quaternary bromide **23** at high dilution in xylene in the presence of phenothiazine. After filtration of the polymer, chromatography yielded 41% of the anti isomer **29** and 4% of the syn isomer **30**. It appears that the higher temperature allowed for formation of the syn isomer **30**; the overall improvement of yields may well have been due to both the lower concentration of water and the more dilute reaction conditions.

A highly efficient synthesis of [2.2](1,4)naphthalenophane **(29)** developed by Brown and Sondheimer,[34] which is a variation of Cram's dimerization procedure, is also shown in Scheme 5. The dihydronaphthalene **31** was reduced to the diol, which was converted to the corresponding ditosylate **32**; the latter, on solvolysis in refluxing pyridine, gave a 90% yield of the cyclophane **29**. The thermal dimerization of **28** to **29** is formally a $(6\pi + 6\pi)$ electrocyclic addition reaction and, as such, is not allowed under Woodward–Hoffmann selection rules. Consequently, the dimerization must involve a multistep process.

Another synthesis of the cyclophane **29** is shown in Scheme 6.[35] This was also a multistep approach based on the [2.2]paracyclophane tetracarboxylate **33** as the starting material. After reduction and bromination to form **34,** debromination with zinc gave a reactive bisdiene, which in the presence of dimethyl acetylenedicarboxylate formed **35**. Subsequent modification led to a low overall yield of **29**. Although the synthesis of **29** was not very efficient, the availability of **34** made it possible to synthesize other cyclophane products by the use of the appropriate dienophile.

The ^{1}H-nmr spectra of the anti and syn isomers **29** and **30** are distinct.[33] The H_a and H_b protons of **29** (Scheme 5) closely resemble those of 1,4-dimethylnaphthalene; however, H_a and H_b in **30** are more highly shielded

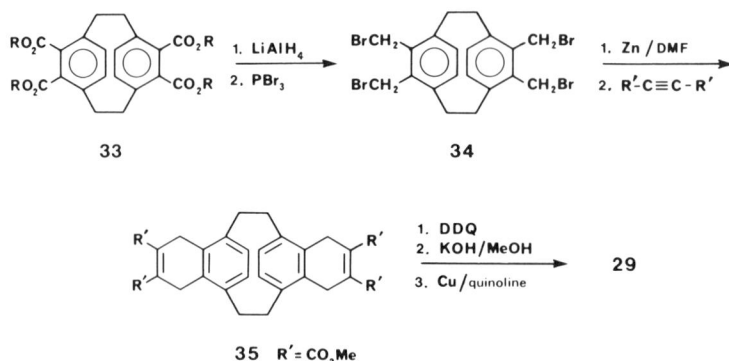

SCHEME 6

and show a shift upfield of 0.35 ppm. Similarly, the H_c protons in **30** show a reverse effect and are shifted 1.00 ppm downfield with respect to those in **29**. Confirmation of the syn structure was obtained by an independent X-ray crystallographic analysis.[36]

The [2.2](1,4)naphthalenophanes **29** and **30** undergo quite remarkable thermal and photochemical reactions. This is due to the strained nature of the bridged aromatic rings and the close proximity of the opposed naphthalene moieties. On being heated above its melting point (243–245°C), the syn isomer **30** resolidifies at 250°C and then remelts at 300–303°C, showing that complete conversion to the anti isomer **29** has occurred.[33] It was proposed that conversion of **30** to **29** occurred by cleavage of the bond between adjacent benzylic carbons to form a diradical **(36)**, which subsequently reclosed to produce the thermodynamically more stable anti isomer. A similar analogy has been found in the chemistry of [2.2]paracyclophanes.[37] The thermochemistry of both isomers **29** and **30** together with that of the photochemical product, dibenzoequinene **(38)**, has been reported.[38]

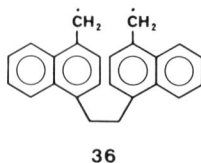

36

The photochemical reactions of cyclophanes **29** and **30** are complex and provide several interesting products, which are shown in Scheme 7. The syn and anti isomers can be interconverted photochemically. Irradiation of the syn isomer **30** in degassed benzene yields predominantly the anti isomer;[33] however, continued irradiation of solutions of the anti isomer **29** leads to the formation of other products.

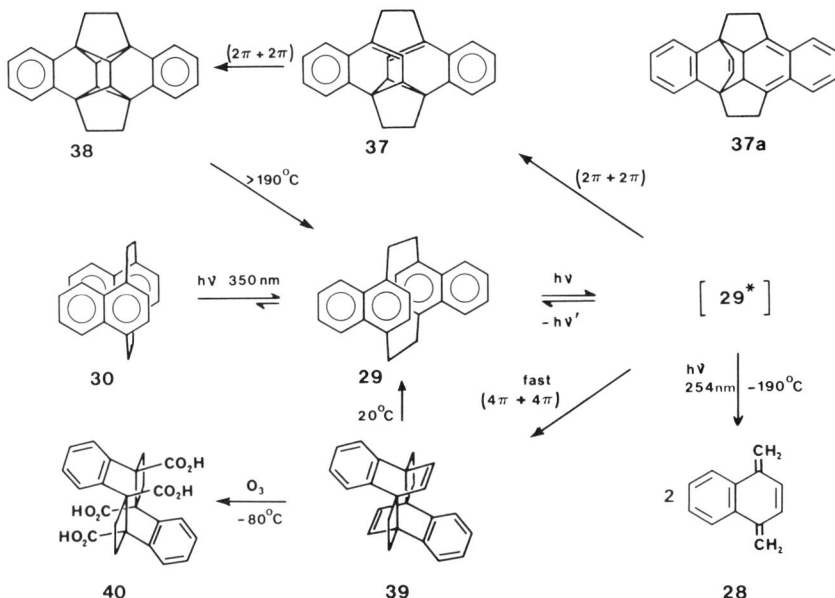

SCHEME 7

On irradiation in cyclohexane solution above 290 mm, **29** forms a thermally labile $(4\pi + 4\pi)$ intramolecular addition product **(39)**.[39] At room temperature, **39** rapidly rearomatizes ($t_{1/2} = 76$ sec at $20°C$). The addition product was characterized by ozonolysis at $-80°C$ with oxidative workup to the tetracarboxylic acid **40**. The $(4\pi + 4\pi)$ addition compound **39** may be regarded as the kinetic product of the reaction, because on prolonged irradiation at room temperature (10 days) the thermodynamically more stable product, dibenzoequinene **(38)**, forms (25–50%).[39,40] This remarkable compound **38** presumably arises from two consecutive $(2\pi + 2\pi)$ additions, the first leading to an intermediate **(37**, as yet uncharacterized) and the second producing the final product **(38)**. An alternative intermediate **(37a)** derived from a $(4\pi + 2\pi)$ addition reaction as proposed in the original paper[39] seems less attractive, because such a photochemical Diels–Alder reaction would have to proceed in a suprafacial–antarafacial sense. The structure for **38** was suggested by ^{1}H-nmr spectral data, which was confirmed by an X-ray crystallographic determination.[41] Irradiation of solutions of **29** in diethyl ether–ethanol at $-190°C$ produces the $(6\pi + 6\pi)$ cleavage product benzo-1,4-xylylene **(28)**, which has enough stability at room temperature (~30% disappearance in 20 hr at $25°C$) to make it a potentially useful synthon.[42] Similar $(6\pi + 6\pi)$ cleavage reactions were reported with [2.2](2,5) furanophane and [2.2]paracyclophane, which produced 2,5-dimethylene-2,5-dihydrofuran and 1,4-xylylene, respectively.[42] It is of interest that these photochemical cleavage reactions must be effected in media that are still "fluid" at $-196°$. Cleavage products formed in a hard glass at low temperatures are unable to diffuse apart.

The strained nature of the naphthalene rings in **29** is reflected by its ready photosensitized autoxidation to the novel polycyclic product **41** (Eq. 1)[43]:

$$\text{(1)}$$

29   41

Irradiation of a dilute solution of **29** in methanol in the presence of a sensitizer and air for 10 days produced a crystalline oxidation product in 20% yield. Spectroscopic and chemical data suggested that this product was the internal Diels–Alder adduct **41,** and this finding was confirmed by single-crystal X-ray analysis.[41] Presumably, the formation of **41** occurs by an initial reaction of **29** with singlet oxygen to produce the transannular peroxide **42.** Naphthalene and benzene do not normally show much reactivity toward singlet oxygen. In the case of **32** the greater reactivity of the dienoid system is due to the strain inherent in the bridged naphthalene ring. A second internal Diels–Alder reaction would convert **42** to **43,** which under solvolysis by methanol produces **41.**

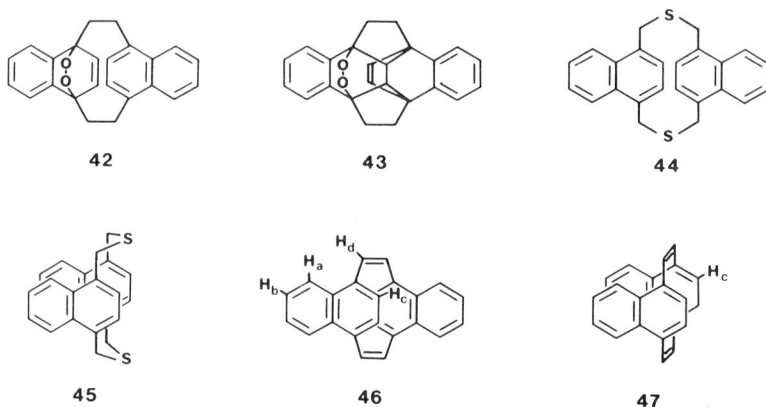

42   43   44

45   46   47

The anti and syn isomers of [2.2](1,4)naphthalenophane-1,13-diene (**46** and **47**) were prepared by Otsubo and Boekelheide.[44] A coupling reaction produced the anti- and syn-[3.3] isomers **44** and **45** in a 5:1 ratio (64% total yield), and these were separated by chromatography. Ring contraction of dithiacyclophanes **44** and **45** with benzyne produced [2.2]cyclophanes, which were oxidized to the corresponding sulfoxides, and these in turn were pyrolyzed to give the dienes **46** and **47** in low yields (4 and 1%, respectively). The ^{1}H-nmr spectra supported the structures of the dienes. In particular, the H_c proton of the anti isomer **46** showed an upfield shift of 1 ppm by comparison with that in **47**, and the vinyl protons (H_d) of both dienes **46** and **47** were observed at δ 7.45 and 7.64. The deshielding effect of the naphthalene ring on the vinyl protons is even greater than that in [2.2]paracyclophane-1,9-diene[45] and [2.2.2](1,3,5)cyclophane-1,9,17-triene.[46]

A number of hydroxy derivatives of the [2.2](1,4)naphthalenophanes have been reported by Staab[47] in connection with a study of charge-transfer interactions in cyclophanes. The anti and syn isomers **49** and **50** were produced in a 7 : 1 ratio (26% total yield) by pyrolysis of the quaternary ammonium hydroxide **48** in xylene. The mixture of isomers **49** and **50** was separated by chromatography, and the ^{1}H-nmr data were in accord with those of the parent cyclophanes. Subsequent demethylation and oxidation produced the corresponding naphthoquinophanes **51** and **52** without isomerization occurring. However, it is interesting that at a higher temperature (230°C, 1 hr, argon) **52** was converted to **51** to give a 1 : 1 mixture of the compounds.

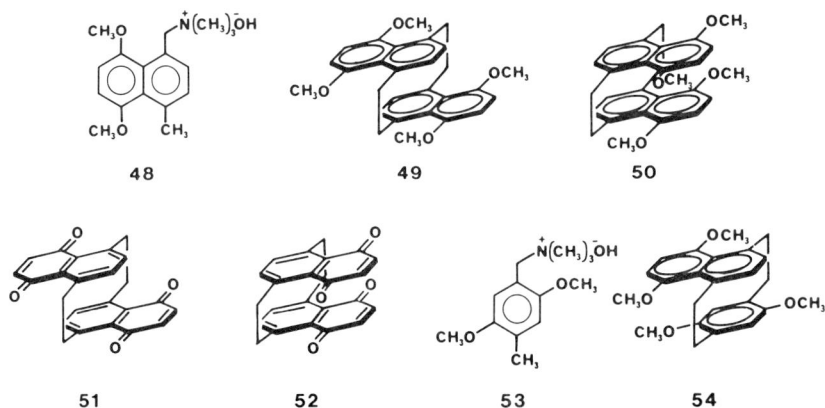

In a similar synthetic sequence [2](1,4)naphthaleno[2]paracyclophane **(54)** was prepared in 5 to 10% yield as a crossed product from the pyrolysis of **48** and **53** in boiling toluene.[48] This reaction also gave the expected stereopairs **49** and **50** and those of the paracyclophane derived solely from **53**. Staab reported the electronic spectra of three quinones derived from **54** together with those of **51** and **52**.[47,48]

Another example of the (6π + 6π) addition of mixed p-xylylenes to produce [2.2]cyclophanes is shown in Scheme 8.[49] In this case quite good

yields were obtained for the four possible products. The [2.2](1,4)naph-
thalenofuranophane **57** was isolated only as the anti form; none of the syn
isomer was observed.

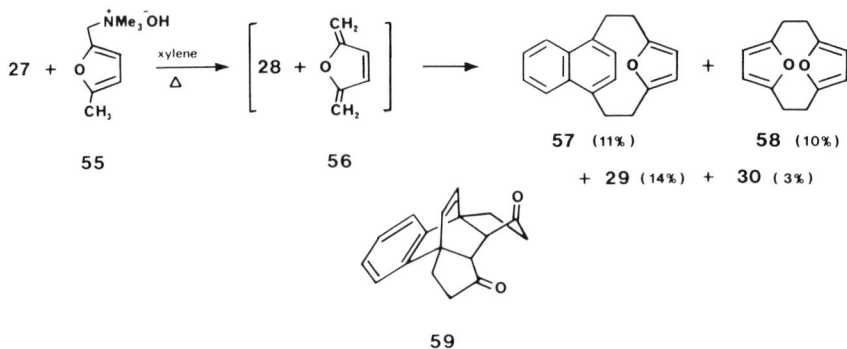

57 (11%)   **58** (10%)

+ **29** (14%) + **30** (3%)

59

SCHEME 8

The naphthalenofuranophane **57** undergoes a rapid oxidation with bromine to produce the diketone **59**.[49] The conversion, in part, involves an
intramolecular Diels–Alder addition. [2.2](2,5)Furanophane (**58**) undergoes a facile Diels–Alder addition with dimethyl acetylenedicarboxylate[50]
to produce **60**. In a similar fashion naphthalenofuranophane **57** gave the
adduct **62**, which is the result of two consecutive $(4\pi + 2\pi)$ additions,[51]
the first addition producing a proposed intermediate compound (**61**). Although several modes of intramolecular addition are possible in the intermediate **61**, only one appears to occur, namely, the addition between the
unsubstituted aromatic ring and the more reactive dienophile (the dimethyl maleate residue).

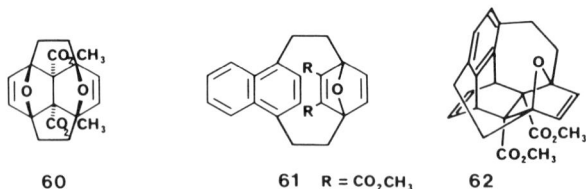

60   **61** R = CO$_2$CH$_3$ **62**

C. Other [2.2 . . .]Cyclophanes from (1,4)Naphthalenes

Boekhelheide[52] reported the synthesis of the cyclophanetriene **64** by the
standard ring contraction methods from the [3.3]dithiacyclophane precur-

sor **63**. The dihydro derivative of **64** has also been reported.[53] Photoclosure of the triene **64** was affected at −70°C in degassed perdeuteriotetrahydrofuran to produce, in high yield, a mixture of the isomers **65** and **66**.

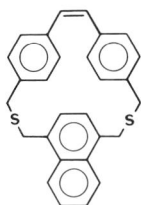 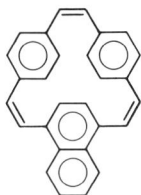 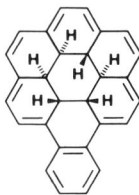 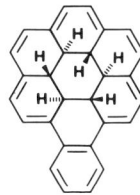

  63   **64**   **65**   **66**

   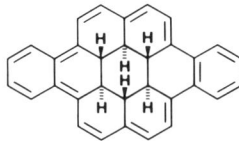

      67   **68**

A similar sequence of reactions led to the diene **67**, and photoclosure produced the dibenzohexahydrocoronene **68** as a mixture of isomers, one of which is shown here. As expected, the internal protons of **65/66** and **68** all show strong upfield shifts ($\delta \simeq -1$ to -5), and the peripheral aromatic protons are observed at low fields due to deshielding ($\delta \simeq 7.4-9.5$). In line with results reported by Mitchell[20–22] on the dihydropyrenes, benzannulation leads to increased ring current within these molecules, and the chemical shift effects are greater in **68** than in **65/66**. The work of both Boekelheide and Mitchell illustrates the usefulness of cyclophanedienes and cyclophanetrienes as precursors of bridged annulenes and other aromatic compounds that would otherwise be formed only with difficulty.

The approach to the synthesis of condensed aromatic cyclophanes developed by Wennerström[54–56] can be illustrated by the preparation of the tetraene **71**, as shown in Scheme 9.[56] The method uses a fourfold Wittig reaction to produce a tetraene directly and, although the yields obtained using this general method are low (1–15%), the procedure has the advantage of being a one-step reaction carried out with readily available materials. Reaction between 1,4-benzenedicarboxaldehyde **(69)** and the bisphosphonium salt **70** gave the tetraene **71** in 1.5% yield. Reaction conditions were chosen to enhance the formation of double bonds with cis geometry, which is essential for macrocycle formation. Cyclophanes of the size of **71** are better regarded as macrocycles in that (*a*) there is fast rotation of the benzene and naphthalene rings on the nmr time scale, (*b*)

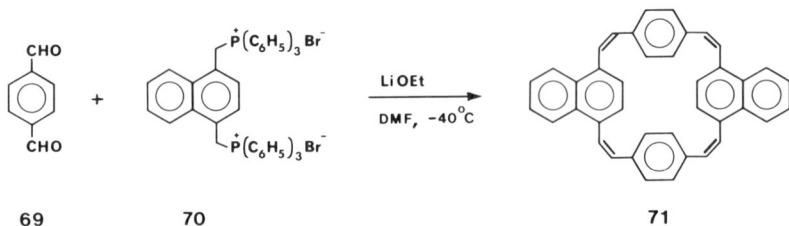

SCHEME 9

CPK models suggest that the central cavity may be large enough to accept host molecules, and (c) the molecule can attain a relatively planar conformation with extensive π-electron overlap, as evidenced by its uv spectrum.

D. [3.2]- and Higher (1,4)Naphthalene Cyclophanes

There are a number of reports concerned with the synthesis and chemistry of [3.2]- and [3.3](1,4)naphthalenophanes. Blank and Haenel[57] converted the dinaphthylpropane 72 to thia[3.3](1,4)naphthalenophane (73), which was obtained solely as the anti isomer. Oxidation to the sulfone 74 followed by pyrolysis at 500°C produced a 3:1 mixture (45% yield) of diastereomers 75 and 76, which were separated by chromatography. In a fashion similar to that of the simpler [2.2] compound 29, isomers 75 and 76 undergo photochemical $(4\pi + 4\pi)$ ring closure to the thermally labile products 77 and 78.[57]

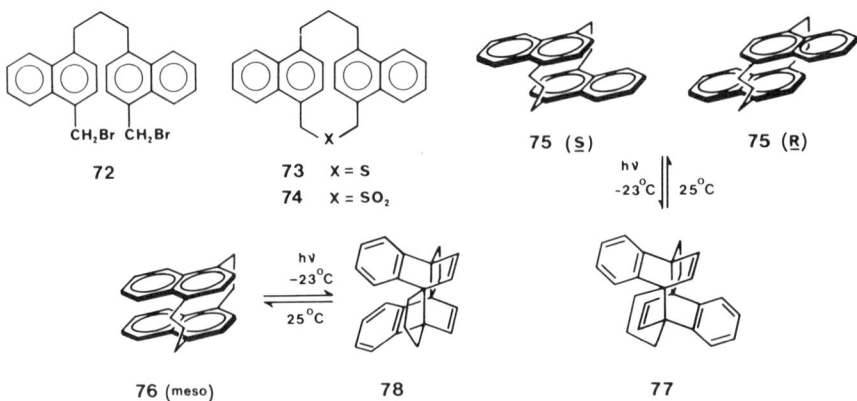

The *anti-* and *syn*-[3.3](1,4)naphthalenophanes **82** and **83** have been prepared.[58,59] Inazu[58] employed a malonic ester method to prepare the syn and anti isomers of **79,** which were reduced to the desired cyclophanes in a three-step sequence. Misumi's approach via the dithia[4.4]cyclophane **80** and pyrolysis of the sulfone **81** gave **82** and **83** (3 : 2) in a 30% yield.[59] Earlier papers[60,61] summarized the scope of both the pyrolytic method and direct photodesulfurization of dithiacyclophanes such as **80** to produce [3.3]cyclophanes. The electronic absorption spectra of **82** and **83** were similar to those of the corresponding anti and syn isomers of [2.2](1,4)naphthalenophane (**29** and **30**), indicating that the spectra depend on the stacking mode of the two naphthalene rings. Both compounds **82** and **83** in particular show long-wavelength absorption maxima at 325 and 330 nm, respectively, which could be ascribed to transannular π-electronic interactions.

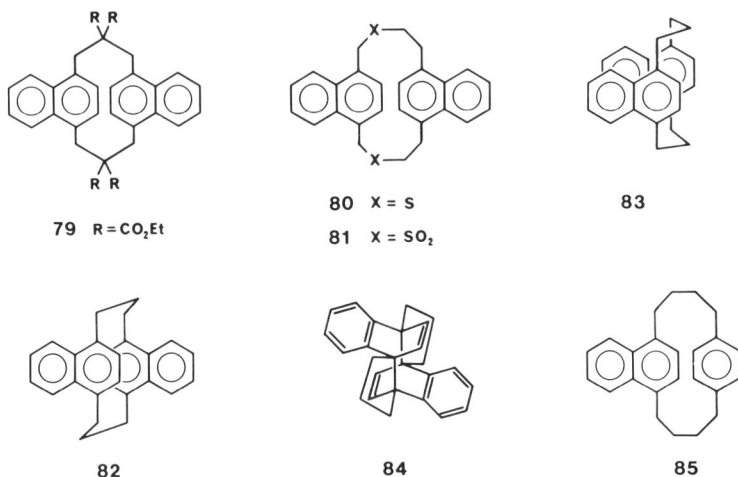

79 R = CO₂Et

80 X = S
81 X = SO₂

83

82   84   85

On irradiation of a solution of **82,** a thermally stable (4π + 4π) intramolecular addition product **(84)** was obtained.[58,59] Similar results were found for **83.**[58]

The [4.4](1,4)naphthalenoparacyclophane **85** was synthesized by a Friedel–Crafts succinoylation, reduction, and dehydrogenation procedure commencing with [4.4]paracyclophane.[62]

A series of [n](1,4)naphthalenophanes (**87:** n = 8, 9, 10, and 14) was prepared as shown in Scheme 10.[63] Acid-catalyzed ring opening of the furanophane **57** produced **86,** and reduction of **86** gave **87** (n = 8). Diazomethane homologation of **86** led to the [9]- and [10]cyclophanes, and the alternative and well-documented[4,64] acyloin condensation of **88** gave moderate yields of the [10]- and [14]cyclophanes.

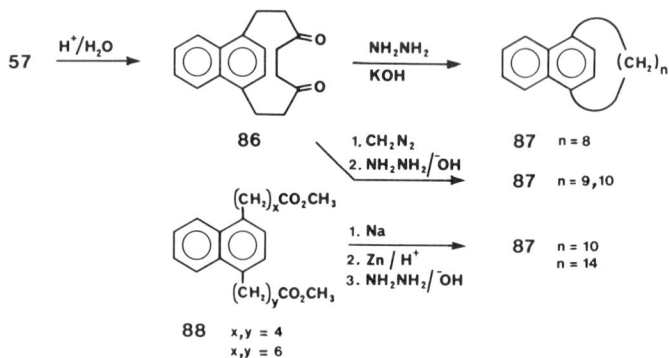

SCHEME 10

The ^{1}H-nmr data for cyclophanes **87** showed that only the [14]cyclophane **87** ($n = 14$) was conformationally mobile on the nmr time scale at ambient temperature. An earlier paper[65] had reported the conformational mobility of the bridging chain in **86** due to flipping of the central four carbon atoms of the chain above the immobile aromatic ring. These results were consistent with those obtained from a number of [8]paracyclophane derivatives.[66,67]

The cyclophanes **87** ($n = 8$, 9, or 10) reacted with dicyanoacetylene at 100°C to produce Diels–Alder addition products **(89)** of the unsubstituted ring. The [14]cyclophane reacted similarly, giving addition products to both rings, namely, **89** ($n = 14$) and **90**, the desired paddlane derivative, in 54 and 7% yield, respectively.[63]

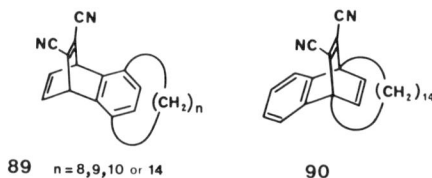

The [12](1,4)naphthalenophane **92** was prepared (as shown in Eq. 2[68]) by an extension of a method developed by Allen and Van Allan[69] from the pentadienone **91.**

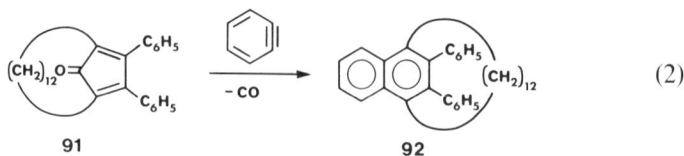

(2)

E. Heterocyclic (1,4)Naphthalene Cyclophanes

A number of heterocyclic [2.2]cyclophanes incorporating naphthalene residues have been examined. These include the furan, thiophene, and pyrrole derivatives **57, 93**, and **94**. The furan derivative **57**, synthesized by

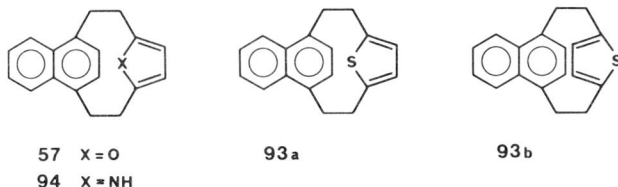

57 X = O
94 X = NH

93a

93b

the cross-breeding method shown in Scheme 8, was obtained[49] only in the anti form. Variable-temperature ^{1}H-nmr studies of **57** showed no significant changes,[70] and the anti conformation in the solid state was confirmed by X-ray crystallography.[71] By contrast, the thiophene analog **93**, prepared by a cross-breeding Hofmann elimination procedure,[72,73] was obtained as both the anti and syn isomers **93a** and **94b** ($\sim$15 : 1), which could be clearly differentiated by their ^{1}H-nmr spectra. Attempts to isomerize the syn form **94b** to the anti form **93a** thermally were unsuccessful, because decomposition occurred.

The furan cyclophane **57** proved to be a useful intermediate in the synthesis of the related pyrrole-derived cyclophanes,[74-76] because the latter compounds could not be easily synthesized by the Hofmann elimination route. Acid hydrolysis of **57** gave the crystalline diketone **86**, which condensed with ammonia (Paal–Knorr reaction) to form the [2.2]pyrrolo(1,4)naphthalenophane **95** as the anti isomer (Eq. 3).

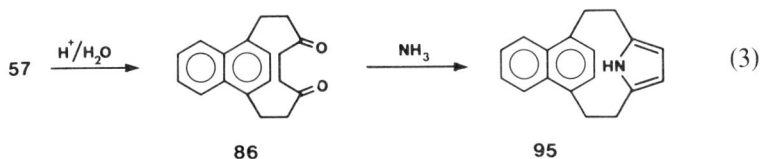

86   **95**

F. (1,5)- (2,6)-, and (2,7)Naphthalene Cyclophanes

The chemistry of the 1,3- and 1,4-disubstituted naphthalene cyclophanes in which a single ring is spanned can be appropriately compared directly with that of the simpler meta- and paracyclophanes derived

from benzene. However, in the 1,5-, 2,6-, and 2,7-disubstituted naphthalenophanes, bridging groups span both rings of the aromatic nucleus,
which opens up new chemical possibilities. Because much of the work that
has been reported on the 1,5, 2,6, and 2,7 series overlaps, all three groups
of compounds are considered here.

A coupling reaction between 1,5-bis(bromomethyl)naphthalene and 1,4-
bis(mercaptomethyl)benzene gave a good yield of the dithiacyclophane
96, and photochemical extrusion of sulfur in triethyl phosphite formed the
1,5-bridged naphthalene **97.**[77] In a similar way the [2.2](1,5)naphtha-
lenophanes, achiral **100** and chiral **101,** were prepared by photolysis of
either of the dithiacyclophane isomers **98** or **99.**[78] The structure assignment[78] was based on the 360-MHz ^{1}H-nmr spectra of **100** and **101** and an
X-ray crystal structure of **100,** which showed that the reverse assignment
originally proposed[79] was incorrect.

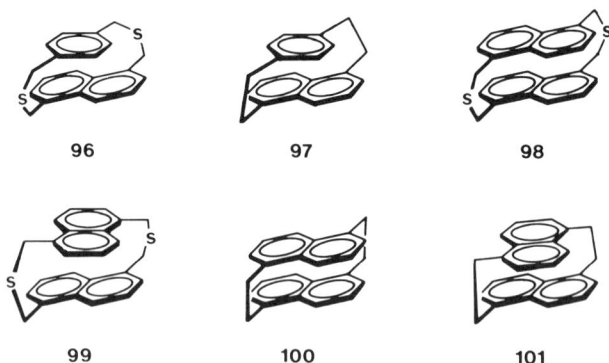

96   97   98

99   100   101

The conformationally mobile pyridine derivative from a 1,5-disubstitu-
ted naphthalene, cyclophane **104,** was prepared by pyrolysis of the
[3.3]disulfone **103** obtained by oxidation of **102.** A ring contraction procedure converted **102** to the diene **105,** which, from ^{1}H-nmr data, appears to
have the orthogonal structure shown.[80]

Among a series of macrocyclic polyethers reported by Sousa *et al.* are
the interesting "rope-skipping" variety, of which cyclophane **106** is an
example.[81,82]

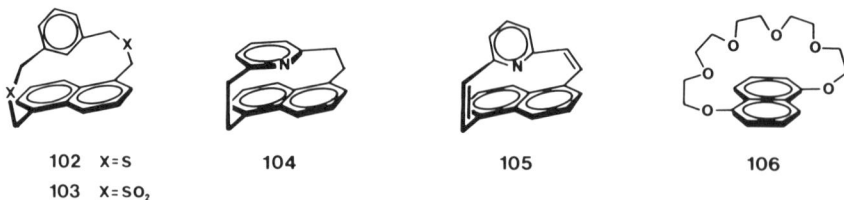

102   X = S   104   105   106
103   X = SO$_2$

Haenel and Staab reported the synthesis of the chiral compounds [2.2](2,6)naphthalenophane **(108)**[83,84] and its diene **(109)**[84] obtained by standard ring contraction and elimination reactions from the [3.3]disulfide **107.** No evidence was obtained for the formation of the achiral isomers **110** and **111.** Photoextrusion of SO_2 from the disulfone corresponding to **107** also produces **108.**[85] The diene **109** could be catalytically hydrogenated to **108,** and resolution of racemic **108** was effected with Newman's reagent (TAPA).[86] The S chirality was proposed for the levorotatory enantiomer $(-)$-**108,** which is shown here.

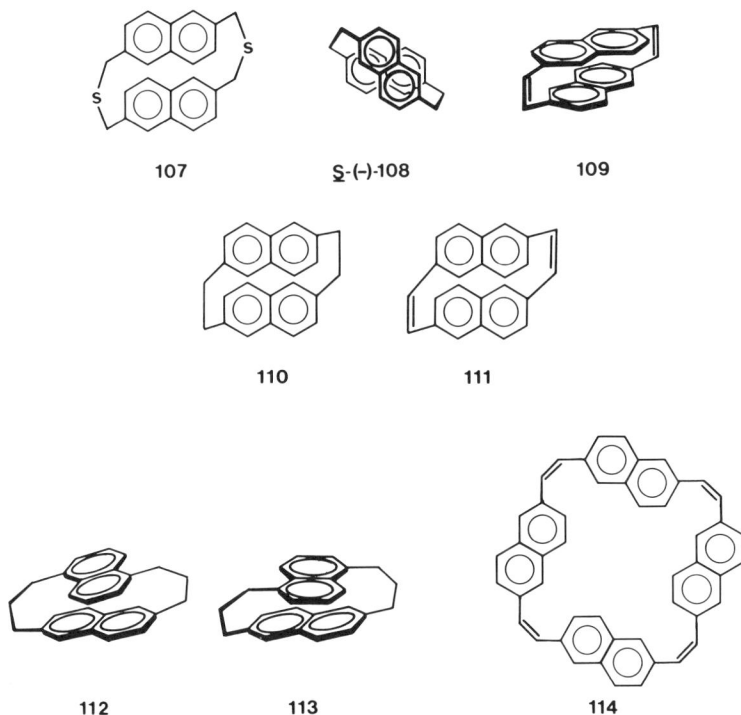

107   S-(–)-108   109

110   111

112   113   114

A number of [3.3](2,6)naphthalenophanes, including [3.3](2,6)naphthalenoparacyclophane,[60,61] a diastereomeric mixture of [3.3](2,6)naphthalenophanes,[87,88] and the [3.3](1,5)(2,6)naphthalenophanes **112** and **113,** have been prepared,[87] all via ring contraction methods from dithiacyclophane precursors. A series of macrocyclic [2.2.2.2]cyclophanes with saturated and unsaturated bridges (e.g., **114**) prepared from 2,6-disubstituted naphthalenes using Wittig reactions has also been reported.[55] Kemp[89] reported macrocyclic [9.6]- and [6.6](2,6)naphthalenophanes for use in encapsulation studies and selective affinity for smaller aromatic species.[89] Whitlock *et al.*[90,91] prepared similar macrocyclic [8.8](1,4)- and [8.8](2,6)naphthalenophanes having diyne bridging functions.

The [2.2](2,6,2',7')naphthalenophane **115**[92,93] and the corresponding diene **116**[93] were both prepared by ring contraction methods, the latter compound in particular by pyrolysis of a disulfoxide. In an earlier attempt[92] to prepare compound **116** by a base-induced elimination of dimethyl sulfide from **117,** a rearrangement apparently occurred to give a low yield of the 2,7 isomer **2** instead.[93] Both compounds **115** and **116** are conformationally mobile and show temperature-dependent ^{1}H-nmr spectra.

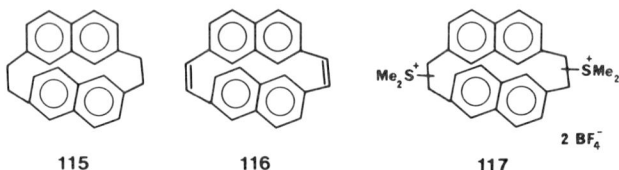

115   **116**   **117**

The nonplanar polycyclic aromatic compound corannulene (**118,** also known as 5-circulene) has inherently interesting chemical and physical properties and is one of a family of cyclic aromatic structures.[94] The only recorded synthesis consists of approximately 17 discrete steps starting from acenaphthene.[95,96] Two other unsuccessful approaches have been recorded, one from a fluoranthenediacetic acid[97] and the other from the [2.2](2,7)naphthalenoparacyclophane-1,11-dienes **119** and **120.**[98,99] Both **119** and **120** and the saturated analog **121** were prepared from dithia[3.3]cyclophane precursors. However, none of these three [2.2]cyclophanes could be dehydrogenated to a corannulene derivative by either photochemical or chemical methods.[98,99] It is possible that the product, if formed, may not have survived the reaction conditions. The ^{1}H-nmr spectra suggested that these naphthalenoparacyclophanes have a stepped geometry, as shown in structure **121.**

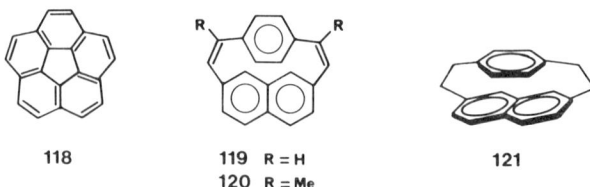

118   **119** R = H   **121**
                     120 R = Me

[2.2](2,7)Naphthalenophane (**5**) was synthesized by a sodium- or phenyllithium-induced benzylic coupling of 2,7-bis(bromomethyl)naphthalene.[6,7] Not surprisingly, **5** on reaction with $AlCl_3/CS_2$ followed by dehydrogenation over palladium yielded coronene (**3**).[6,7] Subsequent work by Sato[100] showed that the benzylic coupling reaction can produce both the anti and syn isomers of **5.** The ^{1}H-nmr spectra were in accord with the assigned structures, and ^{13}C-nmr and uv spectra were also determined.

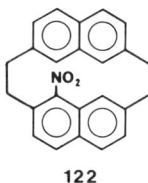

122

Griffin[101] observed the remarkable nitration reaction of compound **5** to produce the derivative **122** as the major product (55%), in complete contrast to the results observed for [2.2]metacyclophane and related compounds. This intraannular reaction appears to be unique in cyclophane chemistry and deserves further investigation.

An early paper[2] reported an unsuccessful attempt to prepare diene **2,** considered to be a good precursor of coronene **(3).** Diene **2** was subsequently synthesized[3,102,103] from a [3.3]dithiacyclophane[104] and was shown to undergo efficient oxidative photochemical ring closure to produce **3.**[3,102] Attempts to identify the anticipated tetrahydrocoronene intermediate **123** were unsuccessful.[52,105] Spectral data again suggested that the diene **2** had an anti-stepped conformation. Electron spin resonance spectra of the radical-anions of cyclophanes **2** and **5** and a dideuterated derivative of **5** were determined,[106] and evidence was obtained for the intermediacy of anions of type **124** (a partially ring-closed product toward coronene, the final compound obtained from the reaction of **2** with potassium). The dithia[2.2]- and tetrathia[2.2]cyclophanes **125**[104] and **126**[107] were both obtained as anti-stepped forms by straightforward preparative methods.

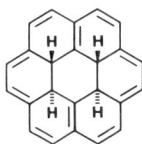 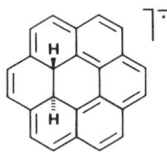 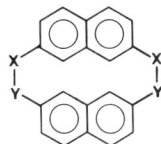

123   **124**   **125** X = CH$_2$; Y = S

                                            126 X = Y = S

1,16-Didehydrohexahelicene (or hexa[7]circulene, **128**) was prepared in high yield by oxidative photochemical ring closure of the biphenylnaphthalenophanediene **127,**[108,109] which was prepared from dithia[3.3]cyclophane precursors. Molecular models of the circulene **128** indicate that the molecule has a saddle-shaped geometry and consequently would be dissymmetric and could potentially be resolved into enantiomers.

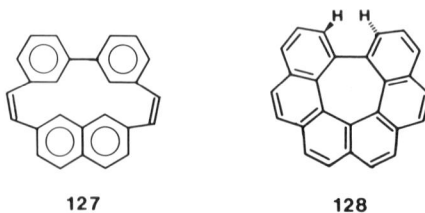

127   **128**

G. (1,7)- and (1,8)Naphthalene Cyclophanes

Several compounds derived from 1,7- and 1,8-disubstituted naphthalenes are worthy of mention. These include the cyclic alkynes **129**[110] and **130,**[111] used in studies of annulated annulenes and prepared by oxidative coupling of alkynes from 1,7-disubstituted naphthalenes. Bieber and Vögtle[112-115] prepared several cyclophanes of type **131,** which contain 1,8-disubstituted naphthalenes orthogonal to other aromatic residues, and the diyne **132** has also been made[116] in low yield. A series of macrocyclic 1,4,5,8-tetrasubstituted naphthalene cyclophanes has been prepared for studies of electron donor–acceptor effects.[117,118]

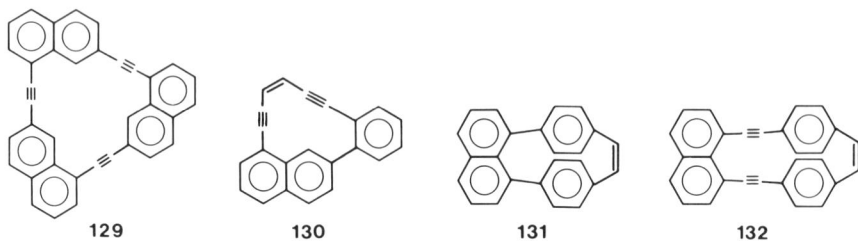

129   **130**   **131**   **132**

IV. ANTHRACENE-DERIVED CYCLOPHANES

Virtually all of the anthracenophanes reported to date have been derived from either 1,4- or 9,10-disubstituted anthracenes. Golden published the first paper on this class of compounds in 1961[119] and showed that the readily prepared [2.2](9,10)anthracenophane **134**[120] underwent a facile, photochemically induced cyclization to the cage compound **135** (Scheme

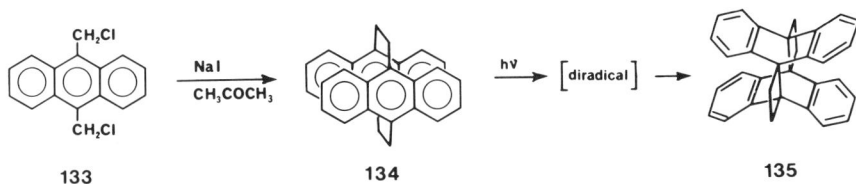

SCHEME 11

11), characteristic of addition reactions across the 9,10 positions of an-thracene. Subsequent investigations by Kaupp,[121,122] in which quantum yields were determined, established the existence of a diradical intermedi-ate in this photochemical reaction.

Many cyclophanes incorporating anthracene, which has a lower ioniza-tion potential than either naphthalene or benzene, have been prepared by Misumi[123] and co-workers in order to study the effect of transannular π-electron interactions in excimer fluorescence, photodimerization, and esr phenomena. The first compounds reported in this series were the anti and syn isomers of [2.2](1,4)anthracenophane (**136** and **137**), formed from the dimerization of 1,4-anthraquinodimethane, in turn derived by a Hofmann elimination procedure.[124] The syn isomer **137** underwent a rapid light-induced cyclization (to the cage structure **138**), which was reversible ei-ther thermally or photochemically. Isomer **137** could also be thermally isomerized to the anti form **136** (Scheme 12). X-ray crystallographic anal-ysis confirmed the structures of both **137**[125] and **138**.[123]

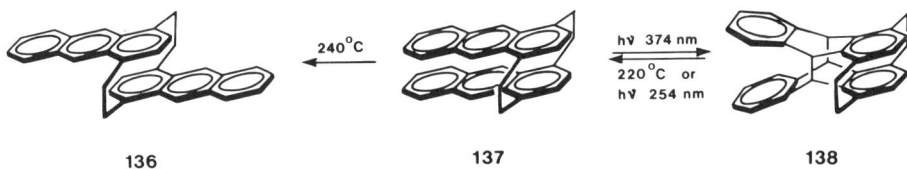

SCHEME 12

Cross-breeding reactions from the Hofmann degradation of quaternary ammonium hydroxides led to a number of related anthracenophanes,[126] including [2.2](1,4)(9,10)anthracenophane **(139)**,[127] *anti-* and *syn-*[2.2](1,4)naphthaleno(1,4)anthracenophanes,[127] [2.2]paracyclo(9,10)an-thracenophane **(140)**[128] (previously reported as an impure compound[28]), [2.2]paracyclo(1,4)anthracenophane,[128] and [2.2](1,4)naphthaleno(9,10)-anthracenophane **(141)**.[128] Dimethyl derivatives of compound **140** and [2.2]paracyclo(1,4)anthracenophane, together with the synthesis of a triple-layered 1,4-anthracenophane, have also been reported.[126] The electronic absorption spectra of these anthracene-derived cyclophanes showed bathochromic shifts of the long-wavelength band in comparison with the acyclic analogs, indicating considerable transannular π-electronic interaction.[126]

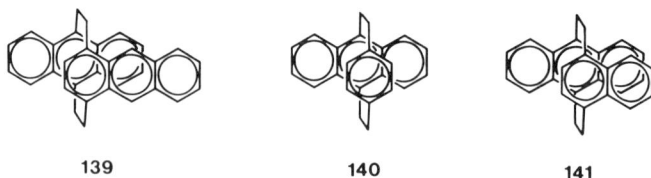

139   140   141

The esr spectrum of the radical-anion derived from 134 suggested that the unpaired electron was delocalized over both anthracene rings. However, the esr spectra of the radical-anions derived from some of the other cyclophanes such as 139 were very complicated.[129] A more recent paper reported ENDOR studies on the radical-anions of the anti and syn isomers 136 and 137, and it was found that the spin density in the radical-anion tends to accumulate on the overlapping part of the molecule.[130] Excimer fluorescence studies were reported on compounds 134, 136, 137, and 139 together with the appropriate acyclic analogs.[131]

A number of [2.2]heteroanthracenophanes have been reported. Wynberg and Helder[132] showed that [2.2](9,10)anthraceno(2,5)furanophane (142) could be prepared in 40% yield by a cross-breeding cycloaddition reaction from the pyrolysis of the appropriate ammonium hydroxides. Minor amounts of the two symmetric addition products, [2.2](2,5)-furanophane (58) and [2.2](9,10)anthracenophane (134), were also observed. Variable-temperature uv and nmr spectroscopy of 142 provided evidence that the aromatic rings are perpendicular to one an-

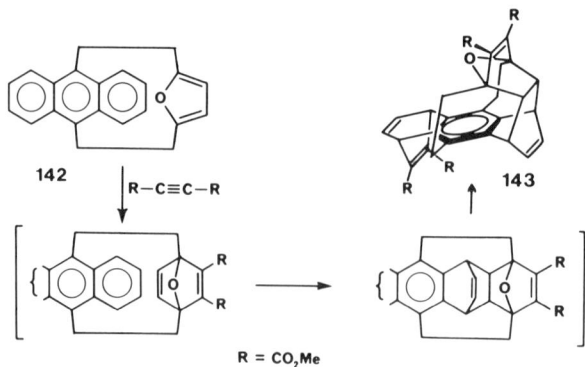

142   R–C≡C–R   143

R = CO₂Me

SCHEME 13

other at low temperature and probably also at room temperature.[70] Compound **142** reacted with oxygen and also with 2 mol of acetylenedicarboxylate to give an adduct of type **143,** which represents one of two possible structures. This adduct formation differed from previously reported reactions of [2.2](2,5)furanophane **(58)**[50] and [2.2](1,4)naphthalenofuranophane **(57)**[51] in that the initial addition of dimethyl acetylenedicarboxylate to the furan ring was followed by an internal Diels–Alder addition of the nonactivated double bond across the 1,4 position of the anthracene ring (Scheme 13).

The syntheses and electronic absorption and emission spectra of the *anti*-[2.2](1,4)anthracenophanes **144** and **145** as well as compound **146** were also reported.[72,73] It is of interest that the cross-breeding reaction used to make **144** and **145** led to none of the syn isomers.

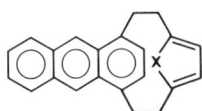
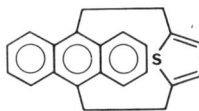

144 X = O
145 X = S

146

Other anthracene cyclophanes that have been investigated include the [4.4.4.4] compound **147,** prepared in low yield by a Wittig reaction[133]; the cyclic hydrazide **148,** which under oxidative conditions in aprotic media shows intramolecular sensitized chemiluminescence approaching the quantum yield of luminol[134,135]; [2.4]- and [2.5](9,10)anthracenophanes, the crystal structures of which have been determined[136]; and [3.3](1,4)naphthaleno(9,10)anthracenophane **149,** which forms the photoisomer **150** on exposure to light.[137] The related [3.3]paracyclo(9,10)anthracenophane undergoes a similar photochemical isomerization and also a thermally induced intramolecular Diels–Alder reaction.[137]

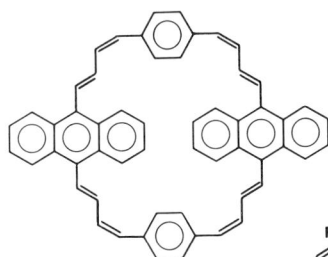

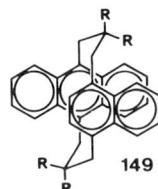
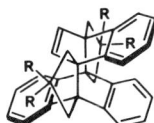
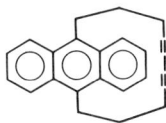
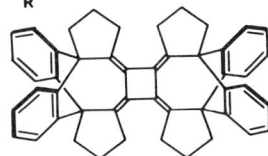

147

148

149

150

151

152

Misumi and co-workers showed that, on exposure to sunlight, [10](9,10)anthracenopha-4,6-diyne (151) in benzene produced the remarkable dimer 152 in quantitative yield[138] and that the intermediate was a highly strained cyclic butatriene.[139] Carbon-13 nmr data for 151 and related compounds have been recorded.[140,141] A photochromic crown ether incorporating anthracene has been synthesized by the same research group.[142] This [2.13]cyclophane (153) forms a 9,10 intramolecular adduct on exposure to light characteristic of cyclophanes similar to 134. The incorporation of residues such as anthracene into a macrocycle offers many possibilities for controlled and reversible modification of the shape of the macrocycle. Another recorded macrocyclic compound is [12](1.4)anthracenophane (154), which was prepared in a manner similar to that for compound 92 (see Eq. 2).[68]

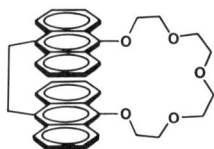 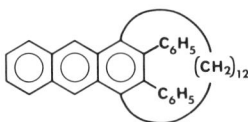 

153   154   155

There are few reports of cyclophanes from 1,8-disubstituted anthracenes. Vögtle *et al.* recorded the synthesis of 155 and the quinone derived from it[143] and a series of 1,8-bridged anthraquinones, including a polyether derivative.[144] A [2.2](1,8)anthracenophanediene has been synthesized by a Wittig reaction, but this compound would be better regarded as an annulene.[145]

A number of research groups have reported their results of esr spectroscopy of radical species from [2.2](9,10)anthracenophane (134).[32,146,147] Together with the studies on the naphthalene cyclophanes already considered[30,32,106] and that of Eargle[148] on the [2.2](1,4)naphthalenophane radical-anion, the papers by Gerson and co-workers[147] and Pearson *et al.*[146] provide information on radical-anions (and radical-cations) of molecules having more than one reducible aromatic ring and present a systematic study of the π-electron density in such polycyclic aromatic species.

V. PHENANTHRENE-DERIVED CYCLOPHANES

Potter and Sutherland[149] showed that a series of [*n*](3,6)phenanthrenophanes (157) could be synthesized efficiently by the oxidative photochemical ring closure[150,151] of a precursor bridged stilbene **(156)**. This ring closure to form the phenanthrene moiety is a good illustration of an approach that has been used a number of times in the cyclophane chemistry of phenanthrene. Apart from a [12](4,5)phenanthrenophane prepared by an acyloin condensation,[152] the other recorded alkenyl-bridged 3,6-phenanthrenes are better regarded as annulene derivatives[10,153–155] and are not discussed here.

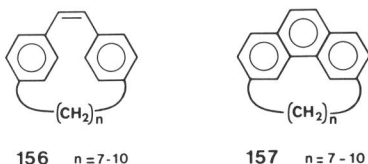

156 n = 7-10   **157** n = 7 - 10

Much of the research on cyclophanes incorporating phenanthrene rings has been done with the aim of producing suitable precursors for circulenes, heterocirculenes, and bridged annulenes. Consequently, many papers could be discussed equally thoroughly from the viewpoint of aromaticity, synthetic strategy, or spectroscopic properties. This section presents the cyclophane chemistry of phenanthrene in order of increasing structural complexity, followed by a discussed of related cyclophanes from dibenzofurans, dibenzothiophenes, and fluorenes.

Photochemical extrusion of sulfur from dithiacyclophane **(158)** gave [2.2](3,6)phenanthroparacyclophane **(161)**, which was dehydrogenated over $AlCl_3$ to coronene **(3)**.[156] Ring contraction of the sulfonium salt **(159)** derived from **148** led to poor yields of the desired compound **162** (Stevens rearrangement conditions), and it was proposed, from the nature of the polymer product obtained, that a competing 1,6-elimination, through the para-disubstituted benzene ring to form a reactive p-xylylene intermediate **(164)**, had occurred.[157] This competing side reaction could be minimized by the use of the dimethylcyclophane **160**, in which the kinetic and thermodynamic acidity of the proton adjacent to the benzene moiety had been reduced with respect to that of the benzylic proton adjacent to the phenanthrene. Ring contraction of **160** gave good yields of the [2.2]cyclophane **163**.[157] Similar observations were made in other paracyclophane systems.[98,99,158]

The diene **165** was made[157] by subjecting **158** to a Wittig rearrangement,[17] oxidizing the resulting product to a disulfoxide, and effecting a thermal elimination of methylsulfenic acid. An alternative route to diene **165**[159] employed the benzyne–Stevens rearrangement procedure[45] and commenced with compound **158**. A degassed solution of diene **165** on exposure to light changed color to deep orange and gave an ¹H-nmr spectrum indicative of the bridged annulene **166** (the shielded internal protons were observed at δ −2.74 to −3.02). In the presence of oxygen[159] or

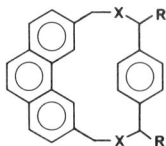

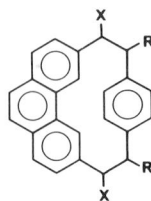

158 X = S; R = H   **161**   **162** X = SMe; R = H
159 X = S⁺Me ⁻BF₄; R = H   **163** X = S Me; R = Me
160 X = S⁺Me ⁻BF₄; R = Me

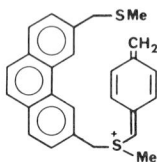

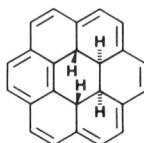

    164   **165**   **166**   **167**

iodine,[157] oxidation proceeded to produce coronene (**3**) quantitatively. Boekelheide and co-workers reported[159] the synthesis of several other related cyclophanedienes, including the dihydrophenanthrene cyclophanediene **167**,[160] all of which cyclized in a manner similar to the cyclization of bridged annulenes. The synthesis of the dibenzothiophene and biphenyl analogs is discussed in Section VII. This work formed part of a study to determine diamagnetic ring currents in a series of bridged annulenes.

The phenanthrene cyclophanedienes **168** and **169** and the corresponding tetrahydro derivatives were prepared by the usual ring contraction methods from dithia[3.3]cyclophane precursors.[161,162] The ¹H-nmr data for both compounds suggested that they had the syn-configuration, as shown in structures **168** and **169**. Attempts to photocyclize the dienes oxidatively to the respective polycyclic aromatic compounds **171** and **172** failed, and this may have been due partly to the stereochemistry of the dienes. Similar observations were made with [2.2](3,6)phenanthranophanediene (**170**),[163] prepared by a Wittig reaction–aryl iodide photolysis sequence, which also failed to ring close to the unrecorded [8]circulene **173**.[164]

Haenel and Staab[83,165] reported the synthesis of [2.2](2,7)phenanthrenophane (**174**) as a mixture of isomers, together with the corresponding compound prepared from a 9,10-dihydrophenanthrene precursor and mixed phenanthrene–biphenyl compounds of the type shown in structure **175**. The synthetic route included pyrolysis of the disulfones prepared from dithia[3.3]cyclophanes.

168

169

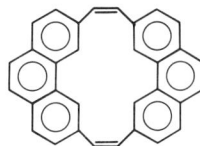

170

171

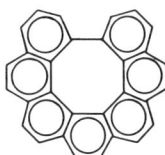

172

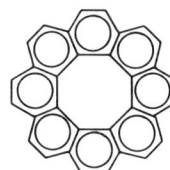

173

174

175

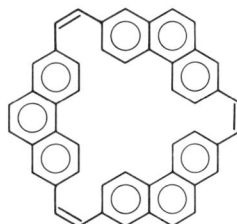

176

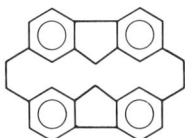

177

178

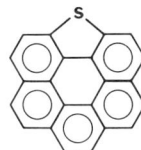

179

The [2.2.2](2,7)phenanthrenophanetriene **176** was prepared by allylic bromination and debromination of a 9,10-dihydrophenanthrene analog, in turn made by a Wurtz coupling reaction. A second product of the coupling reaction was a macrocyclic [2⁵](2,7)phenanthrenophane.[166–168]

Cyclophanes derived from fluorenes, fluorenones, and dibenzothiophenes are briefly mentioned here because of their obvious structural similarity to phenanthrene. Haenel prepared a series of anti and syn isomers from 2,7-disubstituted fluorenes, 9-fluorenones, and 9,9-dideuteriofluorene, for example, compound **177,** using disulfide precursors.[169] On reaction with *n*-butyllithium, [2.2](2,7)fluorenophane **(177)** produced dianions that could be alkylated with methyl iodide at the 9 position of the fluorenyl ring.[170] The cyclophanediene **178,** from dibenzothiophene, photocyclized under anaerobic conditions to a tetrahydrothiacoronene (a bridged annulene) as discussed previously, and on exposure to oxygen it formed thiacornonene **(179).**[159,171] A trisdibenzothiophene cyclophane synthesized by Staab *et al.*[172] is better regarded as a hexa-*m*-phenylene derivative.

VI. CYCLOPHANES FROM POLYCONDENSED AROMATIC COMPOUNDS

A. Pyrenes

[2.2](2,7)Pyrenophane **(183)** was first prepared by Misumi *et al.*[173] in 1975, and this report was followed almost simultaneously by accounts of the preparation of the same compound by independent groups in Victoria, Canada[174] and Heidelberg.[175,176] The synthetic routes to the pyrene moieties were invariably based on transannular reactions of metacyclophanes[177]; for example, [2.2]metacyclophane **(180)** on reaction with bromine, with a small amount of iron powder, gave the tetrahydropyrene derivative **181.** Oxidation with DDQ at a later stage in the synthesis produced the pyrene ring.[173] Standard reactions led to the formation of dithia[3.3]cyclophanes such as **182,** which could be ring-contracted to [2.2](2,7)pyrenophane **(183)** or the corresponding 1,13-diene.[173] Crystal structures have been determined for both compound **183**[176] and its diene.[178] Absorption, fluorescence, and phosphorescence spectra of compound **183,** together with the effects of solvent polarity on excimer formation, have been reported.[179–181] The esr and endor spectra of the radical-anion of **183** have been recorded.[182]

A number of isomers of pyrenophane **183** have been prepared from

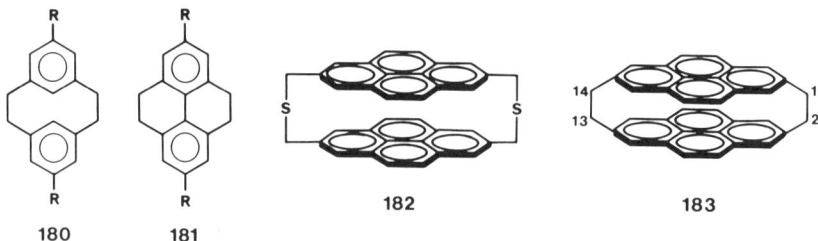

180   181   182   183

dithia intermediates for fluorescence studies,[123] including [2.2](1,6)py-
renophane, [2.2](1,6)(2,7)pyrenophane, [2.2](1,8)pyrenophane,[183] and
[2.2](1,3)pyrenophane **(184)**.[184] Related pyrenophanes incorporating
phenanthrene and naphthalene residues include [2.2](2,7)phenanthro-
(2,7)pyrenophane and [2.2](2,6)naphthaleno(1,6)pyrenophane **185,**[185]
the latter structure representing one of the two stereoisomers constituting
the mixture obtained. Two other interesting cyclophanes prepared in Misumi's laboratories and derived from stepped metacyclophanes are compounds **186** and **187.**[186] A number of other similar stepped cyclophanes
have been made, and these are discussed in Chapter 10.

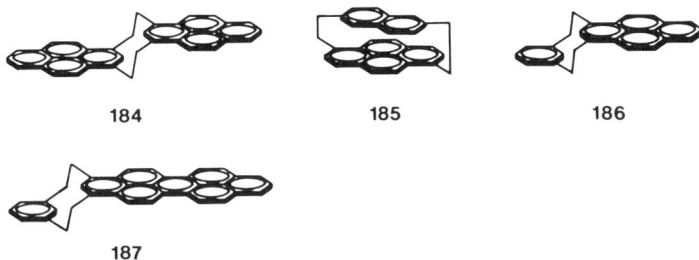

184   185   186

187

B. Benzannulated Anthracenes and Phenanthrenes

Diederich and Staab prepared [2.2](3,11)dibenzo[*aj*]anthracenophane
(189) in a series of reactions from 5,6,8,9-tetrahydrodibenzo[*aj*]anthra-
cene **(188),** again proceeding through dithiacyclophane intermediates.[187]
The 1,17-diene corresponding to **189** was resistant to photocyclodehydro-
genation. However, the octahydroderivative **190** could be easily converted to kekulene **(191)** by a short irradiation in the presence of iodine
followed by dehydrogenation of the product with DDQ.[187,188] The early
attempts[167,168,189] to prepare kekulene were frustrated to a large extent by
the restricted synthetic pathways available at that time. The subsequent
development of the valuable dithiacyclophane intermediates and the
many associated reaction pathways made it possible to complete the synthesis of **191.**

Peter and Jenny[190] reported the preparation of [2.2](3,10)benzo[*c*]-
phenanthrophane **(192)** by a Wurtz coupling reaction from a disubstituted benzo[*c*]phenanthrene. It would appear that **192** was prepared as a
precursor of a polycyclic aromatic structure of the kekulene variety.

188   189   190

191   192

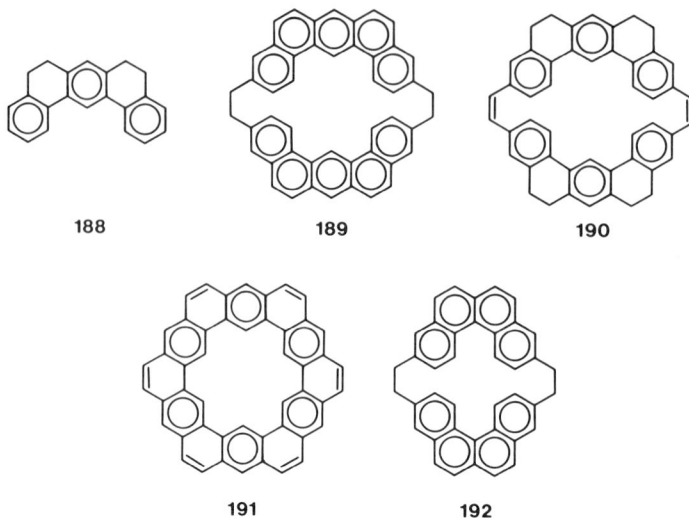

C. Cyclophanes from Helicenes

Propellicene, or bis-2,13-pentahelicenylene **(194)**, a compound that has the geometry of a two-bladed propeller, was made by the photochemical cyclization of the cyclic tetraene **193**.[192,193] Presumably, an intermediate compound in this conversion would have been a phenanthrenophane. Other bridged helicenes that have been reported are the paracyclophano derivative **195**,[194,195] whose absolute configuration as shown was determined to be $(-)$-(M)-**195** and was used to confirm an earlier assignment of

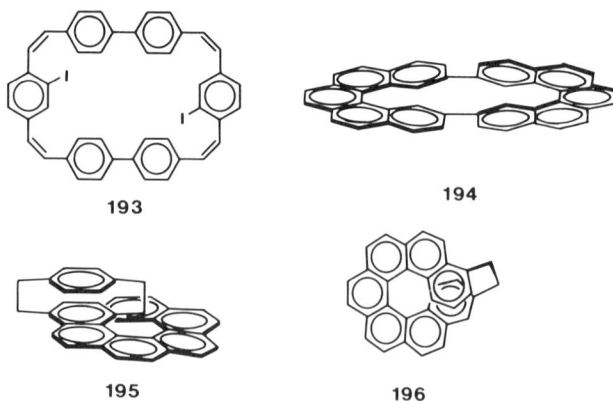

193   194

195   196

a paracyclophane derivative,[196] and also 3,15-ethano[7]helicene **(196)** and its 3,15-(2-oxapropano) derivative. The crystal structures of the latter two compounds were determined and compared with that of 3,15-dimethyl [7]helicene.[197]

VII. CYCLOPHANES FROM BIPHENYLS

Although the biphenyl residue does not strictly belong to the class of condensed benzenoid aromatic compounds, it is worth considering the cyclophanes derived from this ring system in the context of this chapter because many of the compounds are directly related to those derived from phenanthrene in terms of structure and chemistry.

The biphenylmetacyclophane **197** was obtained by pyrolysis of a precursor disulfone.[198] Ring contraction reactions led to the preparation of the biphenylparacyclophanes **198**[158,160] and **199,**[158] which on oxidative photocyclization produced the benzo[*ghi*]perylenes **200** and **201.**[158] The light-induced closure of a degassed solution of **198** at −80°C gave the expected bridged annulene **202.**[159,160]

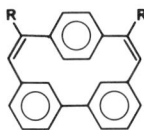

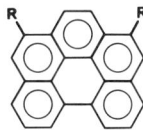

| **197** | **198** R = H | **200** R = H |
| | **199** R = Me | **201** R = Me |

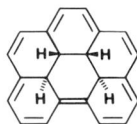

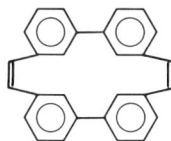

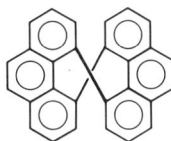

202   **203**   **204**

Two research groups reported the synthesis of the diene **203** (one approach utilized the Wittig reaction,[199] and the other used intermediate sulfur compounds[200]) and its oxidation to dibenzo[*def,pgr*]tetraphenylene **(204).** Compound **204,** with symmetry point group D_2, has a propeller-shaped geometry similar to that of **194.** [2.2](3,3′,4,4′)Biphenylophane **(205)** and the corresponding 1,15-diene showed temperature-dependent nmr spectra, and a scheme of ring flipping involving concerted and two-step conformational processes was proposed.[201,202] Both of these molecules are structurally similar to metaparacyclophane[12-15] and the naphthalenophanes **115** and **116**,[93] all of which are conformationally mobile. [2.2](4,4')Biphenylophane **(206)** was obtained in good yield by pyrolysis of a disulfone.[83,165]

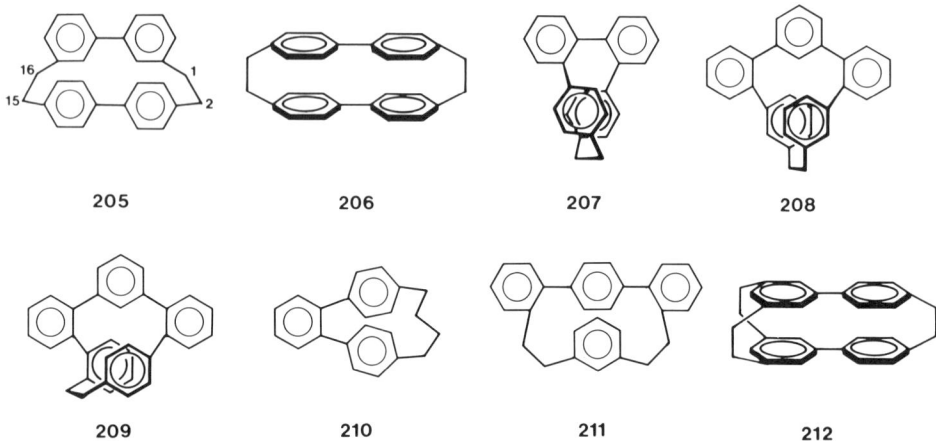

205   206   207   208

209   210   211   212

The helical bridged [2](4,4''')-o-quaterphenyl **207**,[203] and quinquephenyls **208**[204] and **209**[205] have been reported by Vögtle and co-workers.[206] In contrast to that of the conformationally mobile **208,** the nmr spectrum of compound **209** or the ethylenic analog shows no temperature dependence. These reports form part of a large series of papers on cyclophanes from o-terphenyls,[207,208] p-terphenyls,[209] triply clamped biphenyls,[210] and triphenylbenzenes.[211] Generally, the approach to these compounds has been via disulfide intermediates. The structures illustrated, **210**,[207] **211**,[109] and **212**,[210] provide only a brief overview of the scope of this work. Nakazaki *et al.*[212] prepared polymethylene-bridged biphenyls, Doomes and Beard[213] reported a similar bridged 4,4'-biphenyl, and Inazu *et al.*[214] made a [6.6](4,4')biphenylophane, but all of these compounds are rather removed from the main theme of this chapter and are not considered in detail.

VIII. OUTLOOK

The outstanding synthetic challenge in this area of chemistry is the development of suitable cyclophane precursors of [7]- and [8]circulenes

(171 and **173).** Molecular models and calculations[164] support the view that [8]circulene is probably no more distorted than some of the helicenes, and as a consequence both compounds **171** and **173** are worthwhile objectives. The ideal cyclophane precursors of these unknown circulenes should be those derived from 9,10-dihydrophenanthrenes, in much the same way that Staab and co-workers successfully used tetrahydrodibenzo[*aj*]anthracenes in the synthesis of kekulene **(191).**[187,188] The [9]- and [10]circulenes pose greater problems in that the synthesis of suitable aromatic moieties becomes a major problem. However, extrapolation of such methods suggests pathways to aromatic Möbius molecules (bent, battered, and twisted!). The introduction of polar groups into the aromatic cyclophanes to provide solubility in water will probably lead to much productive research. Because many of the cyclophanes reported in recent years are macrocycles capable of encapsulating smaller molecules, the development of synthetic procedures to obtain large cyclic compounds that will partition between aqueous and lipid phases will undoubtedly provide many useful models.

REFERENCES

1. C. J. Brown and A. C. Farthing, *Nature (London)* **164,** 915 (1949).
2. J. H. Wood and J. A. Stanfield, *J. Am. Chem. Soc.* **64,** 2343 (1942).
3. J. R. Davy and J. A. Reiss, *J. Chem. Soc., Chem. Commun.* p. 806 (1973).
4. B. H. Smith, "Bridged Aromatic Compounds." Academic Press, New York, 1964.
5. J. Abell and D. C. Cram, *J. Am. Chem. Soc.* **76,** 4406 (1954).
6. W. Baker, F. Glockling, and J. F. W. McOmie, *J. Chem. Soc.* p. 1118 (1951).
7. W. Baker, J. F. W. McOmie, and W. K. Warbuton, *J. Chem. Soc.* p. 2991 (1952).
8. D. A. Whiting and A. F. Wood, *J. Chem. Soc., Perkin Trans. 1* p. 623 (1980).
9. R. C. Helgeson, J. P. Mazaleyrat, and D. J. Cram, *J. Am. Chem. Soc.* **103,** 3929 (1981), and previous papers in this series.
10. H. A. Staab, U. E. Meissner, and B. Meissner, *Chem. Ber.* **109,** 3875 (1976).
11. H. H. Wasserman and P. M. Keehn, *J. Am. Chem. Soc.* **94,** 298 (1972).
12. S. Akabori, S. Hayashi, M. Nawa, and K. Shiomi, *Tetrahedron Lett.* p. 3727 (1969).
13. F. Vögtle, *Chem. Ber.* **102,** 3077 (1969).
14. D. T. Hefelfinger and D. J. Cram, *J. Am. Chem. Soc.* **92,** 1073 (1970).
15. D. T. Hefelfinger and D. J. Cram, *J. Am. Chem. Soc.* **93,** 4767 (1971).
16. R. H. Mitchell and V. Boekelheide, *J. Am. Chem. Soc.* **92,** 3510 (1970).
17. R. H. Mitchell and V. Boekelheide, *J. Am. Chem. Soc.* **96,** 1547 (1974).
18. R. H. Mitchell, T. Otsubo, and V. Boekelheide, *Tetrahedron Lett.* p. 219 (1975).
19. F. Vögtle and P. Neumann, *Synthesis* p. 85 (1973); R. H. Mitchell, *Heterocycles* **11,** 563 (1978).
20. R. H. Mitchell and R. J. Carruthers, *Tetrahedron Lett.* p. 4331 (1975).
21. R. H. Mitchell and J. S. Yan, *Can. J. Chem.* **55,** 3347 (1977).
22. R. H. Mitchell, R. J. Carruthers, and L. Mazuch, *J. Am. Chem. Soc.* **100,** 1007 (1978).

23. W. E. Parham, D. R. Johnson, C. T. Hughes, M. K. Meilahn, and J. K. Rinehart, *J. Org. Chem.* **35,** 1048 (1970).
24. W. E. Parham, D. C. Egberg, and W. C. Montgomery, *J. Org. Chem.* **38,** 1207 (1973).
25. W. E. Parham and W. C. Montgomery, *J. Org. Chem.* **39,** 3411 (1974).
26. P. Grice and C. B. Reese, *Tetrahedron Lett.* p. 2563 (1979).
27. P. Grice and C. B. Reese, *J. Chem. Soc., Chem. Commun.* p. 424 (1980).
28. D. J. Cram, C. K. Dalton, and G. R. Knox, *J. Am. Chem. Soc.* **85,** 1088 (1963).
29. J. Bruhin and W. Jenny, *Tetrahedron Lett.* p. 1215 (1973).
30. J. Bruhin, F. Gerson, W. B. Martin, and C. Wydler, *Helv. Chim. Acta* **60,** 1915 (1977).
31. E. J. Corey and E. Block, *J. Org. Chem.* **34,** 1233 (1969).
32. F. Gerson, W. B. Martin, and C. Wydler, *J. Am. Chem. Soc.* **98,** 1318 (1976).
33. H. H. Wasserman and P. M. Keehn, *J. Am. Chem. Soc.* **91,** 2374 (1969).
34. G. W. Brown and F. Sondheimer, *J. Am. Chem. Soc.* **89,** 7116 (1967).
35. J. Kleinschroth and H. Hopf, *Tetrahedron Lett.* p. 969 (1978).
36. A. V. Fratini, cited as a footnote in Wasserman and Keehn.[33]
37. H. J. Reich and D. J. Cram, *J. Am. Chem. Soc.* **89,** 3078 (1967).
38. J. M. McBride, P. M. Keehn, and H. H. Wasserman, *Tetrahedron Lett.* p. 4147 (1969).
39. G. Kaupp and I. Zimmerman, *Angew. Chem., Int. Ed. Engl.* **15,** 441 (1976).
40. H. H. Wasserman and P. M. Keehn, *J. Am. Chem. Soc.* **89,** 2770 (1967).
41. A. V. Fratini, *J. Am. Chem. Soc.* **90,** 1688 (1968).
42. G. Kaupp, *Angew. Chem., Int. Ed. Engl.* **15,** 442 (1976).
43. H. H. Wasserman and P. M. Keehn, *J. Am. Chem. Soc.* **88,** 4522 (1966).
44. T. Otsubo and V. Boekelheide, *J. Org. Chem.* **42,** 1085 (1977).
45. T. Otsubo and V. Boekelheide, *Tetrahedron Lett.* p. 3881 (1975).
46. V. Boekelheide and R. A. Hollins, *J. Am. Chem. Soc.* **95,** 3201 (1973).
47. H. A. Staab and C. P. Herz, *Angew. Chem., Int. Ed. Engl.* **16,** 392 (1977).
48. C. P. Herz and H. A. Staab, *Angew. Chem., Int. Ed. Engl.* **16,** 394 (1977).
49. H. H. Wasserman and P. M. Keehn, *Tetrahedron Lett.* p. 3227 (1969).
50. D. J. Cram, C. S. Montgomery, and G. R. Knox, *J. Am. Chem. Soc.* **88,** 515 (1966).
51. H. H. Wasserman and R. Kitzing, *Tetrahedron Lett.* p. 3343 (1969).
52. T. Otsubo, R. Gray, and V. Boekelheide, *J. Am. Chem. Soc.* **100,** 2449 (1978).
53. F. Imashiro, M. Oda, T. Iida, and Z. Yoshida, *Tetrahedron Lett.* p. 371 (1976).
54. B. Thulin and O. Wennerström, *Acta Chem. Scand., Ser. B* **B31,** 135 (1977).
55. H. E. Högberg, B. Thulin, and O. Wennerström, *Tetrahedron Lett.* p. 931 (1977).
56. D. Tanner, B. Thulin, and O. Wennerström, *Acta Chem. Scand., Ser. B* **B33,** 443 (1979).
57. N. E. Blank and M. W. Haenel, *Chem. Ber.* **114,** 1531 (1981).
58. T. Kawabata, T. Shinmyozu, T. Inazu, and T. Yoshimo, *Chem. Lett.* p. 315 (1979).
59. M. Yoshinaga, T. Otsubo, Y. Sakata, and S. Misumi, *Bull. Chem. Soc. Jpn.* **52,** 3759 (1979).
60. T. Otsubo, M. Kitasawa, and S. Misumi, *Chem. Lett.* p. 977 (1977).
61. T. Otsubo, M. Kitasawa, and S. Misumi, *Bull. Chem. Soc. Jpn.* **52,** 1515 (1979).
62. D. J. Cram and R. A. Reeves, *J. Am. Chem. Soc.* **80,** 3094 (1958).
63. K. B. Wiberg and M. J. O'Donnell, *J. Am. Chem. Soc.* **101,** 6660 (1979).
64. D. J. Cram and H. Steinberg, *J. Am. Chem. Soc.* **73,** 5691 (1951).
65. J. F. Haley and P. M. Keehn, *Chem. Lett.* p. 999 (1976).
66. G. M. Whitesides, B. A. Pawson, and A. C. Cope, *J. Am. Chem. Soc.* **90,** 639 (1968).
67. T. Hiyama, S. Hirano, and H. Nozaki, *J. Am. Chem. Soc.* **96,** 5281 (1974).
68. J. C. Dignan and J. B. Miller, *J. Org. Chem.* **32,** 490 (1967).
69. C. F. H. Allen and J. A. Van Allan, *J. Org. Chem.* **18,** 882 (1953).

70. C. B. Shana, S. M. Rosenfeld, and P. M. Keehn, *Tetrahedron* **33**, 1081 (1977).
71. M. Corson, B. M. Foxman, and P. M. Keehn, *Tetrahedron* **34**, 1641 (1978).
72. S. Mizogami, N. Osaka, T. Otsubo, Y. Sakata, and S. Misumi, *Tetrahedron Lett.* p. 799 (1974).
73. T. Otsubo, S. Mizogami, N. Osaka, Y. Sakata, and S. Misumi, *Bull. Chem. Soc. Jpn.* **50**, 1858 (1977).
74. J. F. Haley and P. M. Keehn, *Tetrahedron Lett.* p. 4017 (1973).
75. J. F. Haley, S. M. Rosenfeld, and P. M. Keehn, *J. Org. Chem.* **42**, 1379 (1977).
76. S. Rosenfeld and P. M. Keehn, *Tetrahedron Lett.* p. 4021 (1973).
77. M. W. Haenel, *Tetrahedron Lett.* p. 4191 (1977).
78. M. W. Haenel, *Chem. Ber.* **111**, 1789 (1978).
79. M. W. Haenel, *Tetrahedron Lett.* p. 3053 (1974).
80. M. W. Haenel, *Tetrahedron Lett.* p. 4007 (1978).
81. J. M. Larson and L. R. Sousa, *J. Am. Chem. Soc.* **100**, 1943 (1978).
82. H. S. Brown, C. P. Muenchausen, and L. P. Sousa, *J. Org. Chem.* **45**, 1682 (1980).
83. M. Haenel and H. A. Staab, *Tetrahedron Lett.* p. 3585 (1970).
84. M. Haenel and H. A. Staab, *Chem. Ber.* **106**, 2203 (1973).
85. R. S. Givens and P. L. Wylie, *Tetrahedron Lett.* p. 865 (1978).
86. P. Block and M. S. Newman, *Org. Synth. Collect. Vol.* **5**, 103 (1973).
87. N. E. Blank and M. W. Haenel, *Tetrahedron Lett.* p. 1425 (1978).
88. N. E. Blank and M. W. Haenel, *Chem. Ber.* **114**, 1520 (1981).
89. D. S. Kemp, M. E. Garst, R. W. Harper, D. D. Cox, D. Carlson, and S. Denmark, *J. Org. Chem.* **44**, 4469 (1979).
90. B. J. Whitlock, E. T. Jarvi, and H. W. Whitlock, *J. Org. Chem.* **46**, 1832 (1981).
91. S. P. Adams and H. W. Whitlock, *J. Org. Chem.* **46**, 3474 (1981).
92. V. Boekelheide and C. H. Tsai, *Tetrahedron* **32**, 423 (1976).
93. M. N. Iskander and J. A. Reiss, *Tetrahedron* **34**, 2343 (1978).
94. D. Hellwinkel, *Chem.-Ztg.* **94**, 715 (1970).
95. W. E. Barth and R. G. Lawton, *J. Am. Chem. Soc.* **88**, 380 (1966).
96. W. E. Barth and R. G. Lawton, *J. Am. Chem. Soc.* **93**, 1730 (1971).
97. J. T. Craig and M. D. W. Robins, *Aust. J. Chem.* **21**, 2237 (1968).
98. J. R. Davy, M. N. Iskander, and J. A. Reiss, *Tetrahedron Lett.* p. 4085 (1978).
99. J. R. Davy, M. N. Iskander, and J. A. Reiss, *Aust. J. Chem.* **32**, 1067 (1979).
100. T. Sato, H. Matsui, and R. Komaki, *J. Chem. Soc., Perkin Trans. 1* p. 2051 (1976).
101. R. W. Griffin and N. Orr, *Tetrahedron Lett.* p. 4567 (1969).
102. J. R. Davy and J. A. Reiss, *Aust. J. Chem.* **29**, 163 (1976).
103. J. R. Davy and J. A. Reiss, *Tetrahedron Lett.* p. 3639 (1972).
104. F. Vögtle, R. Schäfer, L. Schunder, and P. Neumann, *Justus Liebigs Ann. Chem.* **734**, 102 (1970).
105. J. R. Davy and J. A. Reiss, unpublished observations.
106. C. Elschenbroich, F. Gerson, and J. A. Reiss, *J. Am. Chem. Soc.* **99**, 60 (1977).
107. F. Bottino and S. Pappalardo, *Heterocycles* **12**, 1331 (1979).
108. P. J. Jessup and J. A. Reiss, *Tetrahedron Lett.* p. 1453 (1975).
109. P. J. Jessup and J. A. Reiss, *Aust. J. Chem.* **29**, 173 (1976).
110. K. Endo, Y. Sakata, and S. Misumi, *Bull. Chem. Soc. Jpn.* **44**, 2465 (1971).
111. H. A. Staab and H. J. Shin, *Chem. Ber.* **110**, 631 (1977).
112. W. Bieber and F. Vögtle, *Angew. Chem., Int. Ed. Engl.* **16**, 175 (1977).
113. W. Bieber and F. Vögtle, *Chem. Ber.* **111**, 1653 (1978).
114. F. Vögtle and R. Wingen, *Tetrahedron Lett.* p. 1459 (1978).
115. W. Bieber and F. Vögtle, *Chem. Ber.* **112**, 1919 (1979).

116. A. Kashahara, T. Izumi, and T. Katou, *Chem. Lett.* p. 1373 (1979).
117. L. G. Schroff, A. J. A. Van der Weerdt, D. J. H. Staalman, J. W. Verhoeven, and T. J. De Boer, *Tetrahedron Lett.* p. 1649 (1973).
118. R. L. J. Zsom, L. G. Schroff, C. J. Bakker, J. W. Verhoeven, T. H. De Boer, J. D. Wright, and H. Kuroda, *Tetrahedron* **34,** 3225 (1978).
119. J. H. Golden, *J. Chem. Soc.* p. 3741 (1961).
120. J. H. Golden, British Patent 1,015,783 (1966); *Chem. Abstr.* **64,** 8110b (1966).
121. G. Kaupp, *Angew. Chem., Int. Ed. Engl.* **11,** 313 (1972).
122. G. Kaupp, *Liebigs Ann. Chem.* p. 844 (1973).
123. S. Misumi, *Mem. Inst. Sci. Ind. Res., Osaka Univ.* **36,** 37 (1979).
124. T. Toyoda, I. Otsubo, T. Otsubo, Y. Sakata, and S. Misumi, *Tetrahedron Lett.* p. 1731 (1972).
125. T. Toyoda and S. Misumi, *Tetrahedron Lett.* p. 1479 (1978).
126. A. Iwama, T. Toyoda, M. Yoshida, T. Otsubo, Y. Sakata, and S. Misumi, *Bull. Chem. Soc. Jpn.* **51,** 2988 (1978).
127. A. Iwama, T. Toyoda, T. Otsubo, and S. Misumi, *Tetrahedron Lett.* p. 1725 (1973).
128. A. Iwama, T. Toyoda, T. Otsubo, and S. Misumi, *Chem. Lett.* p. 587 (1973).
129. T. Hayashi, N. Mataga, Y. Sakata, and S. Misumi, *Bull. Chem. Soc. Jpn.* **48,** 416 (1975).
130. F. Nemoto, K. Ishiza, T. Toyoda, Y. Sakata, and S. Misumi, *J. Am. Chem. Soc.* **102,** 654 (1980).
131. T. Hayashi, N. Mataga, Y. Sakata, S. Misumi, M. Morita, and J. Tanaka, *J. Am. Chem. Soc.* **98,** 5910 (1976).
132. H. Wynberg and R. Helder, *Tetrahedron Lett.* p. 4317 (1971).
133. D. Tanner, B. Thulin, and O. Wennerström, *Acta Chem. Scand., Ser. B* **B33,** 464 (1979).
134. K. D. Gundermann and K. D. Röker, *Angew. Chem., Int. Ed. Engl.* **12,** 425 (1973).
135. K. D. Gundermann and K. D. Röker, *Liebigs Ann. Chem.* p. 140 (1976).
136. A. Dunand, J. Ferguson, M. Puza, and G. B. Robertson, *J. Am. Chem. Soc.* **102,** 3524 (1980).
137. T. Shinmyozu, T. Inazu, and T. Yoshino, *Chem. Lett.* p. 405 (1978).
138. T. Inoue, T. Kaneda, and S. Misumi, *Tetrahedron Lett.* p. 2969 (1974).
139. T. Hayashi, N. Mataga, T. Inoue, T. Kaneda, M. Irie, and S. Misumi, *J. Am. Chem. Soc.* **99,** 523 (1977).
140. T. Kaneda, T. Inoue, Y. Yasufuku, and S. Misumi, *Tetrahedron Lett.* p. 1543 (1975).
141. T. Kaneda, T. Otsubo, H. Horita, and S. Misumi, *Bull. Chem. Soc. Jpn.* **53,** 1015 (1980).
142. I. Yamashita, M. Fujii, T. Kaneda, and S. Misumi, *Tetrahedron Lett.* p. 541 (1980).
143. R. Wingen and F. Vögtle, *Chem. Ber.* **113,** 676 (1980).
144. E. Buhleier and F. Vögtle, **111,** 2729 (1978).
145. S. Akiyama and M. Nakagawa, *Bull. Chem. Soc. Jpn.* **44,** 3158 (1971).
146. J. M. Pearson, D. J. Williams, and M. Levy, *J. Am. Chem. Soc.* **93,** 5478 (1971).
147. F. Gerson, G. Kaupp, H. Ohya-Nishiguchi, *Angew. Chem., Int. Ed. Engl.* **16,** 657 (1977).
148. D. H. Eargle, *J. Magn. Reson.* **2,** 225 (1970).
149. S. E. Potter and I. O. Sutherland, *J. Chem. Soc., Chem. Commun.* p. 520 (1973).
150. F. B. Mallory, C. S. Wood, and J. T. Gordon, *J. Am. Chem. Soc.* **86,** 3094 (1964).
151. E. V. Blackburn and C. J. Timmons, *J. Chem. Soc., Q. Rev.* **23,** 482 (1969).
152. M. S. Rubin and S. Welner, *J. Org. Chem.* **45,** 1847 (1980).
153. U. Meissner, B. Meissner, and H. Staab, *Angew. Chem., Int. Ed. Engl.* **12,** 916 (1973).

154. G. Ege and H. Vogler, *Tetrahedron* **31**, 569 (1975).
155. G. Hallas and R. M. Potts, *J. Chem. Soc., Chem. Commun.* p. 252 (1975).
156. J. T. Craig, B. Halton, and S. F. Lo, *Aust. J. Chem.* **28**, 913 (1975).
157. P. J. Jessup and J. A. Reiss, *Aust. J. Chem.* **30**, 843 (1977).
158. P. J. Jessup and J. A. Reiss, *Aust. J. Chem.* **29**, 1267 (1976).
159. R. B. Du Vernet, O. Wennerström, J. Lawson, T. Otsubo, and V. Boekelheide, *J. Am. Chem. Soc.* **100**, 2457 (1978).
160. R. B. Du Vernet, T. Otsubo, J. A. Lawson, and V. Boekelheide, *J. Am. Chem. Soc.* **97**, 1629 (1975).
161. P. J. Jessup and J. A. Reiss, *Aust. J. Chem.* **30**, 851 (1977).
162. D. N. Leach and J. A. Reiss, *Aust. J. Chem.* **32**, 361 (1979).
163. B. Thulin and O. Wennerström, *Acta Chem. Scand., Ser. B* **B30**, 369 (1976).
164. T. Liljefors and O. Wennerström, *Tetrahedron* **33**, 2999 (1977).
165. H. A. Staab and M. Haenel, *Chem. Ber.* **106**, 2190 (1973).
166. W. Jenny and R. Paioni, *Chimia* **22**, 248 (1968).
167. R. Paioni and W. Jenny, *Helv. Chim. Acta* **53**, 141 (1970).
168. P. Baumgartner, R. Paioni, and W. Jenny, *Helv. Chim. Acta* **54**, 266 (1971).
169. M. W. Haenel, *Tetrahedron Lett.* p. 3121 (1976).
170. M. W. Haenel, *Tetrahedron Lett.* p. 1273 (1977).
171. J. Lawson, R. Du Vernet, and V. Boekelheide, *J. Am. Chem. Soc.* **95**, 956 (1973).
172. F. Binnig, H. Meyer, and H. A. Staab, *Justus Liebigs Ann. Chem.* **724**, 24 (1969).
173. T. Umemoto, S. Satani, Y. Sakata, and S. Misumi, *Tetrahedron Lett.* p. 3159 (1975).
174. R. Mitchell, R. J. Carruthers, and J. C. M. Zwinkels, *Tetrahedron Lett.* p. 2585 (1976).
175. D. Schweitzer, K. H. Hausser, R. G. H. Kirrstetter, and H. A. Staab, *Z. Naturforsch., A* **31A**, 1189 (1976).
176. H. Irngartinger, R. G. H. Kirrstetter, C. Krieger, H. Rodewald, and H. Staab, *Tetrahedron Lett.* p. 1425 (1977).
177. T. Sato, M. Wakabayashi, Y. Okamura, T. Amada, and K. Hata, *Bull. Chem. Soc. Jpn.* **40**, 2363 (1967).
178. Y. Kai, F. Hama, N. Yasuoka, and N. Kasai, *Acta Crystallogr., Sect. B* **B34**, 1263 (1978).
179. T. Hayashi, N. Mataga, Y. Sakata, and S. Misumi, *Chem. Phys. Lett.* **41**, 325 (1976).
180. T. Hayashi, N. Mataga, T. Umemoto, Y. Sakata, and S. Misumi, *J. Phys. Chem.* **81**, 424 (1977).
181. H. Staab and R. G. H. Kirrstetter, *Liebigs Ann. Chem.* p. 886 (1979).
182. K. Ishizu, Y. Sugimoto, T. Umemoto, Y. Sakata, and S. Misumi, *Bull. Chem. Soc. Jpn.* **50**, 2801 (1977).
183. T. Kawashima, T. Otsubo, Y. Sakata, and S. Misumi, *Tetrahedron Lett.* p. 5115 (1978).
184. T. Umemoto, T. Kawashima, Y. Sakata, and S. Misumi, *Chem. Lett.* p. 837 (1975).
185. R. G. H. Kirrstetter and H. A. Staab, *Liebigs Ann. Chem.* p. 899 (1979).
186. T. Umemoto, T. Kawashima, Y. Sakata, and S. Misumi, *Tetrahedron Lett.* p. 463 (1975).
187. F. Diederich and H. A. Staab, *Angew. Chem., Int. Ed. Engl.* **17**, 372 (1978).
188. C. Krieger, F. Diederich, D. Schweitzer, and H. A. Staab, *Angew. Chem., Int. Ed. Engl.* **18**, 699 (1979).
189. F. Vögtle and H. A. Staab, *Chem. Ber.* **101**, 2709 (1968).
190. R. Peter and W. Jenny, *Chimia* **19**, 45 (1965).
191. R. Peter and W. Jenny, *Helv. Chim. Acta* **49**, 2123 (1966).
192. B. Thulin and O. Wennerström, *Acta Chem. Scand., Ser. B* **B30**, 688 (1976).

193. E. M. Kosower, H. Dodiuk, B. Thulin, and O. Wennerström, *Acta Chem. Scand., Ser. B* **B31,** 526 (1977).
194. M. Nakazaki, K. Yamamoto, and M. Maeda, *Chem. Lett.* p. 1553 (1980).
195. M. Nakazaki, K. Yamamoto, and M. Maeda, *J. Org. Chem.* **46,** 1985 (1981).
196. J. Tribout, R. H. Martin, M. Doyle, and H. Wynberg, *Tetrahedron Lett.* p. 2839 (1972).
197. M. Joly, N. Defay, R. H. Martin, J. P. Declerq, G. Germain, B. Soubrier-Paynen, and M. Van Meerssche, *Helv. Chim. Acta* **60,** 537 (1977).
198. F. Vögtle, *Justus Liebigs Ann. Chem.* **728,** 17 (1969).
199. B. Thulin and O. Wennerström, *Tetrahedron Lett.* p. 929 (1977).
200. D. N. Leach and J. A. Reiss, *J. Org. Chem.* **43,** 2484 (1978).
201. D. N. Leach and J. A. Reiss, *Tetrahedron Lett.* p. 4501 (1979).
202. D. N. Leach and J. A. Reiss, *Aust. J. Chem.* **33,** 823 (1980).
203. F. Vögtle, M. Atzmüller, W. Welmer, and J. Grütze, *Angew. Chem., Int. Ed. Engl.* **16,** 325 (1977).
204. F. Vögtle and E. Hammerschmidt, *Angew. Chem., Int. Ed. Engl.* **17,** 268 (1978).
205. E. Hammerschmidt and F. Vögtle, *Chem. Ber.* **112,** 1785 (1979).
206. E. Hammerschmidt and F. Vögtle, *Chem. Ber.* **113,** 3550 (1980).
207. J. Grütze and F. Vögtle, *Chem. Ber.* **110,** 1978 (1977).
208. N. Jacobson and V. Boekelheide, *Angew. Chem., Int. Ed. Engl.* **17,** 46 (1978).
209. K. Böckmann and F. Vögtle, *Liebigs Ann. Chem.* p. 467 (1981).
210. F. Vögtle and G. Steinhagen, *Chem. Ber.* **111,** 205 (1978).
211. G. Hohner and F. Vögtle, *Chem. Ber.* **110,** 3052 (1977).
212. M. Nakazaki, K. Yamamoto, S. Isoe, and M. Kobayasi, *J. Org. Chem.* **44,** 2160 (1979).
213. E. Doomes and R. M. Beard, *Tetrahedron Lett.* p. 1243 (1976).
214. R. Nagano, J. Nishikido, T. Inazu, and T. Yoshino, *Bull. Chem. Soc. Jpn.* **46,** 653 (1973).

CHAPTER 8

Nonbenzenoid Cyclophanes

SHÔ ITÔ,
YUTAKA FUJISE, AND
YOSHIMASA FUKAZAWA

Department of Chemistry
Faculty of Science
Tohoku University
Sendai, Japan

I. INTRODUCTION

In the past three decades the chemistry of cyclophanes has made a large contribution to the field of aromatic chemistry. The fruitfulness and productivity in the field has prompted us to consider phane systems containing nonbenzenoid aromatic rings. Compounds of this type are worthy of investigation because the resonance energy of these rings is in general

ISBN 0-12-403002-5

smaller than that of the corresponding benzenoid rings; the outcome is the pronounced deformation of the former rings in order to release the internal strain when incorporated into phane systems with short bridge(s). Furthermore, the charged or dipole character of nonbenzenoid rings is suitable for the investigation of transannular interaction when they are incorporated into multidecked phane systems. Thus, nonbenzenoid phanes have been the subject of various research groups.[1] This chapter deals with this group of compounds, which are classified by the number of π electrons in the nonbenzenoid aromatic rings, with emphasis on their geometries and physical properties. Because of space limitations, we regretfully omit any discussion of ferrocenophanes. The reader is referred to the appropriate references.[2]

II. SYNTHETIC METHODS

The synthetic methods developed for the benzenoid phanes are generally applicable to the nonbenzenoid systems, provided that the reaction conditions are mild enough to preserve the more sensitive nonbenzenoid aromatic systems. The methods applied so far to this area can be classified in two groups on the basis of the crucial step: unimolecular cyclization or bimolecular coupling of appropriate derivatives (method A) and aromatization of appropriate polycyclic precursors (method B), as schematically shown in Eq. (1), taking a single-decked phane as an example.

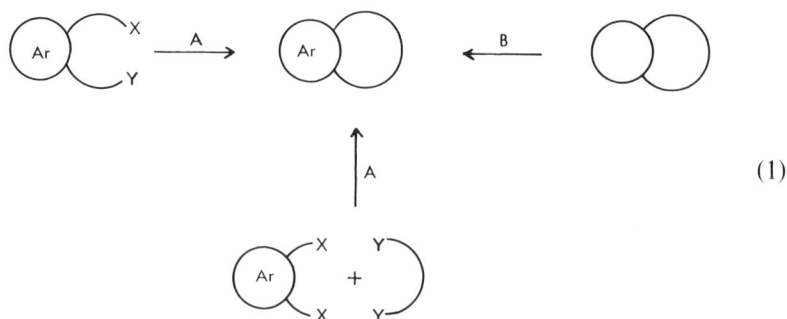

$$\tag{1}$$

Although method A is a direct application of the same reactions used in the benzenoid series, method B is more significant and is much more widely applied to the nonbenzenoid series because the nonbenzenoid aromatic rings are less stable and more reactive than benzenoid rings, and therefore these rings are more conveniently constructed at a later stage.

These general methods, being applicable to single-decked and multi-decked phanes as well as polynuclear phanes, are further classified by the actual methods of bond formation. These methods are shown below with representative examples to illustrate their variety.

A. Cyclization or coupling of aromatic derivatives.
   1. Sulfur route: Consists of coupling a bismercaptoalkyl derivative with compounds having two leaving groups to form dithiaphanes. Two methods of desulfurization have been applied.
      a. Photodesulfurization: This is the method of choice for preparing azulenophanes that are sensitive to acidic oxidation to disulfone (Eq. 2).

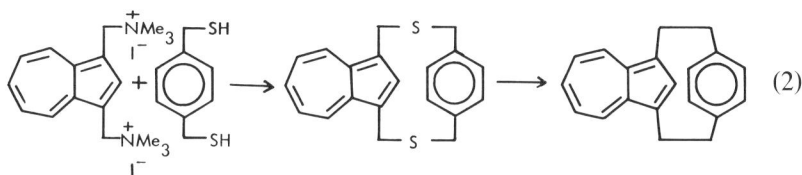

(2)

      b. Thermal desulfurization: Consists of oxidation to disulfone and its thermolysis. The method is always used for photolabile tropolonophanes (Eq. 3).

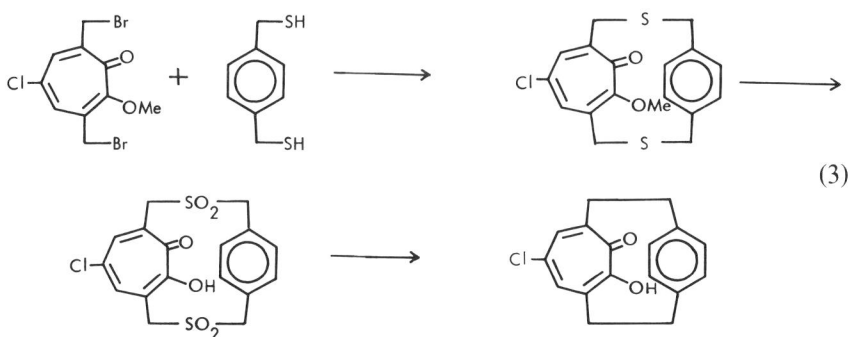

(3)

   2. Coupling reaction: Two methods have been applied.
      a. Hofmann degradation: Although the yield is good generally, the method results in a mixture of orientation isomers whenever possible, and their separation could be tedious (Eq. 4).

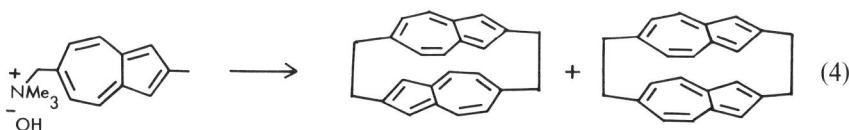

(4)

b. Reductive coupling (Eq. 5):

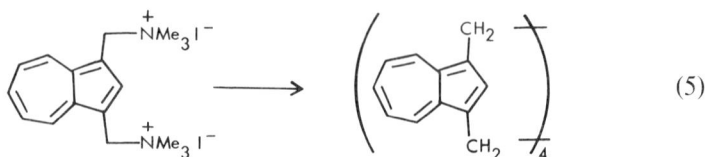

$$(5)$$

3. Cyclization of bridge.
   a. Thorpe–Zeigler reaction: Applicable only to phanes with long bridge(s) (Eq. 6).

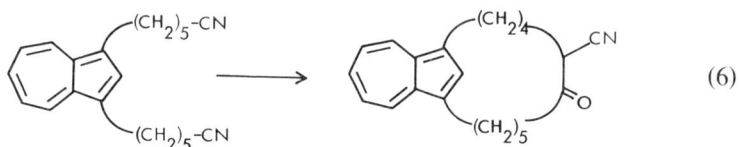

$$(6)$$

B. Aromatization reaction: Includes aromatization in the normal sense and also the direct construction of nonbenzenoid aromatic rings.
   1. Elimination reaction (Eqs. 7, 8, and 9):

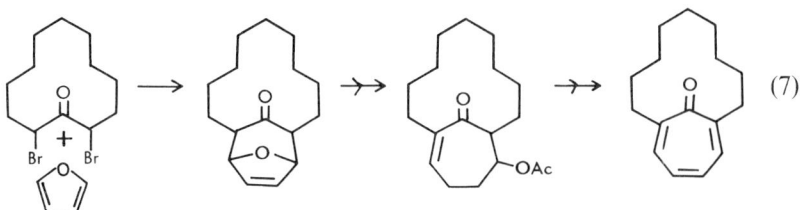

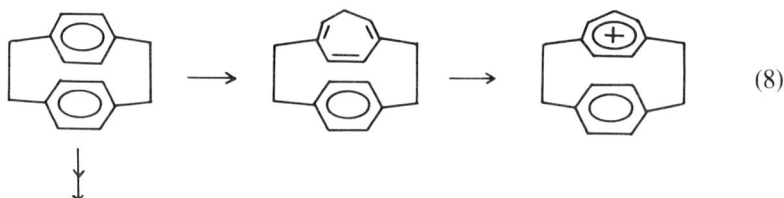

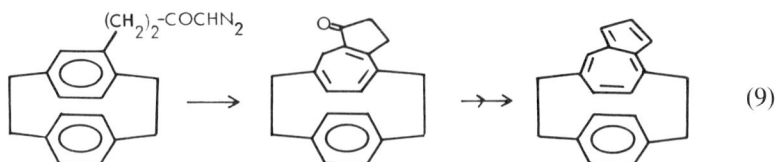

$$(7)$$

$$(8)$$

$$(9)$$

2. Prototropy (Eq. 10):

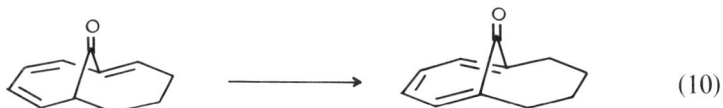
(10)

3. Direct ring construction: Double aldol condensation (Eq. 11) and Hafner's method of azulene synthesis (Eq. 12) have been successful.

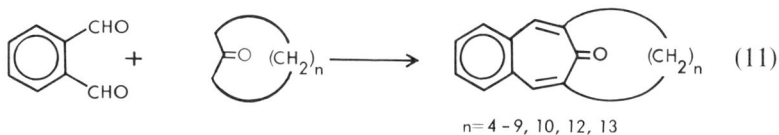
(11)

n = 4 – 9, 10, 12, 13

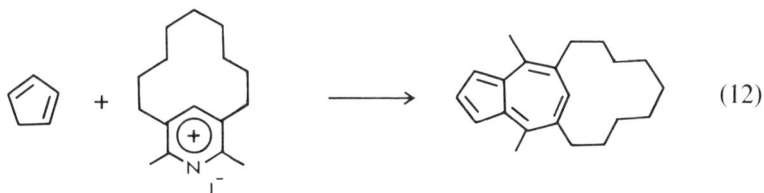
(12)

Some of these nonbenzenoid phanes are known to undergo further reactions as in the unbridged ring systems. By means of these reactions a variety of nonbenzenoid phanes have been synthesized.

III. PHANES CONTAINING NONBENZENOID 6π SYSTEMS

A. Cyclopentadienidophanes

1. [9](1,3)CYCLOPENTADIENIDOPHANE (1)[3]

The first and only example of a phane containing a simple cyclopentadienide anion (1) was synthesized in 1971 by method B,1. The presence of

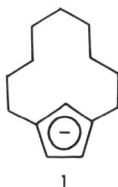

1

a diamagnetic ring current was evidenced by the ^{1}H-nmr spectrum, which exhibits highly shielded methylene protons at 0.55 ppm (2H,m), the rest appearing at 1.0 to 1.6 ppm. Ring protons appear at 5.33 (1H,t) and 5.02 ppm (2H,d), the positions being very similar to that (5.4 ppm) of cyclopentadienide.

2. syn- AND anti-[2.2](2,7)FLUORENIDOPHANES (2 AND 3)[4]

These compounds are only double-decked phanes having cyclopentadienide moieties. They were obtained as dark red solutions via the deprotonation by base of the fluorenophane precursors, which in turn were synthesized by the sulfur route (method A,1,b).[5]

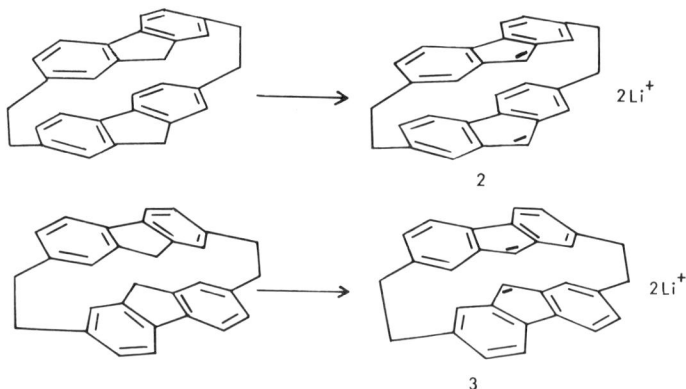

Protons H-1 (and H-8) and H-9 (fluorene numbering) appear at considerably higher field (6.09 and 5.04 ppm, respectively), and H-3 (and H-6) and H-4 (and H-5) at lower fields (6.21 and 7.43 ppm, respectively) in 3 compared with those in 2 (6.75, 5.43, 5.76, and 7.11 ppm, respectively). The chemical shifts are consistent with the assigned orientation of fluorenide moieties in these dianions.

B. Troponophanes

1. [n](2,7)TROPONOPHANES

Single-decked troponophanes have been studied more systematically.

a. Synthesis
Aromatization (method B) of bicyclic precursors has always been used as the synthetic strategy,[6-9] although the method of construction of the

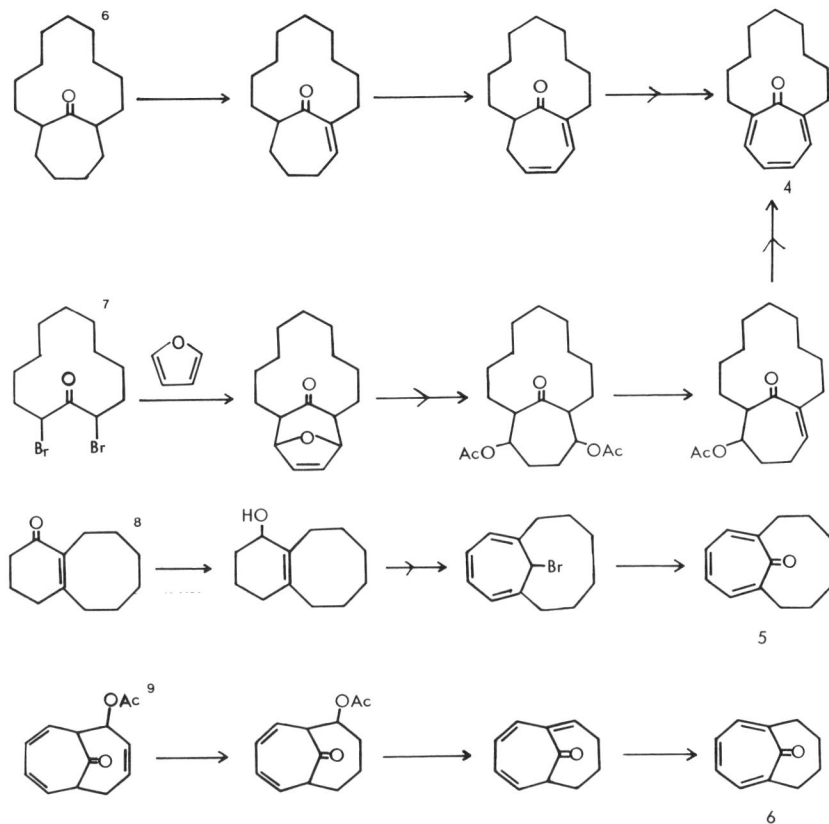

precursors differs from one member of this class to another. The base-catalyzed prototropy used for the synthesis of **6** was applied to the tetraene **7** in the synthesis of dehydro compound **8**.[9]

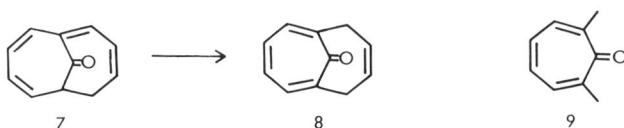

b. Geometry and Physical Properties

All the bridges in compounds **4** to **6** and **8** are fixed in certain geometries, as is demonstrated by the nonequivalence of the "benzylic" protons in their ¹H-nmr spectra. Although the ir spectrum of **4** indicates that the tropone ring is as planer and as polar as that in 2,7-dimethyltropone **(9)**, the carbonyl bands are shifted to higher frequencies on going to **6** through **5**, as shown in Table I, revealing smaller carbonyl bond angles and implying reduced planarity.

TABLE I

Spectral Properties of Bridged Tropones

Compound	ir ($\nu_{C=O}^{KBr}$, cm^{-1})	uv [$\lambda_{max}^{cyclohexane}$, nm (log ε)]
9	1561	232 (4.38)
		311 (3.81)
4	1620	238 (4.41)
		313 (3.90)
5	1656	202 (4.01)^a
		238 (4.00)^a
		284 (3.62)^a
6	1718	238 (3.81)
		247 (3.79)
		~310 (240 sh)

^{a} n-Hexane.

The geometry of **6** as determined by X-ray crystallography is shown in Fig. 1.[9] The tropone ring is deeply bent in a boat form. Furthermore, the ring shows more pronounced bond alternation and a shorter C=O bond than that in tropone itself,[10] suggesting reduced conjugation.

The reduced conjugation is clearly demonstrated in various spectra. Proton nmr and ^{13}C-nmr chemical shifts change regularly on going from planar **4** to **6**.[9] Those of $C_{C=O}$ and C_β and H_β are especially instructive, as shown in Fig. 2. The carbonyl carbon shows a downfield shift to the position normal for the saturated carbonyl carbon; the shift must be due, at least in part, to the net positive charge on the carbon. In accord with this, H_β and C_β shift upfield, becoming very similar to those of tropylidinophane **10**.[11]

The uv spectra of these compounds also reveal reduced conjugation (Table I). The strong and broad band of planar tropones at around 310 nm

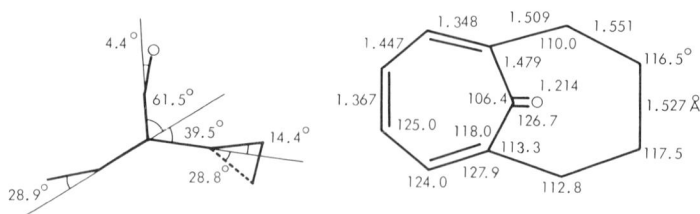

Fig. 1. Molecular structure of **6**.[9] (Reproduced by courtesy of Pergamon Press.)

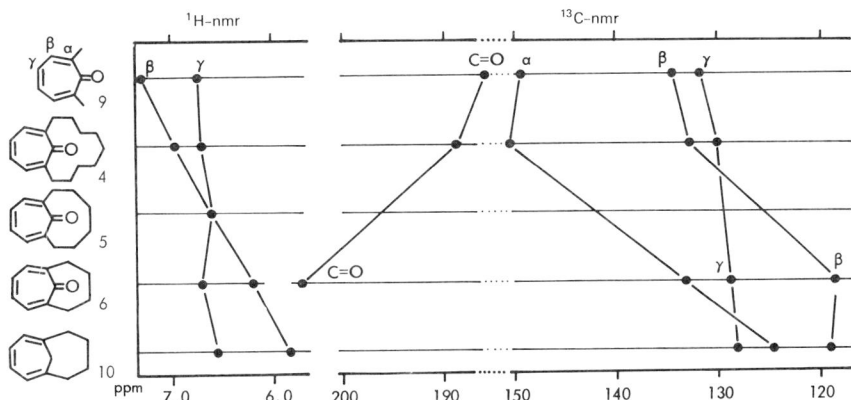

Fig. 2. ^{1}H-nmr and ^{13}C-nmr chemical shifts of [n](2,7)troponophanes.

is greatly reduced to a weak shoulder in **6,** and the general shape of the spectrum of **6** resembles that of **10.**

2. [n]Benzotroponophanes

A series of [n]benzo[d]troponophanes (**11,** n = 4–10, 12, 13) has been synthesized by the double aldol condensation (method B,3) of cycloalkanones (C_7–C_{16}, except C_{14}) with phthalaldehyde,[12-14] and their physical properties have been studied extensively. Some of their physical constants as well as those of dimethylbenzotropone **(12),** a reference compound, are listed in Table II. Although a small effect of bridge formation is seen (compare **12** and the [13] or [12] homolog), a more dramatic change is observed in carbonyl frequency,[12] half-wave reduction potential ($-E_{1/2}$),[15] and chemical shift (δ) of tropone protons[13] when the bridge becomes shorter than —$(CH_2)_7$—. Similar changes occur in dipole moment (μ)[16] and exaltation of molecular refraction ($\Delta[R]_D$).[16]

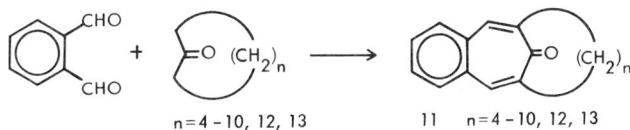

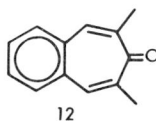

TABLE II

Physical Properties of [n](2,7)Benzo[d]troponophanes (11)

Compounds [n]	$\nu_{C=O}^{Nujol}$ [12] (cm^{-1})	δ_H [13] (ppm)	μ [16] (D)	$-E_{1/2}$ [15] (V)	$\Delta[R]_D$ [16]
Dimethyl (12)	1596	—	3.84	0.679	4.0
13	1602	—	—	0.793	—
12	1590	7.52	3.46	0.828	3.6
10[13]	1610	7.27	—	—	—
9	1610	7.31	—	0.842	—
8	1604	—	—	0.883	—
7	1609	7.2	—	0.925	—
6	1651	—	—	0.977	—
5	1679	6.78	3.09	1.124	2.1
4	1724	—	—	1.142	—

The uv spectra also change regularly. Whereas those of **12** and the [13] homologs consist of four absorption maxima (at ~358, 340, ~272, and ~236 nm) and two shoulders (at ~325 and ~304 nm), the [5] and [6] homologs exhibit only the last two maxima (~280 and ~232 nm) and a shoulder (~330 nm), and the [4] homolog shows only one maximum at 226 nm.[12] All of these changes suggest the decreased bond angle of the carbonyl group and its reduced conjugation with the rest of the π system because of displacement of the carbonyl carbon from the ring plane in the lower homologs. This was verified by X-ray analyses of the two lowest homologs.[17] The tropone ring in the [5] and [4] compounds takes deeply bent tub forms with bending angles of 51.3 and 60.1° (carbonyl side), respectively, and 24.3 and 30.1° (benzene side), respectively. The values are comparable with those of [4](2,7)troponophanes (see Section III,B,1).[9] The delocalization energy in the [5] compound was calculated to be 48.7 kcal/mol, about 35 kcal/mol less than that of **12**. This means that the compound shows very little delocalization in addition to that of benzene.[18]

A series of thienotroponophanes (**13** and **14**, n = 5–9 and 12) were prepared in a similar way and their ^{1}H-nmr spectra examined.[19] No definite conclusion was drawn about their planarity.

In a mechanistic study [4](4,6)benzo[b]troponophanes **15** and **16** were proposed as reaction intermediates.[20] Of particular interest in these species is that, unlike those described above, these troponophanes necessarily possess a double bond of E configuration in one of the seven-membered rings, and both of them seem to behave as discrete species.

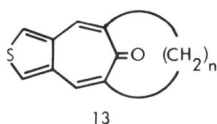

13

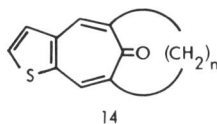

14

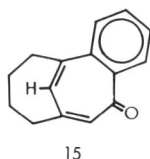

15

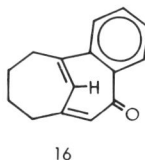

16

When **17** and **18** were treated with tetraethylammonium acetate in acetone, two acetates (**19** and **20**, respectively) were obtained, with complete retention of configuration. From a kinetic study (both bimolecular, and **17** → **19** is faster than **18** → **20**) and studies on compounds with slightly modified structures[20] (sharp rate deceleration), troponophanes **15** and **16** were proposed as reaction intermediates. π-SCF molecular mechanics calculations disclosed a high barrier in their mutual transformation.[21]

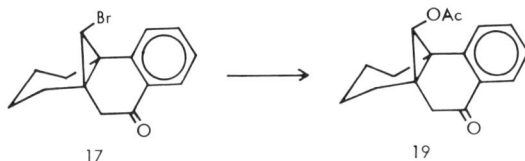

17   19

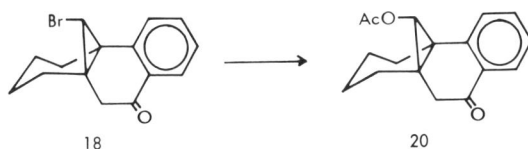

18   20

C. Tropyliophanes

1. [n]METATROPYLIOPHANES

The ^{1}H-nmr spectrum of [9]troponophane **4** in acidic media [~1 equivalent of trifluoroacetic acid (TFA)] is different from the neutral spectrum, tropone protons and benzylic protons exhibiting downfield shifts by 0.9

and 0.1–0.4 ppm, respectively, and some of the bridge protons exhibiting upfield shifts. These changes are consistent with the formation of the corresponding hydroxytropylium ion **21**.[6] In contrast, [6]troponophane **(5)** does not show any of these changes under the same acidic conditions,[8] and [4]troponophane **(6)** shows very small downfield shifts (~0.14 ppm) of vinyl protons, even in 100% TFA.[22] These observations clearly demonstrate that the hydroxytropylium ion is stable only when the seven-membered ring can assume at least near-planar geometry and that 2-hydroxy-[6](1,3)tropyliophane **(22)** is not planar enough to delocalize the positive charge.

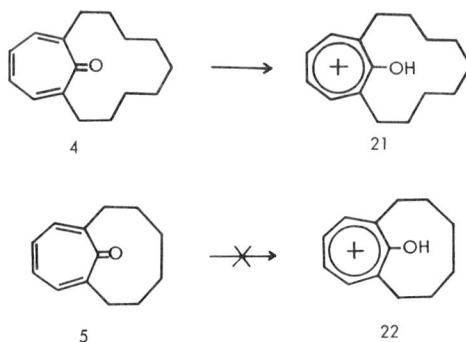

The same problem was investigated with [*n*]benzotroponophanes **11** utilizing the uv spectra. However, not only the [5] compound but also the [13] homolog showed little change in 60% formic acid, whereas **12,** the reference compound, changed to the corresponding hydroxytropylium ion. The difference was attributed to the steric hindrance in protonation of carbonyl groups in these phanes.[12]

Apart from the hydroxylated series, metatropyliophane is known only in benzolog series. Compounds **23** with 10- and 12-methylene bridges were synthesized in the sequence shown and exhibit ¹H-nmr spectra (two tropylium protons resonate at 9.2 ppm) that are similar to those of the hydroxy counterparts.[13] Hydroxy[9]benzohomotropyliophane **(24)** was derived in a similar way.[14,23]

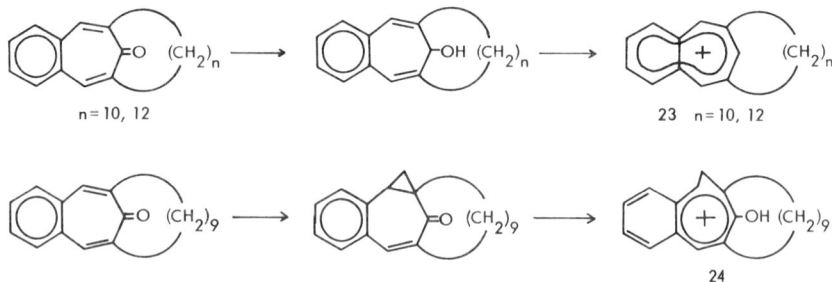

2. TROPYLIOCYCLOPHANES

A variety of double-decked phanes containing a tropylium ion moiety but differing in number, length, and position of bridges have been synthesized as models for studies on charge-transfer (CT) interaction. Currently, compounds 25–33 are the only tropyliophanes known. The method used

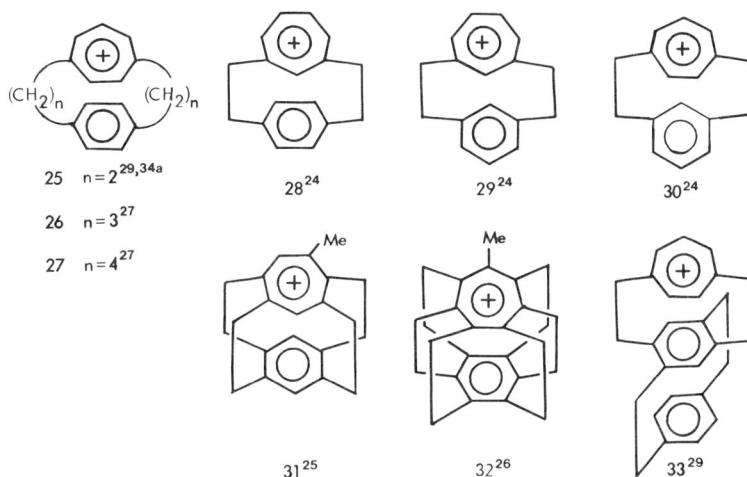

25 n = 2[29,34a]

26 n = 3[27]

27 n = 4[27]

28[24]

29[24]

30[24]

31[25]

32[26]

33[29]

for their synthesis is the ring expansion of the corresponding cyclophanes followed by hydride abstraction (method B,1), as exemplified here.

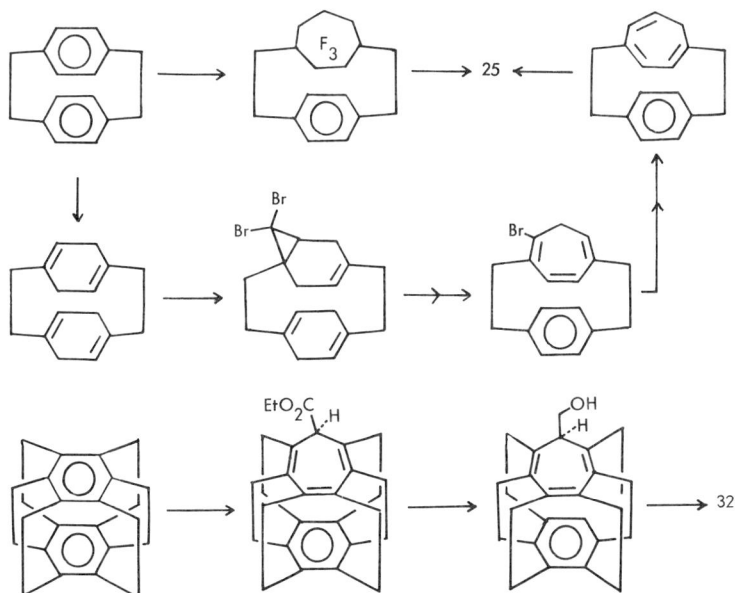

a. Geometry

Although none of the tropyliophanes has been analyzed by X-ray crystallography, it is clear from the molecular model that those with more than three bridges have fixed geometries and that two rings are facing one another. However, those with two bridges have questions as to the relative position of two rings and their flipping, depending on the positions and length of bridges.

Compound **31** with four bridges would provide a good model for [2.2]tropyliocyclophanes in which two aromatic rings are aligned in syn fashion. Chemical shifts of all ring protons in CD$_3$CN are assigned in comparison with that of **25.** All of the tropylium protons of **31** appear at

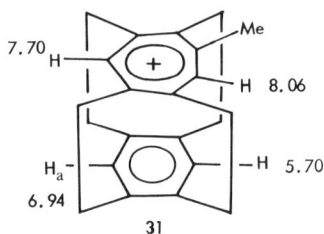

somewhat higher field (up to 1.22 ppm) than do those in tropylium ion (9.28 ppm). The upfield shift is the result of the additive contribution of the diamagnetic ring current effect of the benzene ring and the increased electron density due to CT interaction. (The contribution of CT interaction is discussed in Section III,C,2,b). Ring protons in **25** and **28–30** are

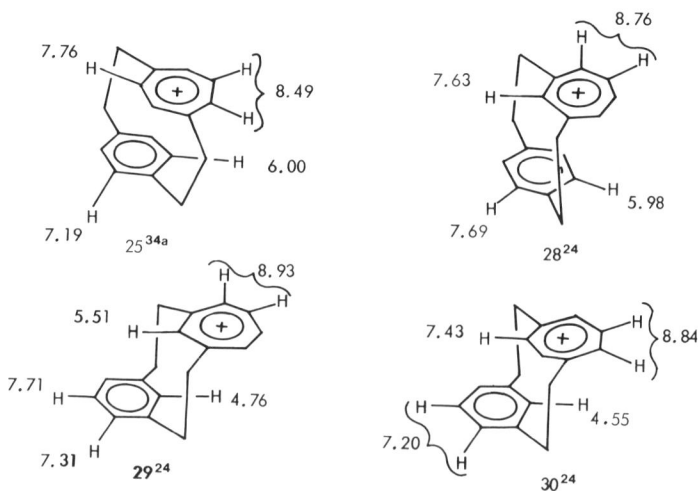

reasonably assigned as shown on the basis of signal intensity and consideration of the ring current effect and charge distribution.

The fact that the singlet benzene protons in **29** and **30** appear in much higher (~2.2 ppm) field and three other benzene protons are in lower field than in **31** suggests the preferred stepped-anti conformations for these compounds at room temperature. For compounds **25** and **28,** nonequivalence of benzene protons reveals fixed geometries in the nmr time scale at room temperature.

Whereas compound **30** shows no sign of ring flipping up to 120°C, **28** shows flipping between the equivalent conformers. The energy barrier of this dynamic process was calculated to be 16 kcal/mol, ~5 kcal/mol lower than that of [2.2]metaparacyclophane, indicating the decrease of strain in the rotational transition state associated with the change in ring size.[24]

The [3.3]cyclophane **26** still exhibits two separate benzene proton signals and shows most of its aromatic protons at lower field than in **25**.

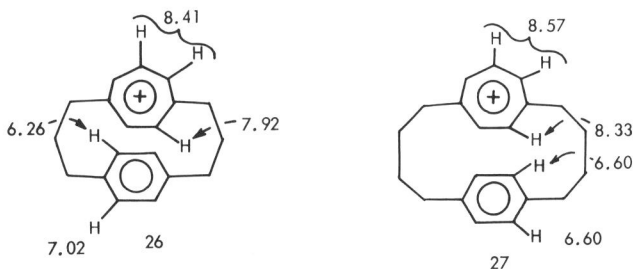

These features of **26** suggest the absence of ring flipping and a longer interlayer distance than in **25**. However, the lower benzene signal appears in still higher field, and the difference in chemical shift of two benzene signals is smaller than in **25,** implying the presence of slow ring flipping and/or a sliding motion of two aromatic rings. In the [4.4]cyclophane **27** the equivalence of the benzene ring protons indicates the presence of ring flipping of the tropylium part and/or rotation of the benzene ring, its activation energy being estimated to be 8.9 kcal/mol.[27]

b. CT Interaction

In addition to the ring current effect of the opposite ring, an important factor affecting the chemical shift of the ring protons in tropyliocyclophanes is CT interaction, which depletes the electron density in the benzene ring.

Because one of the benzene ring protons (H_a) in **31** has about the same relative position with respect to the opposite ring as that in [2.2.2.2](1,2,4,5)cyclophane **(34)** and therefore the ring current effect can

be assumed to be about the same for the protons in both compounds, their chemical shift difference, which amounts to 0.98 ppm, corresponds to the effect of CT interaction in **31**. Similarly, the chemical shift difference (0.69 ppm) between the benzene protons of **25** and those of [2.2]paracyclophane (**35**) corresponds to the CT interaction effect in **25**. The magnitude of the CT interaction is consistent with the interplanar distances in these two compounds. Tropylium ring protons in **25** and **31** also appear at higher field than those in tropylium fluoroborate. As stated earlier, this is the combined effect of the ring current of the facing benzene ring and CT interaction. Without a proper reference compound, these effects cannot be estimated separately, but a larger upfield shift for the protons in **31** than for those in **25** is consistent with the result obtained for the benzene protons.[25]

Electronic spectra have been used more widely for the detection of CT interaction. However, because the difference in experimental conditions applied by each author makes a systematic survey impossible, an attempt is made here to compare the strength of CT interaction within closely related series.

The effect of interdeck distance and skewing of the two aromatic rings was beautifully demonstrated by Boekelheide.[25,26] A regular bathochromic shift of the CT bands is observed, as shown in Table III, on increasing the number (2 to 6) of C_2 bridges by which both interdeck distance and skewing are decreased.

These observations justify qualitatively the conclusion that CT interac-

TABLE III

Charge-Transfer Bands and Probable Interplanar Distances of Tropyliocyclophanes

Compound	Distance (Å)	CT bands [$\lambda_{max}^{CH_3CN}$ (ε)]		
25	3.1[34]	323 (2230)	353 (1590)	400 (524 sh)
31	2.7[29a]	351 (2000)	400 (750)	460 (400 sh)
32	2.6[26]	375 (2990)	431 (1440)	490 (919 sh)

tions decrease when the distance of two aromatic planes becomes longer by the elongation of bridges. However, in many cases the elongation would at the same time make two aromatic rings more conformationally flexible (skewing, sliding, and even ring flipping), and these movements would make the detailed analysis of the spectral change more difficult. The electronic spectra of [2.2]-, [3.3]-, and [4.4]tropyliocyclophanes (**25, 26,** and **27**) have been discussed,[27] and it has been concluded that CT interaction is in the decreasing order **26 > 25 > 27**. The order is in agreement with that in the paracyclophanequinone series.[28]

The effect of the relative positions of two aromatic rings has been disclosed by the spectra of isomeric [2.2]tropyliocyclophanes **25, 28, 29,** and **30**. In **25, 28,** and **30,** in which two rings overlap more extensively, weak CT bands extend to ~450 nm, whereas in **29** of the metacyclophane type the strong band appears at ~410 nm.[24] The triple-decked compound **33** exhibits a huge broad band at 434 nm, which extends to 600 nm.[29]

c. Chemical Reactions

Very little is known about the chemical behavior of tropyliocyclophanes, although an unusual chemistry might emerge because of their novel structures. An interesting reaction was observed when **32** was exposed to moisture or treated with aqueous methanol.[26] The compound underwent an extrusion reaction to give superphane (**36**) in good yield. A possible mechanism was proposed. The reaction should be associated with the release of the strain that the original **32** has.

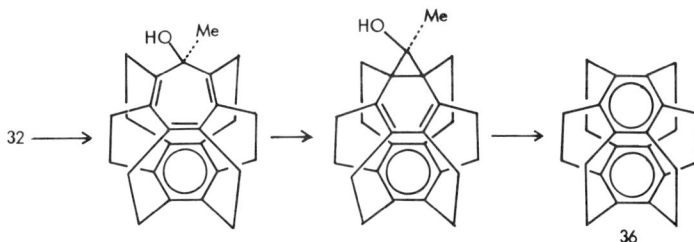

3. [2.2]TROPYLIOCYCLOHEPTATRIENOPHANE AND [2.2](1,4)TROPYLIOPHANE

[2.2](1,4)Tropyliophane ditetrafluoroborate (**37**) and the corresponding tropyliocycloheptatrienophane tetrafluoroborate (**38**) have been synthesized by the scheme shown here.[30]

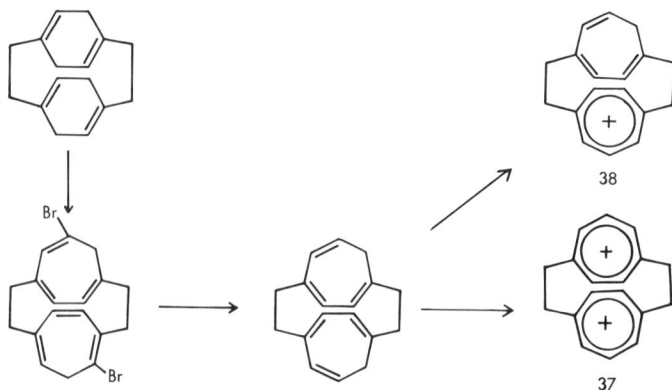

The electronic spectrum of **38,** having a peak at 313 nm and shoulders at 360 and 423 nm, is similar to that of [2.2](1,4)tropylioparacyclophane **(25)** and demonstrates the capacity of cycloheptatriene as an electron donor. Tropyliophane **37** apparently was detected by ^{1}H-nmr, but was too unstable for further investigation. However, [4.4](4,6)benzotropyliophane **(39),** a benzolog and higher homolog of **37,** forms stable perchlorate.[14,31] Compound **39** was derived from the [4.4]benzotroponophane **40,** which in turn was synthesized with its higher ([5.5] and [8.8]) homologs by method B,3 and possesses planar tropone rings.

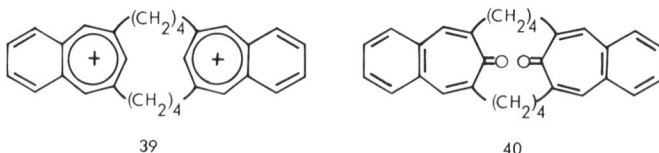

D. Tropolonophanes

Interesting questions concerning short-bridged tropolonophanes include (1) whether the tropolone ring exists in the enolone form (A) despite the strain or in one of the diketo forms (B or C) to release the strain, (2) if it exists as the enolone whether or not the ring shows some enhanced

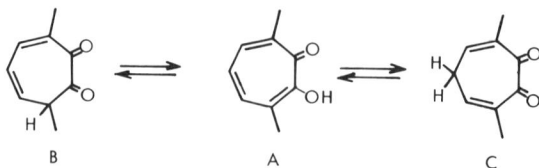

tendency to ketonize, and (3) whether the effect of CT interaction can be observed. How the internal strain is reflected in the shape of two aromatic rings is also of general interest.

Only three [2.2]tropolonophanes are known: [2]paracyclo[2](3,7)tropolonophane (41) and its halo derivatives (42 and 44).[32] They were synthesized in a long reaction sequence utilizing the synthetic method A,1,b. Thermal elimination of SO₂ was the method of choice of desulfurization, the photochemical method giving no definite result.

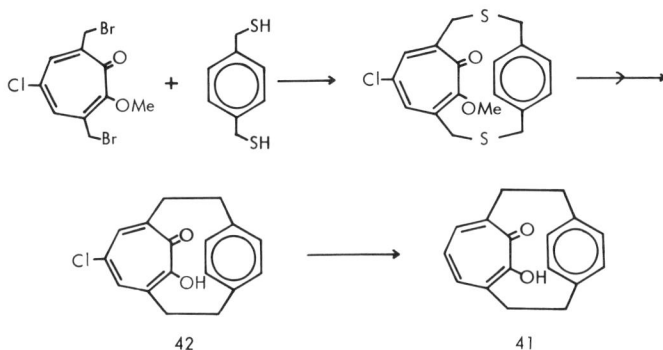

1. STRUCTURE

X-ray crystallographic analysis of 41 revealed that (a) unlike tropolone,[33] 41 has a monomeric structure with no sign of intermolecular hydrogen bonding; (b) the interplanar distance is similar to that of [2.2]paracyclophane[34]; (c) the out-of-plane deformation of the tropolone ring is larger than that of the benzene ring, reflecting the relative extent of delocalization; and (d) despite the distortion, the seven-membered ring still exists in the tropolone form (A) but not in the keto forms (B or C) (Fig. 3).

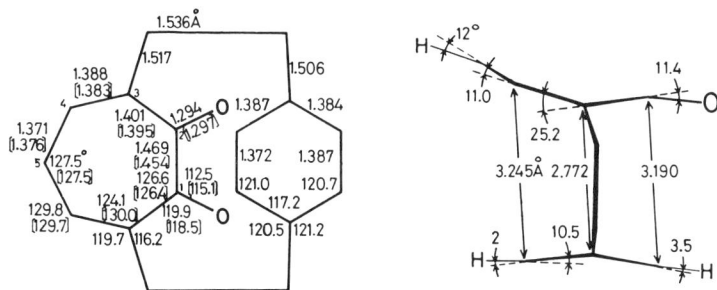

Fig. 3. Molecular structure of 41.[32] Numbers in brackets indicate averaged values of tropolone. (Reproduced by courtesy of Pergamon Press.)

Compounds **41** and **42** exist in the enolone form in solution also and have no tendency to ketonize, as shown by the positive $FeCl_3$ color reaction and their methyl ether formation, both characteristic of tropolones, and resistance to form quinoxaline derivatives. Their ir spectra (no strong carbonyl band higher than 1600 cm^{-1}) and ^{1}H-nmr spectra [AB$_2$ pattern at 6.84 (2H) and 6.49 ppm (1H)] are consistent with the conclusion. However, when brominated, **41** gave exclusively the bromo diketone **43**, which on heating changed to **41** and bromotropolone **44**. Furthermore, on

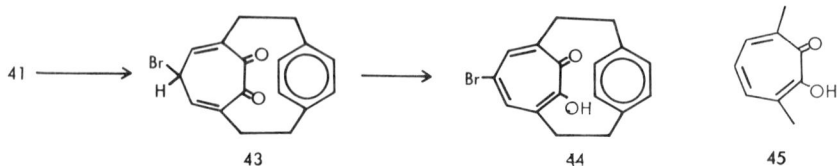

catalytic (Pd–C) hydrogenation **42** yielded a large amount of overreduction products. This behavior, differing from that of 3,7-dimethyltropolone **(45),** may reflect the destabilization of the tropolone ring by the strain.

2. CT INTERACTION

The electronic spectra of **41** and **42** exhibit general features similar to those of the electronic spectrum of **45,** although broadening and some red shift are also noted. This would indicate the presence of only weak CT interaction. However, when the spectra were measured in acidic solvent, a weak but clear shoulder appeared at ~420 nm. This corresponds to the CT absorption (400–430 nm) in tropylioparacyclophane **(25)**[24,27,29,34a] and suggests the contribution of the dihydroxytropylium ion **46**.

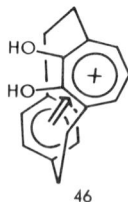

Charge-transfer interaction plays an important role in a reaction. Acetate **47** undergoes acetotropy, as does tropolone acetate.[35] However, the free energy of activation for the process ($\Delta G^{\ddagger} \approx 8.7$ kcal/mol) is considerably lower than in the latter ($\Delta G^{\ddagger} \approx 10.8$ kcal/mol). The difference was accounted for by assuming stabilization in the transition state **48** by CT interaction.

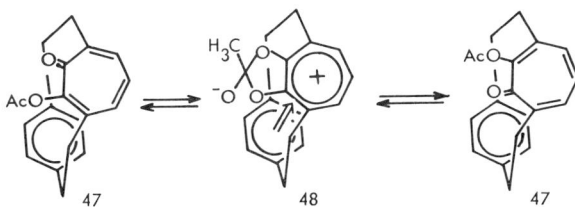

47   48   47

E. Tropoquinonophanes

A series of cyclophanes **(49–52)** containing the *p*-tropoquinone ring were synthesized,[36,37,37a] and their properties were discussed. The synthe-

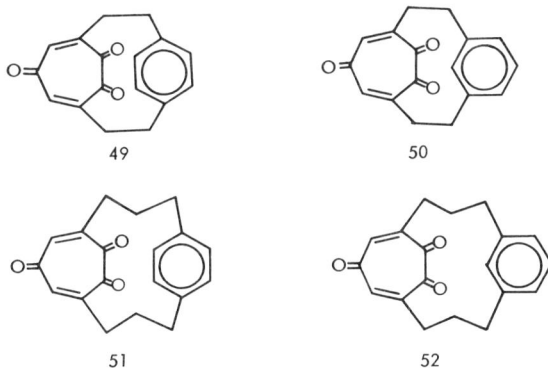

49   50

51   52

sis utilize the sulfur route (method A,1,b) starting from 5-hydroxytropo-lone derivatives, as the following example shows.

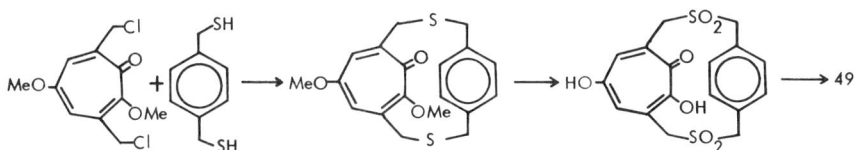

1. STRUCTURE

X-ray crystallographic analysis of **49** revealed that the internal strain caused by the phane structure was reflected in the deformation of the unsaturated rings. However, the bending of the benzene ring is smaller in **49** than in [2.2]paracyclophane **(35),**[34] and the *p*-tropoquinone ring is bent into a deep boat form (Fig. 4). Furthermore, the diketone part of the *p*-

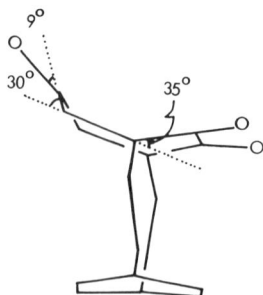

Fig. 4. Molecular structure of **49.**

tropoquinone ring is nearly parallel to the mean plane of the benzene ring, whereas the dienone part of the ring is away from it. These features also appear in structure **51,** but to a lesser extent.[37]

The strain and deformation are reflected in other spectral properties. Whereas the mass spectra of the [3.3]cyclophanes **51** and **52** show strong fragment ions due to three consecutive losses of CO, as in *p*-tropo-quinones in general,[38] those of the more strained **49** and **50** exhibit only two losses of CO, indicating that the elimination of the third CO requires higher energy. In ir spectra, tropoquinones exhibit two groups of complex carbonyl bands of nearly equal frequencies, at $\sim$1670 and $\sim$1605 cm^{-1}.[36] However, on going through **51** and **52** to **49** and **50,** the intensity of the latter band decreases to culminate in a single absorption at $\sim$1670 cm^{-1} for **50,** in which the tropoquinone ring is presumably most strained.

Methylene signals in the ^{1}H-nmr spectra of **51** and **52** are consistent with larger conformational flexibility in these compounds compared with **49** and **50.** The ring protons and carbons, especially those of the quinone ring, showed the following structural features in solution. (1) Whereas compound **50** exists in a stepped-anti form, **52** adopts an overlapped syn conformation (upfield shift values of quinone protons). (2) In accord with X-ray results, which revealed the eclipsed disposition of the α-diketone carbonyls, the reduced polarization of the α-diketone is clearly indicated for **49** and **50** (large upfield shift of the carbonyl carbons). (3) This effect is reduced somewhat in **51,** in which the longer bridge gives the ring more flexibility, and greatly reduced in **52,** in which practically no steric factor keeps the carbonyl eclipsed in the syn form.

2. CT Interaction

A slight enhancement of intensity or the appearance of a shoulder was detected at 350 to 400 nm in the electronic spectra of **49** to **52,** and, from their bathochromic shifts in polar solvent, they were attributed to weak and broad CT bands as in benzoquinonophanes.[37] First half-wave reduction potentials $E_{1/2}$ of **49** (−0.64 V), **50** (−0.64 V), **51** (−0.53 V), and **52** (−0.52 V) disclosed a considerable increase in LUMO energy compared with that of 3,7-dimethyl-*p*-tropoquinone (−0.35 V). Because **51** and **52** would have more nearly planar quinone rings, the difference [(−0.52) − (−0.35) = −0.17 V] in LUMO energy would be largely due to the CT interaction.[39] A further increase in the energy in **49** and **50** would be a reflection of bending of the quinone rings.

IV. PHANES CONTAINING NONBENZENOID 10π SYSTEMS

A. Single-Decked Phanes

1,6-Methano[10]annulene **(53)** can be considered to be the lowest homolog of [10]annulenophanes. Its synthesis by Vögel[40] opened up a new era of 10π aromatic chemistry, which expanded rapidly to include hetero analogs **(54)**[41,42] and ionic species **(55**[43] and **56**[44]**)**. The synthesis of 1,5-methano[10]annulene **(57)**,[45] an isomer of **53**, has contributed further to the chemistry of annulenes.

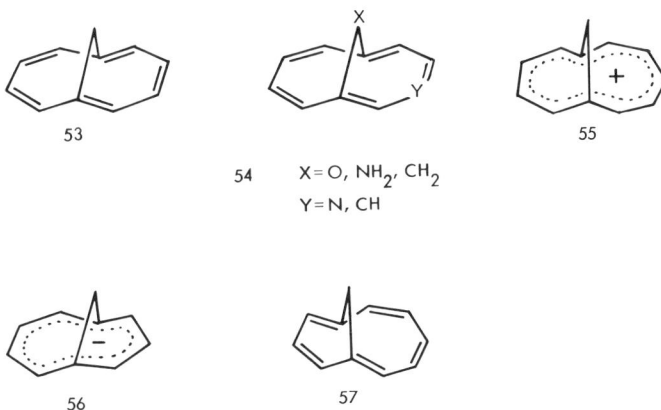

53

54   X = O, NH₂, CH₂
         Y = N, CH

55

56

57

However, the main interest in these compounds is directed toward the chemistry of the unsaturated ring itself, that is, nonbenzenoid aromatic chemistry. Therefore, although the chemistry of these compounds is very stimulating, it is not discussed further here. The same applies to the other bridged annulenes with 12-membered and larger rings.

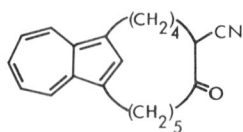

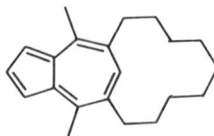

58   59

Apart from these annulenes, [11](1,3)azulenophane **(58)**[46] and [9](5,7)azulenophane **(59)**[47] are the only single-decked 10π phanes known. The former was synthesized by the Thorpe–Ziegler cyclization (method A,3,a) of azulene-1,3-bis(valeronitrile), and the latter by Hafner's azulene synthesis[48] (method B,3) starting from a pyridinophane (see Section II).

B. Double-Decked Azulenophanes

1. [2]AZULENO[2]CYCLOPHANES AND [2.2]AZULENOPHANES

A variety of [2.2]cyclophanes **(60–66)** containing one or two azulene rings have been synthesized by the sulfur route (method A,1,a) starting from azulene bis(methyltrimethylammonium) diiodide. The choice of the starting materials was based on the instability of the halomethylazulenes and their precursors.[49] The azulenocyclophanes **64** and **67** have been syn-

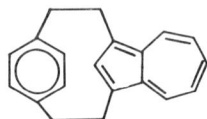

60[67]

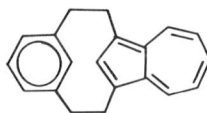

61[62,65]

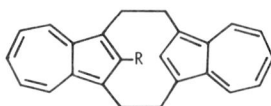

62   R = H[68]

63   R = Me[62]

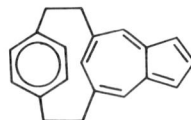

64[50,58]

65[58]

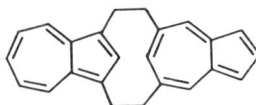

66[62]

thesized from [2.2]metaparacyclophane and [2.2]paracyclophane[50] by the intramolecular carbene insertion reaction[51] (method B,1).

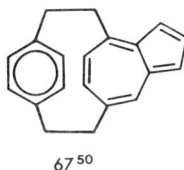

67 [50]

The synthesis of azulenophanes **68** and **69** involves Hofmann-type degradation (method A,2,a)[52] and separation by chromatography on H_3PO_4-impregnated silica gel.[53] Polynuclear [2.2.2.2](1,3)azulenophane **(70)** has been synthesized by reductive coupling (method A,2,b).[54]

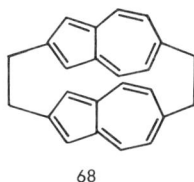

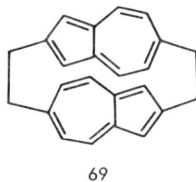

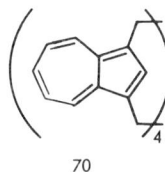

68   69   70

a. Geometry

X-ray crystallographic analysis of **60, 62,** and **64** (Fig. 5) revealed deformation around the bridging sites. Vertical deformation (up to 9°) of the azulene rings in all three, elongation of C-9—C-10 (azulene numbering) in the 1,3-bridged **60** and **62,** and elongation of C-4—C-5 and C-5—C-6 and angular deformations at C-4 and C-5 in the 5,7-bridged **64** are all clearly seen. These types of deformation are also observed for [2.2]cyclophanes.[34,55,55a] Deformation of the benzene rings in **60** and **64** is very similar to that in [2.2]metaparacyclophane.[55]

The geometry of these compounds in solution is established by their ¹H-nmr spectra. The pattern of the bridge proton signals suggests the rigidity of their conformations at room temperature. The relative geometries of two aromatic rings are easily deduced from the large upfield shift of the inner aryl protons and small downfield shifts of almost all the outer aryl protons compared with those in reference compounds, that is, substituted azulenes and xylenes. Thus, it was concluded that all compounds with two meta bridges (i.e., **61, 62, 63, 65,** and **66**) have stepped-anti conformations.[56] Flipping of the azulene ring was observed for **64** at higher temperature. The energy barrier of this dynamic process ($\Delta G^{\ddagger}$ = 16.2 kcal/mol, T_c = 70°C) is smaller than that of [2.2]metaparacyclophane (20.6 kcal/mol, T_c = 146°C).[55] In contrast, **60** showed no indication of such a process up to 190°C, from which the energy barrier of the flipping

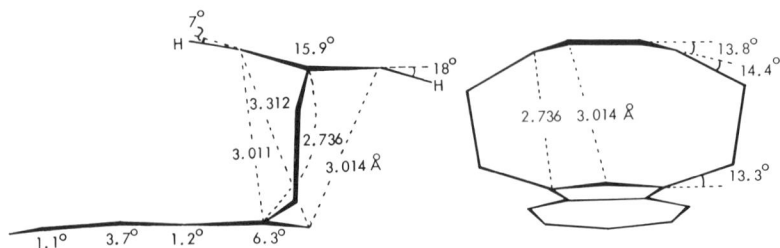

60

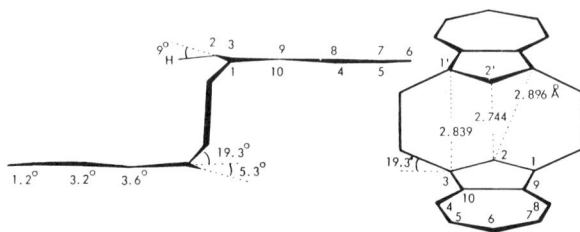

62

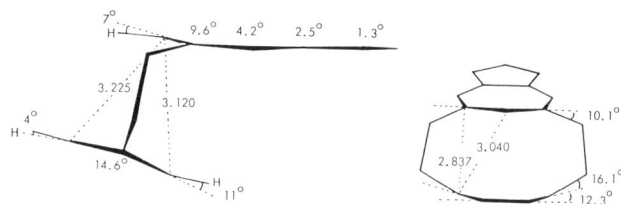

64

Fig. 5. Molecular structures of **60,**[67] **62,**[68] and **64.**[58] (Reproduced by courtesy of Pergamon Press.)

was estimated to be at least 22 kcal/mol. Apparently, the differences reflect the size of steric repulsion between the inner hydrogen and the facing benzene ring in their rotational transition state. The distances between the hydrogen and the benzene ring in the transition state were calculated from X-ray data to be ~1.2, ~1.5, and 1.75Å for **60,** [2.2]meta-paracyclophane, and **64,** respectively.

The (2,6)-bridged azulenophanes **68** and **69** may have some gliding movement at room temperature, as indicated by the singlet and AA′BB′ patterns, respectively, of bridge protons in the ir and ¹H-nmr spectra.

These compounds seem to have barrel shapes on the basis of the ^{1}H-nmr chemical shift of azulene protons, as shown below (smaller upfield shift of H-4 compared with those of the others; reference compound, 2,6-dimethylazulene) and preliminary X-ray results. However, the precise geometry of these compounds could not be deduced because of the extensive disorder of their crystals.[53]

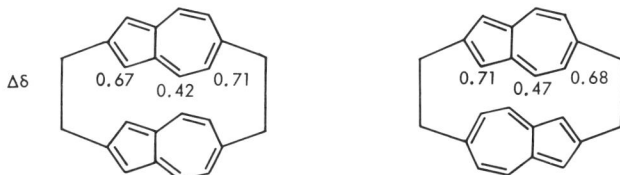

Δδ   0.67   0.71   0.71   0.68
              0.42   0.47

b. Transannular Interactions

The three main electronic absorption bands in azulene (^{1}Lb ~670 nm, ^{1}La ~400 nm, ^{1}Bb ~270 nm)[57] show some bathochromic shifts and broadening (loss of fine structure) when the ring is incorporated in a double-decked [2.2]phane system as is common in [2.2]cyclophanes having weak CT interaction.

In the [2]azuleno[2]cyclophane group, the ^{1}Bb band of metacyclophanes **61** and **65** exhibits a larger bathochromic shift (~2500 cm^{-1}) than those of paracyclophanes **60** and **64** (~300 cm^{-1}). The difference can be explained qualitatively by consideration of the molecular orbitals of azulene and xylenes.[58] As shown on the left side of Fig. 6, the next HOMO of azulene, which is responsible for the ^{1}Bb band, does not interact efficiently with that of *p*-xylene in **60** and **64** because of their geometries. Therefore, the next HOMO of these compounds does not differ greatly from that of azulene. However, the geometries of **61** and **65** allow sufficient overlap of the orbital coefficient of the inner carbons to cause

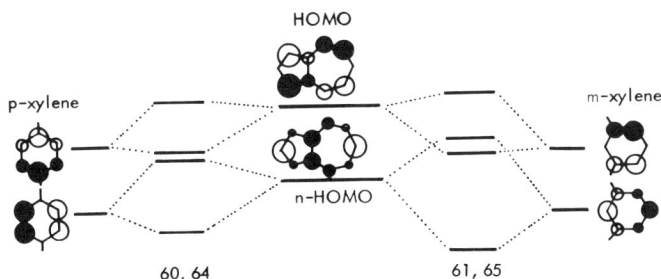

HOMO

p-xylene   m-xylene

n-HOMO

60, 64   61, 65

Fig. 6. Orbital interaction of azulene and xylenes.

the large bathochromic shift, as indicated on the right side of Fig. 6. The azulenophane **67,** with large overlap in asymmetric fashion, shows a rather large bathochromic shift in both ^{1}Bb and ^{1}La bands.

In the [2.2]azulenophane group, somewhat stronger CT interaction is expected for **66** and **69,** as judged from the sizable dipolar character of azulene. In fact, however, no clear absorption band due to the interaction has been observed except extensive broadening in the electronic spectra of all members of this group. The orthogonal nature of the coefficient in the HOMO and LUMO of azulene seems responsible for the absence of a strong CT band in these compounds.[52]

The electronic[53] and photoelectron spectra[60] and magnetic circular dichroism[59] of **68** are very similar to those of **69,** indicating the similarity of the orbital interaction of the two azulene rings in these compounds. The absence of the large orientation effect corroborates the representation of azulene by the perturbed [10]annulene structure. Compounds **68** and **69** emit fluorescence from the third excited singlet states.[59] This anomaly appears to originate from a large gap between the S_2 and S_3 states as in the case of azulene, in which the gap between S_1 and S_2 is concerned.[61]

In contrast to the neutral species described, the monoprotonated cation of all these compounds produces clear CT bands at 400 to 500 nm.[53,62] From the acid concentration required to reach maximum intensity of the CT bands, syn isomer **68** was estimated to be a slightly stronger base than the anti isomer **69,** indicating a small orientation effect, probably due to the dipole–dipole interaction of two azulene rings.[53]

Transannular interactions in anion-radicals corresponding to **60, 62, 68,** and **69** were investigated.[63] The azulene rings of these compounds show a strong tendency to form ion pairs with countercations.

c. Transannular Reactions

Only a few reactions of [2.2]azulenophanes have been investigated so far. Following the precedence of [2.2]metacyclophane,[64] photochemical transannular reactions of **61** and **62** were examined and found to culminate in the synthesis of azulenophenalene **(71)**[65] and naphthodiazulene **(72)**[66] nonalternant isomers of benzopyrene and dibenzopyrene, respectively. A similar transannular reaction of **61** took place under acidic conditions to give the tropylium ion **73.**[62]

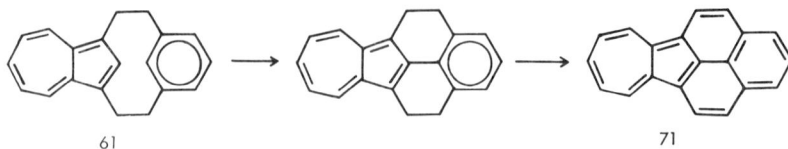

61   71

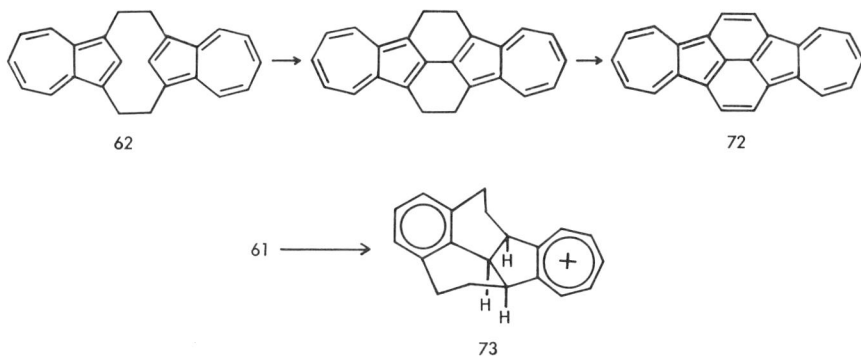

62 → → 72

61 → 73

2. DITHIA[3.3]AZULENOPHANES AND DITHIA[3]AZULENO[3]CYCLOPHANES

In the course of the synthesis of azulenophanes **60–66,** the corresponding dithia[3.3]analogs **(74–81)** were also formed.[58,62,65,67,68] Their geometries and conformational changes, as observed in their [1]H-nmr spectra, are discussed in the following paragraphs.[1]

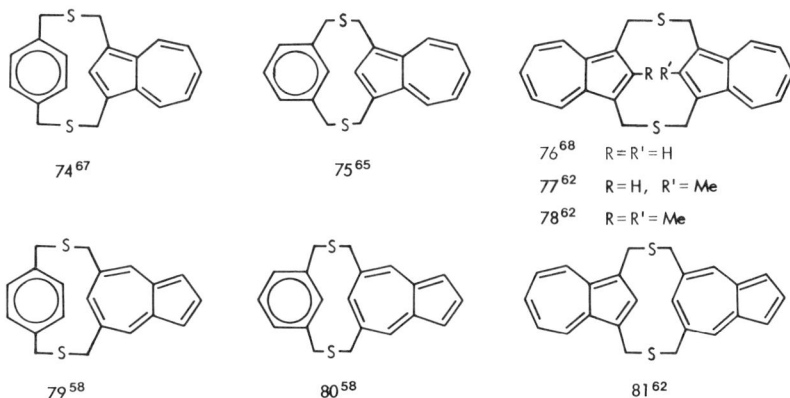

74[67]   75[65]   76[68] R = R' = H
                    77[62] R = H, R' = Me
                    78[62] R = R' = Me

79[58]   80[58]   81[62]

All of the compounds in this group except **78** exhibit rapid dynamic processes at room temperature, as in [3.3]metacyclophane[69] and its dithia analog.[70,70a] This is shown by one or two sharp singlets due to methylene protons and involves both flipping of aromatic rings and conformational change in the bridge. By a lowering of the temperature, the former process can be frozen by −135°C in most cases, whereas the latter process is still operating in some cases. Proton nmr experiments also revealed two important conformational features of these compounds: a preference for

the syn conformation and a difference in the energy barrier for ring flipping.

The activation energy ($\Delta G^{\ddagger}$) of ring flipping is 7.0 kcal/mol ($T_c = -125°$
C) for **74** and less than 7.0 kcal/mol (no sign of broadening at $-110°C$) for
79. The fact that the activation energy for **74** is larger than that for **79,**
being in accord with the results for [2]azuleno[2]paracyclophanes **60** and
64 (Section IV,B,1,a), should originate from the difference in the steric
compression in their respective rotational transition states.

The inner azulene protons of **74** and **79,** which would receive a time-
averaged diamagnetic ring current effect, show large upfield shifts when
compared with the unbridged disubstituted azulenes. The values are comparable with that for dithia[3.3]metaparacyclophane **(82).**[71] The inner protons in **75** and **80,** however, show only small upfield shifts or a downfield

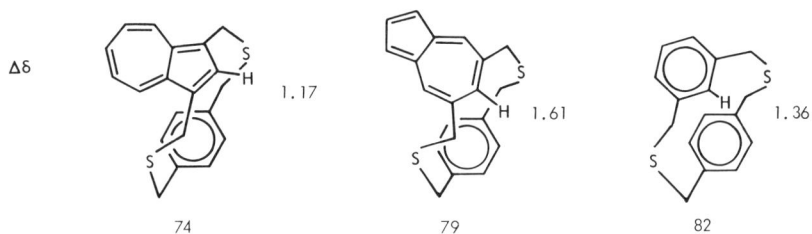

shift. These values, comparable with that for dithia[3.3]metacyclophane
(83),[70a] indicate the predominant contribution of the syn conformation at
room temperature in these azulenocyclophanes. This was verified by low-
temperature ^{1}H-nmr spectroscopy. Both **75** and **80** exist entirely in the syn
conformation, implying that it is at least 1.5 kcal/mol more stable than the
anti conformation. The activation energy of ring flipping was estimated to
be ~10 kcal/mol for **75** and ~10.5 kcal/mol for **80.**

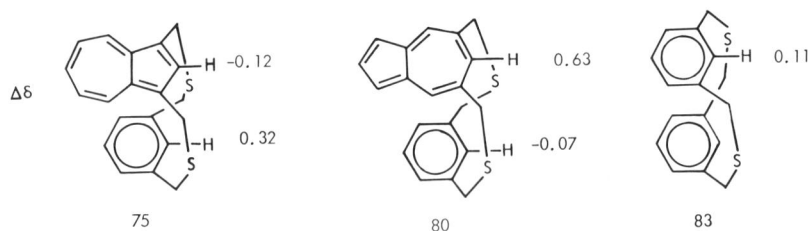

Dimethylazulenophane **(78)** exists in two discrete anti and syn isomers.
Although attempts to separate these isomers were unsuccessful, the
chemical shift of their azulenic protons can be established beyond any
doubt by taking advantage of their unequal composition. The shift values,
upfield for the syn and downfield for the anti isomer, from the unbridged

78 (syn)   78 (anti)

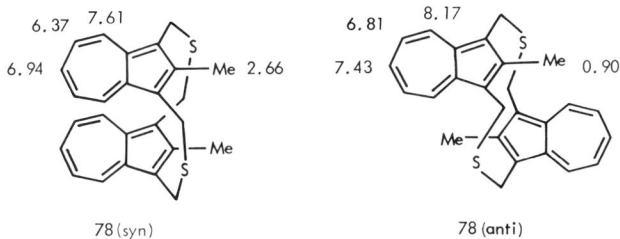

azulene protons would be useful for suggesting the predominant conformer in other azulenophanes **(76, 77,** and **81)** that undergo ring flipping
at room temperature. Most of the azulenic protons in these compounds
show a small upfield shift, as exemplified by **76** and **81,** and therefore the
preferred syn conformation was suggested.

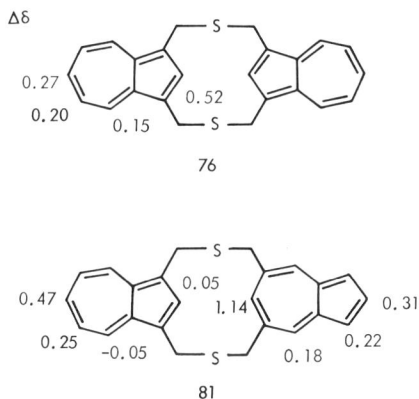

76

81

The low-temperature ^{1}H-nmr spectra supported this view and revealed
the following additional features. Compound **76** exists as a 7 : 3 mixture of
syn and anti conformers at $-120°$C. Therefore, the syn conformer is ~0.3
kcal/mol more stable than the anti form. The activation energy $\Delta G^{\ddagger}$ of the
flipping was estimated to be ~9 kcal/mol. Compound **81** exists only in the
syn conformation at $-70°$C. The activation energy of the flipping was
found to be 12 kcal/mol. The reason for the preferred syn conformation in
these compounds as well as in dithia[3.3]metacyclophane[70a] is believed to
be the nonbonded repulsive interaction in the bridge part. In the syn
conformers all the bonds and two lone pair orbitals on sulfur are always
staggered, whereas in the anti conformers a bond and two lone pair orbitals on sulfur are always eclipsed with three bonds on either one of the
adjacent carbons.

The difference in the activation energy for ring flipping most likely
originates from the interaction of two dipolar azulene rings in the syn form

of **76** and **81.** Because the bridges are long enough and the rotational transition states therefore have nearly equal potential energy, small but clear differences between $\Delta G^{\ddagger}$ (~9 kcal/mol for **76,** 10–10.5 kcal/mol for **75** and **80,** and 12 kcal/mol for **81**) should reflect relative stability of the syn conformers in each compound. Whereas **75** and **80,** in which a benzene ring is facing an azulene, have no particular interaction in their syn forms, **76** and **81,** in which two azulene rings are facing each other, have dipole–dipole interaction, one repulsive and the other attractive, in their syn forms. The interaction stabilizes the syn form of **81** and destabilizes that of **76.**

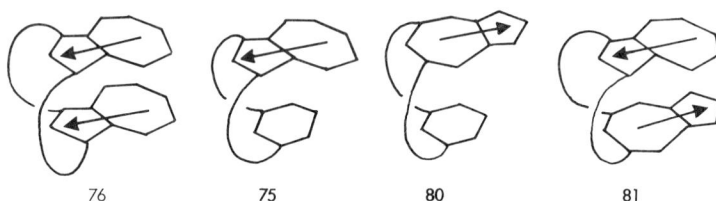

C. 1,6-Methano[10]annulenophanes

Unique cyclophanes **(84–87)** containing a 1,6-methano[10]annulene[40] moiety have been synthesized[72] by the sulfur route (method A,1,a).

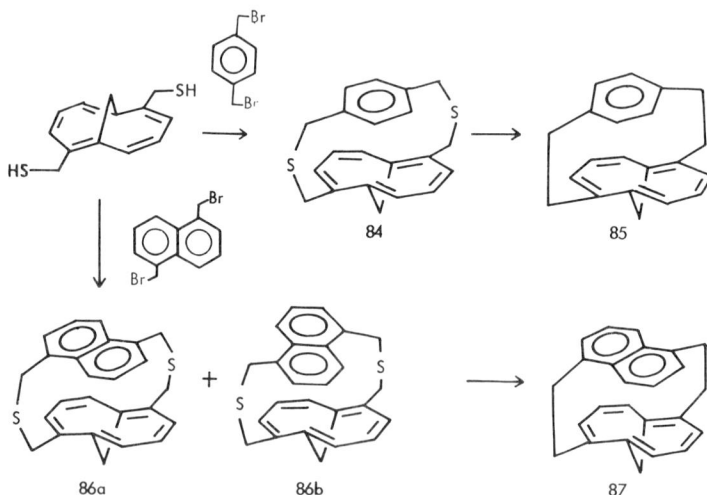

Whereas dithia[3.3]phane **(86)** has two isomers [**86a** (major) and **86b** (minor)] due to the relative orientation of two aromatic ring systems, only the parallel conformation is known for **87.** On the basis of their electronic spectra, it was concluded that transannular interaction is more significant in **87** and very small in **85.**[72]

V. CONCLUSIONS

We have discussed the chemistry of cyclophanes incorporating 6π and 10π nonbenzenoid aromatic rings. Thus far, they have been investigated by a rather small number of groups, and most of the investigations have focused on their physical properties. However, because nonbenzenoid ring systems undergo various chemical reactions, their chemical behavior in strained phane systems would also be fascinating and culminate in a deeper understanding of nonbenzenoid ring systems and eventually of aromatic systems.

Acknowledgments

The authors are grateful to all of their co-workers at Tohoku University. Particular thanks are due to Dr. Nobuo Kato, Dr. Akira Kawamata, Mitsuharu Fujii, Hideshi Saito, Muneo Aoyagi, Hiroshi Matsunaga, Masao Sobukawa, Toshihiro Shiokawa, Yasuhiro Mazaki, and Mrs. Sachiko Hanzawa (née Oeda).

REFERENCES

1. A part of the work done by us has been summarized. S. Itô, *Pure Appl. Chem.* **54,** 957–974 (1982).
2. For reviews, see W. E. Watts, *Organomet. Chem. Rev.* **2,** 231–254 (1967); D. E. Bublitz and K. L. Rinehart, Jr., *Org. React.* **17,** 1–154 (1969).
3. S. Bradamante, A. Marchesini, and G. Pagani, *Tetrahedron Lett.* pp. 4621–4622 (1971).
4. M. W. Haenel, *Tetrahedron Lett.* pp. 1273–1276 (1977).
5. M. W. Haenel, *Tetrahedron Lett.* pp. 3121–3124 (1976).
6. T. Hiyama, Y. Ozaki, and H. Nozaki, *Chem. Lett.* pp. 963–964 (1972); *Tetrahedron* **30,** 2661–2668 (1974).
7. R. Noyori, S. Makino, and H. Takaya, *Tetrahedron Lett.* pp. 1745–1746 (1973).
8. S. Hirano, T. Hiyama, and H. Nozaki, *Tetrahedron Lett.* pp. 1331–1332 (1973); *Tetrahedron* **32,** 2381–2383 (1976).
9. Y. Fujise, T. Shiokawa, Y. Mazaki, Y. Fukazawa, M. Fujii, and S. Itô, *Tetrahedron Lett.* **23,** 1601–1604 (1982).

10. M. J. Barrow and O. S. Mills, *J. Chem. Soc., Chem. Commun.* pp. 66–67 (1973).
11. E. Vogel, W. Wiedemann, H. D. Roth, J. Eimer, and H. Günther, *Justus Liebigs Ann. Chem.* **759,** 1–36 (1972); D. B. Ledlie and L. Bowers, *J. Org. Chem.* **40,** 792–793 (1975).
12. E. Kloster-Jensen, N. Tarköy, A. Eschenmoser, and E. Heilbronner, *Helv. Chim. Acta* **39,** 786–805 (1956).
13. R. E. Harmon, R. Suder, and S. K. Gupta, *Can. J. Chem.* **48,** 195–196 (1970).
14. R. E. Harmon, R. Suder, and S. K. Gupta, *J. Chem. Soc., Perkin Trans. 1* pp. 1746–1749 (1972).
15. R. W. Schmid and E. Heilbronner, *Helv. Chim. Acta* **40,** 950–956 (1957).
16. T. Gäumann, R. W. Schmid, and E. Heilbronner, *Helv. Chim. Acta* **39,** 1985–1993 (1956).
17. K. Ibata, H. Shimanouchi, and Y. Sasada, *Acta Crystallogr., Sect. B* **B31,** 482–489 (1975); Y. Sasada, H. Shimanouchi, and K. Ibata, *Abstr. Natl. Symp. Basic Org. Chem., 3rd, 1976* pp. 143–146.
18. R. W. Schmid, E. Kloster-Jensen, E. Kováts, and E. Heilbronner, *Helv. Chim. Acta* **39,** 806–812 (1956).
19. R. Guilard, P. Fournari, and M. Foutesse, *Bull. Soc. Chim. Fr.* pp. 4349–4356 (1972).
20. R. W. Gray, C. B. Chapleo, T. Vergnani, A. S. Dreiding, M. Liesner, and D. Seebach, *Helv. Chim. Acta* **59,** 1547–1552 (1976), and references cited therein.
21. H. J. Lindner, B. Kitschke, M. Liesner, and D. Seebach, *Helv. Chim. Acta* **60,** 1151–1154 (1977).
22. Y. Fujise, Y. Mazaki, and S. Itô, unpublished result.
23. R. E. Harmon, R. Suder, and S. K. Gupta, *J. Chem. Soc., Chem. Commun.* pp. 472–473 (1972).
24. H. Horita, T. Otsubo, and S. Misumi, *Chem. Lett.* pp. 1309–1312 (1977).
25. R. Gray and V. Boekelheide, *J. Am. Chem. Soc.* **101,** 2128–2136 (1979).
26. Y. Sekine and V. Boekelheide, *J. Am. Chem. Soc.* **103,** 1777–1785 (1981).
27. H. Horita, T. Otsubo, and S. Misumi, *Chem. Lett.* pp. 807–810 (1978).
28. T. Shinmyozu, T. Inazu, and T. Yoshino, *Chem. Lett.* pp. 1347–1350 (1977).
29. H. Horita, T. Otsubo, Y. Sakata, and S. Misumi, *Tetrahedron Lett.* pp. 3899–3902 (1976).
29a. A. W. Hanson, *Acta Crystallogr., Sect. B* **B33,** 2003–2007 (1977).
30. J. G. O'Connor and P. M. Keehn, *Tetrahedron Lett.* pp. 3711–3714 (1977).
31. R. E. Harmon, R. Suder, and S. K. Gupta, *J. Chem. Soc., Chem. Commun.* pp. 1170–1171 (1969).
32. N. Kato, Y. Fukazawa, and S. Itô, *Tetrahedron Lett.* pp. 1113–1116 (1979).
33. H. Shimanouchi and Y. Sasada, *Acta Crystallogr., Sect. B* **B29,** 81–90 (1973).
34. H. Hope, J. Bernstein, and K. N. Trueblood, *Acta Crystallogr., Sect. B* **B28,** 1733–1743 (1972).
34a. J. G. O'Connor and P. M. Keehn, *J. Am. Chem. Soc.* **98,** 8446–8450 (1976).
35. S. Masamune, A. V. Kemp-Jones, J. Green, D. L. Rabenstein, M. Yasunami, K. Takase, and T. Nozoe, *J. Chem. Soc., Chem. Commun.* pp. 283–284 (1973).
36. A. Kawamata, Y. Fukazawa, Y. Fujise, and S. Itô, *Tetrahedron Lett.* **23,** 1083–1086 (1982).
37. A. Kawamata, Y. Fujise, Y. Fukazawa, and S. Itô, to be published.
37a. A. Kawamata, Y. Fukazawa, Y. Fujise, and S. Itô, *Tetrahedron Lett.* **23,** 4955–4958 (1982).
38. S. Itô, Y. Shoji, H. Takeshita, M. Hirama, and K. Takahashi, *Tetrahedron Lett.* pp. 1075–1078 (1975).
39. Very similar differences in the reduction potentials were observed in [2]cyclo[2]- and

[3]cyclo[3](1,3)-*p*-benzoquinonophanes, the spectral data for which indicate nearly flat quinone rings.[37]

40. E. Vogel and H. D. Roth, *Angew. Chem., Int. Ed. Engl.* **3**, 228–229 (1964).
41. F. Sondheimer and A. Shani, *J. Am. Chem. Soc.* **86**, 3168–3169 (1964).
42. E. Vogel, M. Biskup, W. Pretzer, and W. A. Böll, *Angew. Chem.* **76**, 785 (1964); M. Schäfer-Ridder, A. Wagner, M. Schwamborn, H. Schreiner, E. Devrout, and E. Vogel, *Angew. Chem., Int. Ed. Engl.* **17**, 853–855 (1978).
43. W. Grimme, H. Hoffman, and E. Vogel, *Angew. Chem.* **77**, 348–349 (1965); E. Vogel, R. Feldmann, and H. Düwel, *Tetrahedron Lett.* pp. 1941–1944 (1970).
44. W. Grimme, M. Kaufhold, U. Dettmeier, and E. Vogel, *Angew. Chem.* **78**, 643–644 (1966); P. Radlick and W. Rosen, *J. Am. Chem. Soc.* **88**, 3461–3462 (1966).
45. S. Masamune and D. W. Brooks, *Tetrahedron Lett.* pp. 3239–3240 (1977); L. T. Scott, W. R. Brunsvold, M. A. Kirms, and I. Erden, *Angew. Chem., Int. Ed. Engl.* **20**, 274–279 (1981).
46. A. G. Anderson, Jr. and R. D. Breazeale, *J. Org. Chem.* **34**, 2375–2384 (1969).
47. S. Kurokawa and A. G. Anderson, Jr., *Bull. Chem. Soc. Jpn.* **52**, 257–258 (1979).
48. K. Hafner, *Justus Liebigs Ann. Chem.* **606**, 79–89 (1957).
49. K. Hafner and C. Bernhard, *Justus Liebigs Ann. Chem.* **625**, 108–123 (1959); W. Treibs, *Chem. Ber.* **92**, 2152–2163 (1959).
50. T. Kawashima, T. Otsubo, Y. Sakata, and S. Misumi, *Tetrahedron Lett.* pp. 1063–1066 (1978).
51. A. Costantino, G. Linstrumelle, and S. Julia, *Bull. Soc. Chim. Fr.* pp. 907–912, 912–920 (1970); L. T. Scott, *J. Chem. Soc., Chem. Commun.* pp. 882–883 (1973); L. T. Scott, M. A. Minton, and M. A. Kirms, *J. Am. Chem. Soc.* **102**, 6311–6314 (1980).
52. R. Luhowy and P. M. Keehn, *Tetrahedron Lett.* pp. 1043–1046 (1976); *J. Am. Chem. Soc.* **99**, 3797–3805 (1977); N. Kato, Y. Fukazawa, and S. Itô, *Tetrahedron Lett.* pp. 2045–2048 (1976).
53. N. Kato, H. Matsunaga, S. Oeda, Y. Fukazawa, and S. Itô, *Tetrahedron Lett.* pp. 2419–2422 (1979).
54. M. Fujimura, T. Nakazawa, and I. Murata, *Tetrahedron Lett.* pp. 825–828 (1979).
55. D. T. Hefflinger and D. J. Cram, *J. Am. Chem. Soc.* **93**, 4754–4767 (1971).
55a. Y. Kai, N. Yasuoka, and N. Kasai, *Acta Crystallogr., Sect. B* **B33**, 754–762 (1977).
56. Precise geometries of these compounds in solution can be calculated using the line-current approach [C. A. Coulson, J. A. N. F. Gomes, and R. B. Mallion, *Mol. Phys.* **30**, 713–732 (1975)]. To be published by Y. Fukazawa and S. Itô (cf. Itô[1]).
57. E. Heilbronner, *in* "Non-Benzenoid Aromatic Compounds" (D. Ginsburg, ed.), p. 219. Wiley (Interscience), New York, 1959.
58. Y. Fukazawa, M. Sobukawa, and S. Itô, *Tetrahedron Lett.* **23**, 2129–2132 (1982).
59. H. Yamaguchi, K. Ninomiya, M. Fukuda, S. Itô, N. Kato, and Y. Fukazawa, *Chem. Phys. Lett.* **72**, 297–300 (1980).
60. B. Kovač, M. Mohraz, E. Heilbronner, S. Itô, Y. Fukazawa, and P. M. Keehn, *J. Electron. Spectrosc. Relat. Phenom.* **22**, 327–332 (1981).
61. N. J. Turro, V. Ramamurthy, W. Cherry, and W. Farneth, *Chem. Rev.* **78**, 125–145 (1978); S. Murata, C. Iwanaga, T. Toda, and H. Kokubun, *Chem. Phys. Lett.* **13**, 101–104 (1972); *Ber. Bunsenges. Phys. Chem.* **76**, 1176–1183 (1972).
62. Y. Fukazawa, M. Sobukawa, and S. Itô, to be published.
63. M. Iwaizumi, Y. Fukazawa, N. Kato, and S. Itô, *Bull. Chem. Soc. Jpn.* **54**, 1299–1304 (1981).
64. N. L. Allinger, M. A. DaRooge, and R. B. Hermann, *J. Am. Chem. Soc.* **83**, 1974–1978 (1961); N. L. Allinger, B. J. Gorden, S.-E. Hu, and R. A. Ford, *J. Org. Chem.* **32**, 2272–

520 Shô Itô, Yutaka Fujise, and Yoshimasa Fukazawa

2278 (1969); M. Fujimoto, T. Sato, and K. Hata, *Bull. Chem. Soc. Jpn.* **40,** 600–605 (1967); T. Sato, M. Wakabayashi, Y. Okamura, T. Amada, and K. Hata, *ibid.* pp. 2363–2365 (1967); T. Sato, S. Akabori, S. Muto, and K. Hata, *Tetrahedron* **24,** 5557–5567 (1968); T. Sato, M. Wakabayashi, S. Hayashi, and K. Hata, *Bull. Chem. Soc. Jpn.* **42,** 773–776 (1969); T. Sato, K. Nishiyama, S. Shimada, and K. Hata, *ibid.* **44,** 2858–2859 (1971).

65. Y. Nesumi, T. Nakazawa, and I. Murata, *Chem. Lett.* pp. 771–774 (1979).
66. Y. Fukazawa, M. Aoyagi, and S. Itô, *Tetrahedron Lett.* **22,** 3879–3882 (1981).
67. Y. Fukazawa, M. Aoyagi, and S. Itô, *Tetrahedron Lett.* pp. 1067–1070 (1978).
68. Y. Fukazawa, M. Aoyagi, and S. Itô, *Tetrahedron Lett.* pp. 1055–1058 (1979).
69. T. Otsubo, M. Kitasawa, and S. Misumi, *Bull. Chem. Soc. Jpn.* **52,** 1515–1520 (1979).
70. F. Vögtle and L. Schunder, *Chem. Ber.* **102,** 2677–2683 (1969); F. Vögtle and P. Neumann, *Tetrahedron* **26,** 5299–5318 (1970); F. Vögtle, R. Shäfer, L. Schunder, and P. Neumann, *Justus Liebigs Ann. Chem.* **734,** 102–105 (1970); H. Föster and F. Vögtle, *Angew. Chem., Int. Ed. Engl.* **16,** 429–441 (1977); T. Sato, M. Wakabayashi, K. Hata, and M. Kainosho, *Tetrahedron* **27,** 2737–2755 (1971).
70a. W. Anker, G. W. Bushnell, and R. H. Mitchell, *Can. J. Chem.* **57,** 3080–3087 (1979).
71. V. Boekelheide, P. H. Anderson, and T. A. Hylton, *J. Am. Chem. Soc.* **96,** 1558–1564 (1974).
72. M. Matsumoto, T. Otsubo, Y. Sakata, and S. Misumi, *Tetrahedron Lett.* pp. 4425–4428 (1977).

CHAPTER **9**

Multibridged Cyclophanes

HENNING HOPF

Technische Universität Braunschweig
Institute of Organic Chemistry
Braunschweig, Federal Republic of Germany

I. INTRODUCTION

As abundantly illustrated in almost every chapter of this volume, [2.2]paracyclophane (**1**, tricyclo[8.2.2.2^{4,7}]hexadeca-4,6,10,12,13,15-hexaene) has been studied more extensively than any other phane molecule prepared so far. Although itself not a multibridged cyclophane (*vide infra*), **1** may nevertheless be regarded as the parent molecule of this class of aromatic compounds, because by the introduction of additional bridges the title compounds can at least in principle be prepared from it. If these added bridges are also of the ethano type, one obtains the [2$_n$]cyclophanes

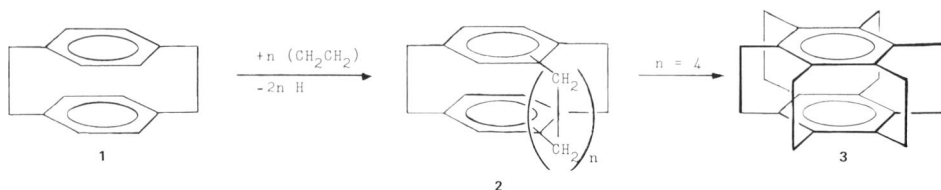

(2), a series of homologous hydrocarbons ending with the celebrated superphane **(3),**[1] which has been described as the "ultimate achievement of work in the cyclophane field."[2]

As will be seen throughout this chapter (which summarizes the literature to the end of 1981) most of the research effort in this area of phane chemistry has been devoted to molecules possessing the general structure **2;** as a result, the terms *multibridged* and [2$_n$]*cyclophanes* are currently employed more or less synonymously. The use of this nomenclature artificially limits the scope of this much broader field.

First, a multibridged cyclophane may not even contain two benzene rings. In a formal sense both benzodicyclobutene **(4)**[3] and tricyclobutabenzene **(5)**[4] as well as their higher and lower homologs are multibridged phanes. However, like dibenzocyclooctadiene, which formally is a [2.2]orthocyclophane, these hydrocarbons are normally not considered phanes, presumably for historic reasons. However, the unknown triply bridged mononuclear molecule **6** definitely is a multibridged phane.

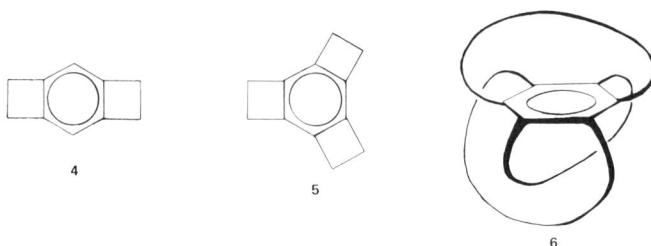

Second, the connecting structural elements need not be saturated or unfunctionalized nor of the same length. They also may be anchored to noncorresponding positions of the various rings, giving rise to multibridged cyclophanes "with nonparallel bridges"[5] or "skewed" cyclophanes. In fact, the first members of this subgroup have been prepared, as described in Section III,B.

Another obvious extension deals with the aromatic unit, which in principle may be polynuclear, carry all types of functional groups, be charged, be homoaromatic, etc. Of particular interest in this context are multibridged phanes containing anti-aromatic building blocks. It has been suggested, for example, that [2$_4$]cyclobutadienophane **(7)** (see Section II for an explanation of this nomenclature) should be stabilized by intraannular electron transfer, making both subunits aromatic, as indicated by the dipolar resonance structure **8**.[6]

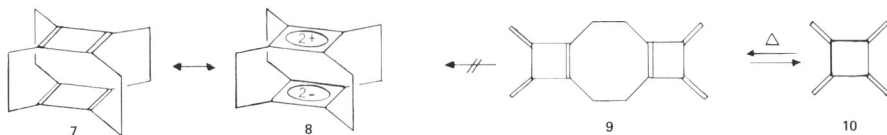

An attempt to synthesize **7/8** by flash-vacuum prolysis of **9** met with failure, the only isolable product being [4]radialene **(10)**, which had served as the starting material for **9**.[7] Although multibridged phanes incorporating a cyclooctatetraene ring have been prepared (see Section IV,C,1) there are as yet no reports of successful attempts to connect, for example, two cyclooctatetraene units by three or more bridges.

II. NOMENCLATURE

To qualify as a phane, an [n]phane must possess at least one molecular bridge, and an [$m.n$]phane at least two. It therefore seems practical to call all phanes multibridged when the number of bridges exceeds that of the aromatic/anti-aromatic nuclei by 1.

As the IUPAC notation for **1** illustrates, the application of this unambiguous nomenclature system is too cumbersome for daily, and even scientific, use. For multibridged phanes, especially if they are polyfunctionalized, the names become increasingly complicated, with the result that virtually no authors in the field use the IUPAC notation for these molecules.

It is therefore not surprising that various nomenclatures have been proposed for phanes,[8–8b] the most comprehensive and self-consistent being that of Vögtle and Neumann.[8b] For multibridged phanes these authors suggest first indicating the type and number of the bridges and subsequently their anchoring points (bridgeheads). The former information is given in brackets, and the latter is written in parentheses. For example, according to these rules cyclophane **3** is [2_6](1,2,3,4,5,6)cyclophane, with the first 2 denoting the ethano (C_2) bridge. Originally, the number and subscript in brackets were separated by a decimal point; this proposal did not prevail, however.

If other aromatic/anti-aromatic subunits are multibridged, as in the case of **7**, for example, the designation *cyclophane* is replaced by the name(s) of the appropriate aromatic/anti-aromatic nucleus/nuclei. In a strict sense the term *cyclophane* is hence reserved for molecules containing benzene rings.

This nomenclature is readily extended to "skew" or "unsymmetric" molecules. For example, the unknown hydrocarbon **11** would be called [2_4](1,2,3,4)(1,2,3,5)cyclophane, the bridgehead carbon atoms being counted in the same direction when the molecule is viewed perpendicularly from the "front" benzene ring (heavy lines).

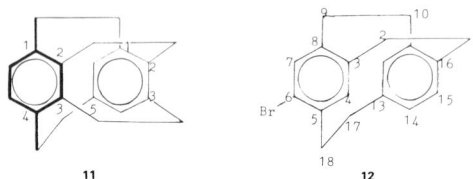

For substituted multibridged cyclophanes these rules have been extended by Kleinschroth.[9] In these cases the numbering begins at the bridge carbon atom that is bound to the less substituted ring. Whenever a branching point, that is, a bridgehead, is reached, the numbering is continued toward the unsubstituted ring atom. The bromide **12,** for example, according to this proposal is 6-bromo[2_3](1,2,4)(1,2,5)cyclophane.

III. PREPARATION OF MULTIBRIDGED CYCLOPHANES

A. [2_n]Cyclophanes with Parallel Bridges

The hydrocarbons represented by the general formula **2** are the multibridged cyclophanes *par excellence.* More importantly, the entire series beginning with $n = 1$ and ending with superphane (**3,** $n = 4$) has been prepared during the past few years, and their structural, spectroscopic, and chemical properties have been investigated. The ethano bridges in **2** end in identical positions of the two benzene rings; that is, the "clamps" are parallel to each other, whereas in the molecules dealt with in Section III,B they are bound to unlike bridgeheads. The latter hydrocarbons have therefore been termed *skewed cyclophanes.*[10]

A historical account of the preparation of the homologous series **2** has been published by the major contributor to the field.[11] The experimental material is presented somewhat differently here, the aim being to show that "epoch-making"[12] syntheses like that of **3** require novel approaches rather than an extension of more or less established methodology. In fact, it will be shown that **3** at first eluded preparation by conventional techniques.

If [2.2]paracyclophane **(1)** is regarded as the parent molecule of the symmetric [2$_n$]cyclophanes, one strategy for the synthesis of these hydrocarbons that immediately comes to mind is to introduce additional bridges into **1** (a commercial product[13]) or, more generally, to start with existing derivatives of **1**. This approach was first taken by Cram and Truesdale,[14] who prepared [2$_3$](1,2,4)cyclophane **(15)** from **1**. Friedel–Crafts acylation

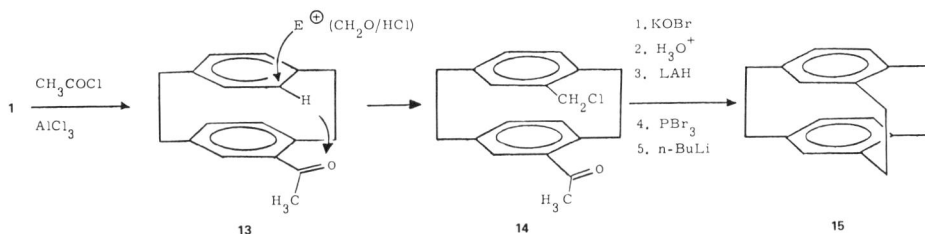

provided the expected methyl ketone **13**, which on chloromethylation exclusively yielded the pseudo-geminally substituted derivative **14**. Transannular directing effects are a well-known phenomenon in cyclophane chemistry[15] (*vide infra*). In the present case the high selectivity is caused by the basic carbonyl group, which is optimally positioned to accept intramolecularly the pseudo-geminal proton coming from the more reactive upper benzene ring and being replaced by the externally attacking electrophile. By a series of oxidations, substitutions, and reductions **14** was subsequently converted to the target molecule **15**. The relatively large effort required to prepare the penultimate precursor of **15** is a disadvantage in this synthesis. The number of steps was reduced, however, when it became possible to obtain the pseudo-geminal diester **18** directly by the addition of 1,2,4,5-hexatetraene **(16)** to methyl propiolate **(17)**.[16,16a]

Besides **18**, the Diels–Alder addition between **16** and **17** also provided its isomeric diesters **19–21**, which are ideally suited for conversion to [2$_4$](1,2,4,5)-,[17] [2$_4$](1,2,3,4)-,[18] and [2$_4$](1,2,3,5)cyclophanes[19] **(22, 23,** and **24,** respectively). As shown in the scheme, in all three syntheses the directing influence of a carbonyl-containing functional group is exploited. The ethano bridges are joined reductively, as in the case of **15,** or via carbene intermediates generated from the corresponding bistosylhydrazones by Bamford–Stevens decomposition. The C—C bond-forming step had been used earlier by both Boekelheide in the first synthesis of **22** (*vide infra*) and Cram in an alternate route to **15**.[14]

Although useful for the preparation of functionalized [2.2]paracyclophanes[16,16a,20] that can serve as precursors for multibridged cy-

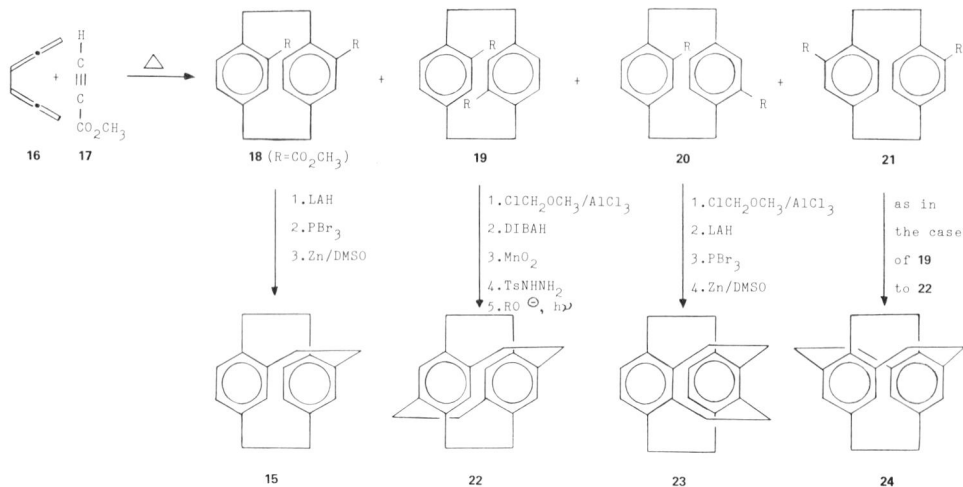

16   17   18 (R=CO₂CH₃)   19   20   21

1.LAH   1.ClCH₂OCH₃/AlCl₃   1.ClCH₂OCH₃/AlCl₃   as in
2.PBr₃   2.DIBAH   2.LAH   the case
3.Zn/DMSO   3.MnO₂   3.PBr₃   of 19
              4.TsNHNH₂   4.Zn/DMSO   to 22
              5.RO⊖, hν

15   22   23   24

clophanes, the one-pot reaction between the bisallene **16** and various activated triple-bond dienophiles generally cannot compete in breadth of application with the currently most important preparative method in phane chemistry: sulfone pyrolysis (and conceptually related approaches; *vide infra*).[21]

The first multibridged cyclophane to be prepared, [2₃](1,3,5)cyclophane **(27),** was indeed synthesized by this method.[22] The ingenuity of this approach lies in the initial construction of a relatively strain-free precursor that already is a phane molecule (in this case **26**), which is subjected to a multifold ring contraction in the last step to provide the highly strained product (e.g., **27**).

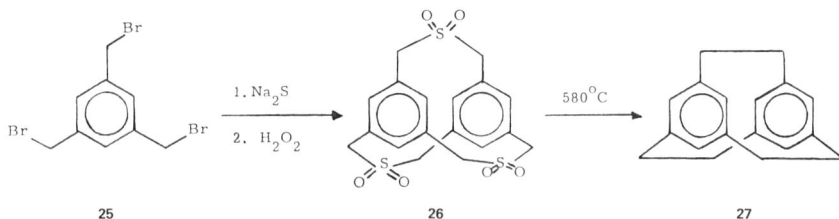

25   26   27

More recently, the sulfone pyrolysis method has been applied to another synthesis of [2₃](1,2,4)cyclophane **(15).**[23] The transannular directing effect of an ester group and the sulfur extrusion process were combined in the first synthesis of a tetra-bridged cyclophane, the (1,2,4,5) isomer **22** by Boekelheide and Gray.[24] Here, a dithia[3.3]paracyclophane **(30)** was prepared by coupling of the dibromide **28** with the bisthiol **29** but, rather than

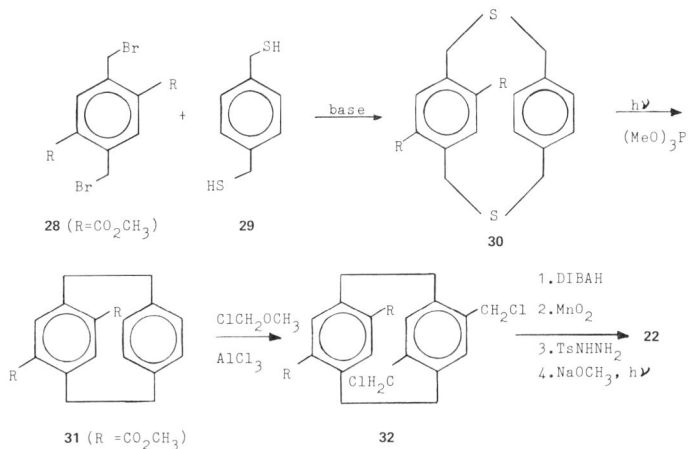

28 (R=CO$_2$CH$_3$) 29 30

31 (R =CO$_2$CH$_3$) 32

oxidizing **30** to the bissulfone, the sulfur was removed by irradiation in trimethyl phosphite.[25] As expected,[15,26] chloromethylation with chloromethyl methyl ether–aluminum chloride led to the pseudo-geminal substitution product **32,** which is readily cyclized to **22.**

The real preparative value of this approach comes to the fore with highly or unusually substituted cyclophanes. For example, the bis-thiaphane **33** on irradiation in trimethyl phosphite is desulfurized to **34,** which according to established methodology[24] is converted to the tetra-bridged tetraether **35.**[27] Derivative **35** then serves as the starting material for the bisquinone **36** and the [2$_4$](1,2,4,5)cyclophanequinhydrone **37.**[27]

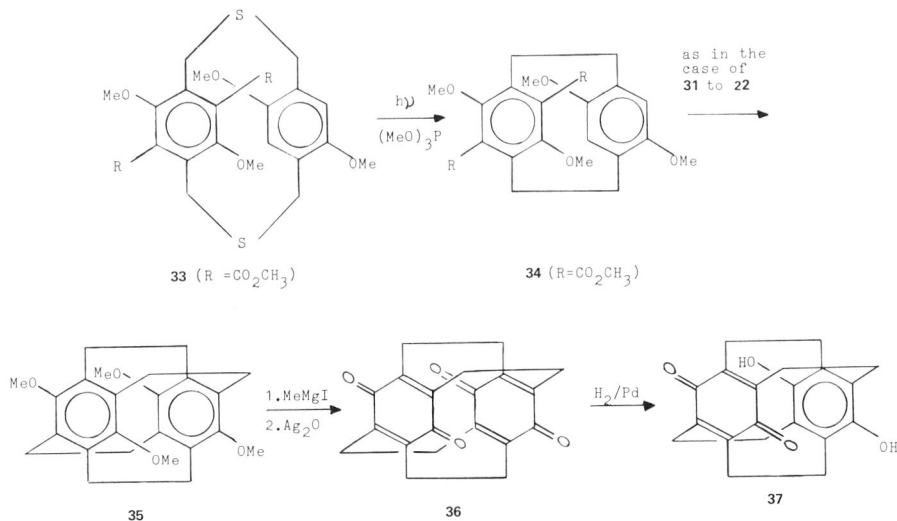

33 (R =CO$_2$CH$_3$) 34 (R=CO$_2$CH$_3$)

35 36 37

Because bridging by carbene insertion has been used successfully several times, and especially because a precursor carrying four "half-bridges" (the tetramethyl[2.2]paracyclophane **38**) can be prepared in multigram quantities,[28] this process is attractive for the preparation of superphane **(3)**.

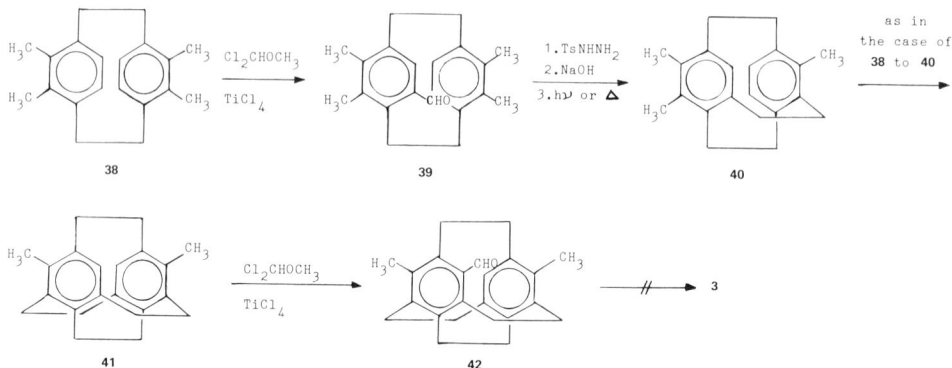

Formylation according to Rieche[29] with dichloromethyl methyl ether–titanium tetrachloride provides the aldehyde **39** in excellent yield (92%), and the bridging step to the [2$_3$]cyclophane **40** according to the principles discussed above proceeds well (60%). However, when this cycle is repeated, a pronounced drop in yield is noted (40% of **41** being isolated) and, when the cycle is carried out a third time, the aldehyde **42** is produced only in 20% yield, not enough to proceed further. The main product of this step is polymeric.[30]

Although this example demonstrated the difficulties of using the one-bridge-at-a-time approach, another attempt to synthesize **3** was undertaken with [2$_4$](1,2,3,5)cyclophane **(24)**, which can be prepared readily and in a sufficient amount by the procedure presented above, serving as substrate. Chloromethylation led to a complicated mixture of products, from which small amounts of the hexachloride **43** (an ipso substitution product; see Section IV,D,1) as well as the two isomeric dialdehydes **44** and **45** (combined yield 2.3%) could be isolated.[31] Ring closure of **44/45** via carbene insertion so far has been unsuccessful.[31] It is evident that this

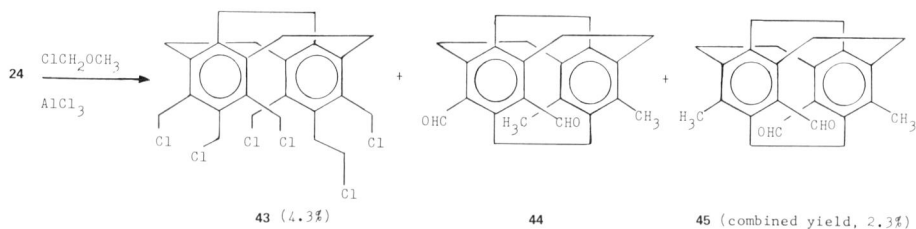

stepwise procedure, even if ultimately successful, would be slow and very tedious; furthermore, it would probably not provide enough material for chemical investigation.[30]

To synthesize the missing members of the [2_n]cyclophane family, Boekelheide and co-workers developed a new approach that capitalizes on the capacity of *o*-xylylene derivatives to undergo [4 + 4] cycloaddition to, *inter alia*, [2.2]orthocyclophanes (dibenzocyclooctadienes),[32,32a] as well as the observation that *o*-chloromethyltoluenes may lose hydrogen chloride when heated to form benzocyclobutenes.[33] Because the latter ring-open to *o*-xylylenes under high-temperature pyrolysis conditions,[32a] a direct method for converting the aromatic halide to these reactive intermediates is available. Furthermore, provided that the dimerization can be carried out intramolecularly, a method for producing two ethano bridges in one step is at hand. This concept was first realized for **22,** which was prepared from 2,5-dimethylbenzyl chloride **(46).**[34]

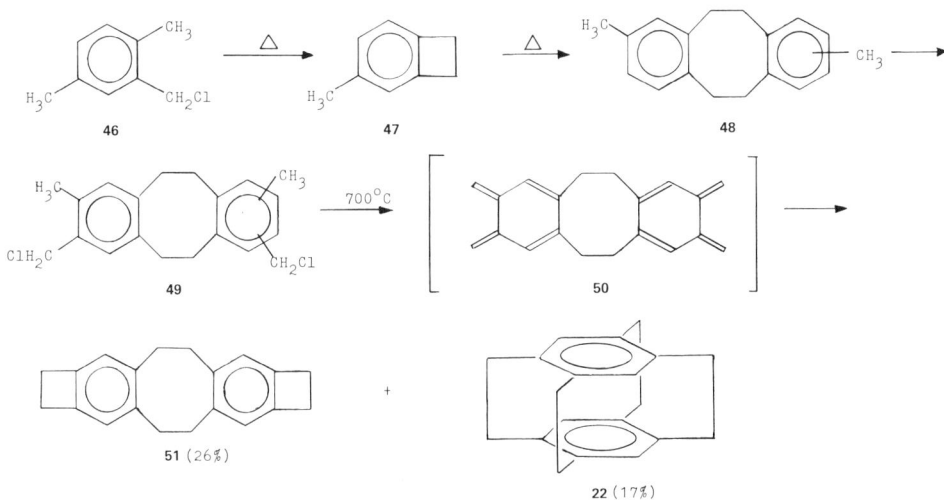

Thermal dehydrochlorination of **46** leads to a mixture of isomeric dimethyl[2.2]orthocyclophanes **(48),** the benzocyclobutene **47** being passed *en route.* By chloromethylation of **48** a substitution pattern corresponding to that of the starting material **46** is created and, when the dichlorides **49** are pyrolyzed at 700°C, a mixture of **51** and **22** is produced. Recycling of **51** through the pyrolysis step converts it to **22** as well, making the overall yield of the process satisfactory. That the benzocyclobutene dimerization method is indeed general was established by preparing [2_3](1,2,4)cyclophane **(15)** from **54**[35,35a] and the then unknown triply bridged hydrocarbon [2_3](1,2,3)cyclophane **(58:** R = H) from **57.**[36]

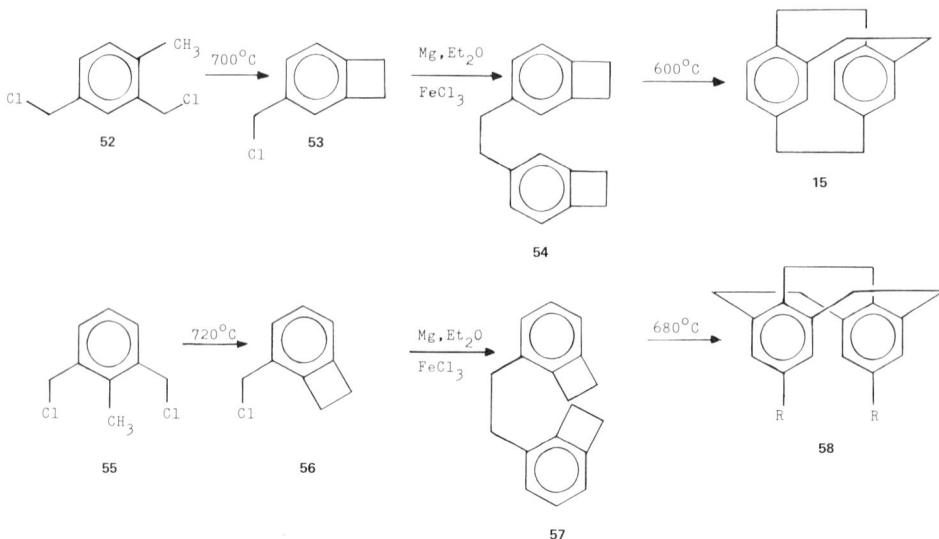

The isomeric precursors **54** and **57,** respectively, were prepared from the chloromethylbenzocyclobutenes **53** and **56,** which in turn were obtained from **52** and **55** by thermal dehydrochlorination.

The (1,2,3)-bridged system is the only one that has also been synthesized by deliberate destruction of one ethano bridge of a higher bridged cyclophane: pyrolysis of [2$_4$](1,2,3,5)cyclophane **(24),** as described in Section IV,B,1, leads to the dimethyl derivative **58** (R = CH$_3$).[37]

The *o*-xylylene dimerization reaction proved its ultimate worth in the brilliant syntheses of [2$_5$](1,2,3,4,5)cyclophane **(65)**[38] and superphane **(3).**[1] The crucial intermediate **64** was obtained by Grignard coupling of **63,** which was synthesized by what at that point amounted to routine methods from **59.** To the extent that organic chemistry is the art of joining carbon atoms, the last step of this sequence, in which four single bonds are formed in one transformation (and in excellent yield), must be regarded as a synthetic masterpiece.

An extraordinary feature of the synthesis of superphane **(3)**[1] is the consistent application of a single concept throughout the sequence. Often in hydrocarbon chemistry, and especially when highly symmetric target molecules are to be prepared, cycloadditions play an important and even central role in the initial stages of the synthesis. Later steps commonly suffer from the fact that it takes several synthetic transformations to produce just one bond, making the whole scheme very involved (and the yield minute). The superphane synthesis avoids this pitfall because it

59   60 (96%)   61 (R=CO$_2$CH$_3$:36%)
                    62 (R=CH$_2$OH:96%)
                    63 (R=CH$_2$BR:99%)

64 (88%)

65

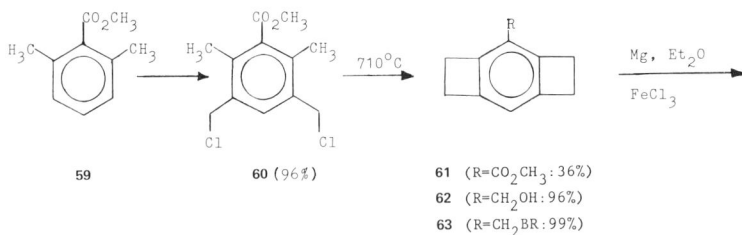

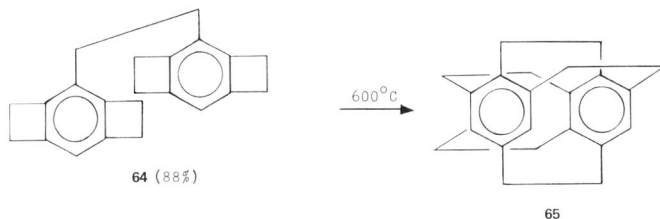

consists basically of a threefold repetition of the *o*-xylylene dimerization. That it furthermore begins with an inexpensive commercial product, the chlorodurene **66,** makes it even more attractive.

Functionalization of the [2$_n$]cyclophanes **(2)** has been extensively studied and is dealt with in detail in Section IV.

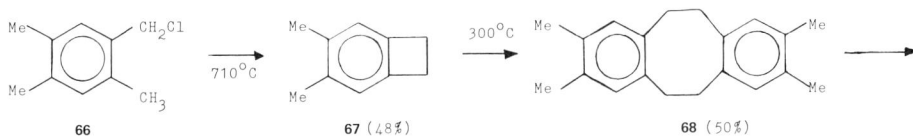

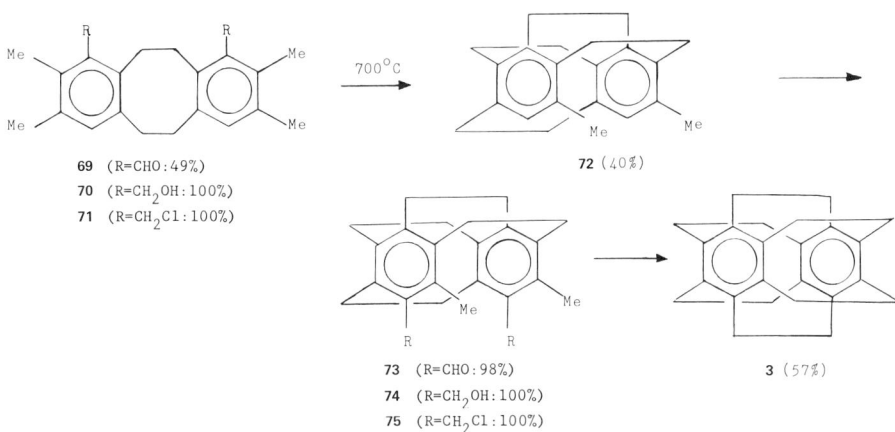

66   67 (48%)   68 (50%)

69 (R=CHO:49%)   72 (40%)
70 (R=CH$_2$OH:100%)
71 (R=CH$_2$Cl:100%)

73 (R=CHO:98%)   3 (57%)
74 (R=CH$_2$OH:100%)
75 (R=CH$_2$Cl:100%)

B. Skewed [2ₙ]Cyclophanes

[2_n]Cyclophanes in which the ethano bridges end in unlike bridgeheads of the two benzene rings have been termed "skewed cyclophanes,"[10] "cyclophanes with nonparallel bridges,"[5] or "unsymmetric cyclophanes."[10,11]

In principle these hydrocarbons can be prepared by the same methods used for the synthesis of their "parallel" isomers **2** (Section III,A). However, whereas sulfone pyrolysis of **76** and of **78** has been applied successfully to prepare both [2_3](1,2,4)(1,3,5)- and [2_3](1,2,4)(1,2,5)cyclophanes (**78** and **79**, respectively),[10] neither the reductive connection of appropriately functionalized carbon atoms, as in **80**,[39] nor carbene dimerization, as

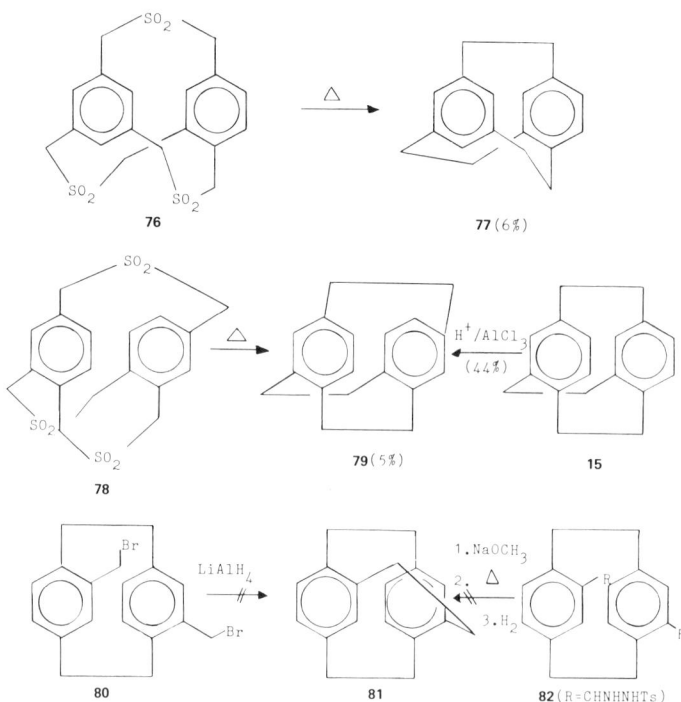

intended for **82,** leads to [2_3](1,2,4)(1,3,4)cyclophane (**81**) or its monoene.[39] Evidently, the distance between the reactive centers in these derivatives is too large to allow bond formation.

That no general methodology to prepare skewed [2_n]cyclophanes has evolved is illustrated by the fact that, although **15** on treatment with hydrochloric acid–aluminum chloride rearranges to the skewed phane **79**,[37] neither **23** nor **24** isomerizes to an unsymmetric [2₄]cyclophane under comparable conditions[37] (see Section IV,B,2). The Lewis acid-catalyzed rearrangement of [2.2]paracyclophane **(1)** to [2.2]metaparacyclophane has been known for several years.[40]

C. Other Multibridged Cyclophanes

1. ALL-CARBON BRIDGES

Although the [2ₙ]cyclophanes, whether symmetric or not, are currently in the center of chemical activity in this area (see Section IV), they are by no means the only multibridged cyclophanes known. Many other multibridged cyclophanes, shown on the next pages, may be regarded as derivatives of the [2ₙ] systems in the sense that they are formally derived from molecules discussed in Sections III,A and III,B by replacement of ethano bridges by other, sometimes quite complicated bridging units (*vide infra*). The family relationship is particularly obvious for [3.2.2](1,2,4)cyclophane **(84)** prepared from **14** via ketone **83** by Cram and Truesdale during their synthesis of [2₃](1,2,4)cyclophane **(15)**.[14]

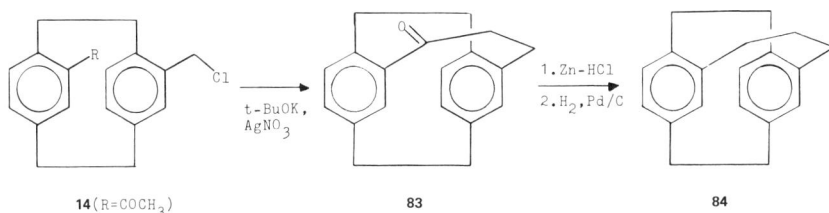

14(R=COCH₃)   83   84

An interesting, although structurally limited approach for preparing [n₃](1,3,5)cyclophanes **(86)** with n exceeding 2 has been reported by Hubert and Dale.[41,42] Treatment of diacetylenes such as **85** with Ziegler catalysts yields the cyclophanes **86** in variable yields (up to 50%) by cyclic trimerization. The method fails with 1,5-hexadiyne; that is, [2₃]-(1,3,5)cyclophane **(27)** cannot be prepared by this route.

When applied to 1,7-cyclododecadiyne **(87)**, this approach does not lead to the fully bridged "percyclophane-4" **89**, [4₆](1,2,3,4,5,6)cyclophane, as originally thought,[43] but rather to the aromatic triyne **88**.[44] In another unsuccessful experiment, **43** failed to provide homosuperphane on reductive cyclization.[45]

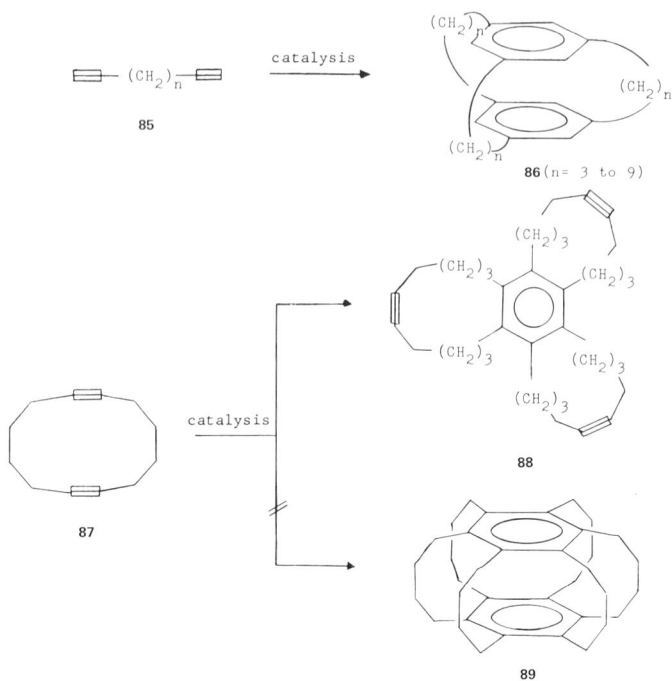

85

86 (n= 3 to 9)

catalysis

87

catalysis

88

89

2. MOLECULAR BRIDGES CONTAINING HETEROATOMS

In the next logical step toward functionalized multibridged cyclophanes the methylene groups of the various parent systems are replaced by heteroatoms. In fact, trissulfones such as **26, 76,** and **78** and their corresponding sulfide precursors belong to this category of compounds.[10,22,46,47] The surprising formation of the thiasuperphane **90** from **3** is described in Section IV,B,2.

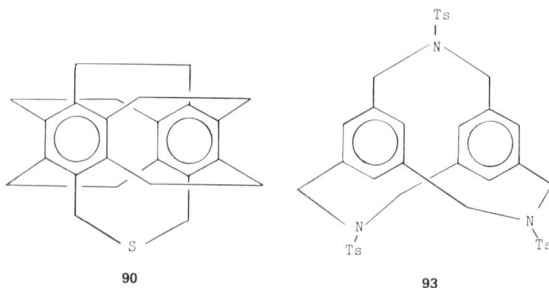

90   93

91   92 (m = 1-3)

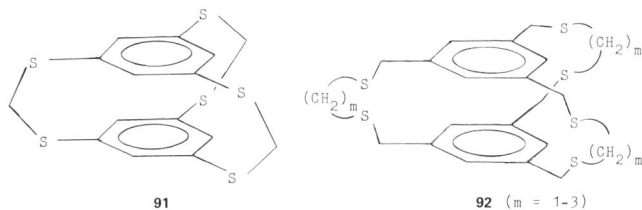

The relatively flexible hexathiaphanes **91**[48] and **92**[49] have been synthe-sized by using special dilution techniques. The ¹H-nmr spectra of these compounds are "normal"; that is, there is no longer a transannular aniso-tropic effect.

In addition to sulfur, oxygen and nitrogen (or functional groups contain-ing these heteroatoms) have been used as bridging atoms most often. Vögtle and co-workers prepared the trisazaphanes **93**,[50] and the same group reported that reduction of the *syn*-dinitro compound **94** with lithium aluminum hydride provides the [3.2.3](1,2,3)phane **95** possessing both sulfur- and nitrogen-containing bridges.[51]

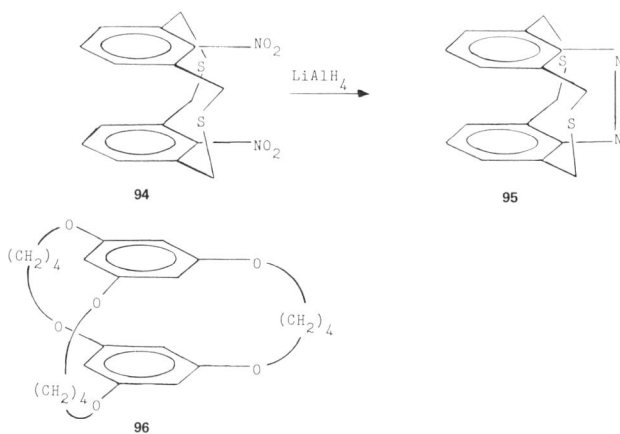

94   95

96

The 1,3,5 substitution pattern is again encountered in various oxa-phanes, interesting hybrids between crown ethers and cyclophanes. For example, the hexaoxa[6₃](1,3,5)cyclophane **96** has been prepared from phloroglucinol and 1,4-dibromo-*n*-butane.[52] Preparative techniques de-veloped for the synthesis of crown ethers have also been exploited for the preparation of a series of macropolycyclic cryptand systems with large, flexible, and partially endolipophilic cavities. A case in point is the cryptand **103** prepared from methyl 3,5-dihydroxybenzoate **(97)** and 8-chloro-3,6-dioxaoctan-1-ol **(98)** via **99** and its ditosylate **100**.[53]

CH$_3$O$_2$C

OH

OH

97

+

Cl

O

O

OH

98

HO$^-$ →

HO$_2$C

O

O

O

OR

O

O

O

OR

99 (R=H)
100 (R=Ts)

+ **97** →

RO$_2$C

O

O

O

O

O

O

O

O

CO$_2$R

101 (R=H)
102 (R=Cl)

+ various diamines →

O

N

O

O

O

N

O

O

O

O

O

O

O

O

O

O

O

103

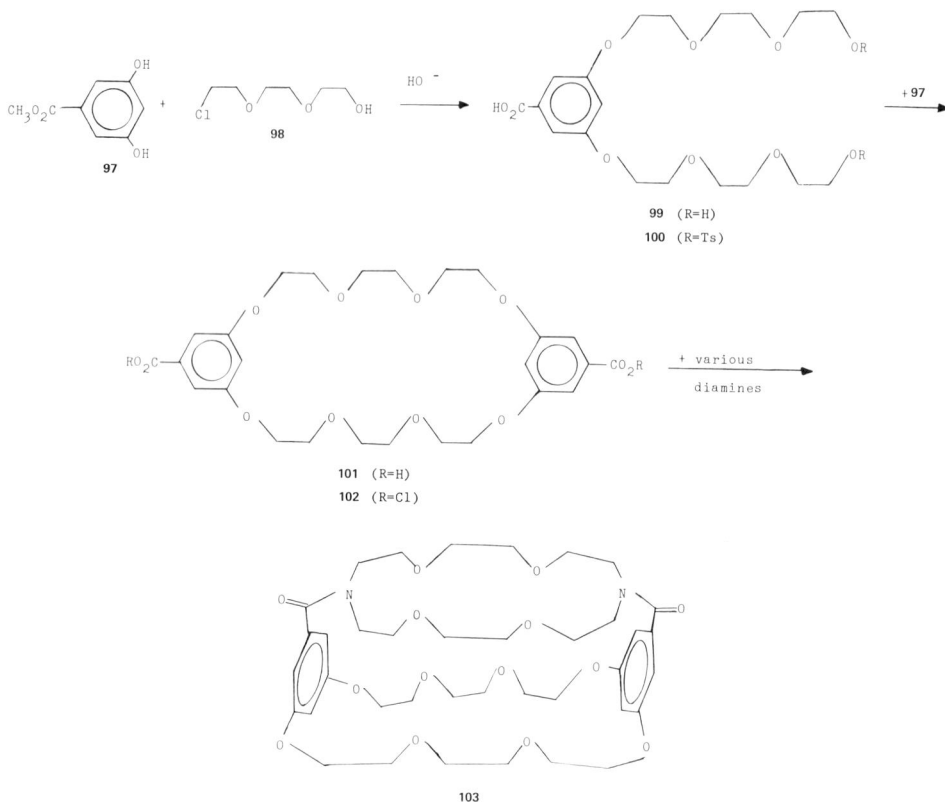

In the second part of the synthesis, **100** is treated with another equivalent of **97** to provide the crown compound **101**, which, after conversion to the bis(acid chloride), is ready to undergo reaction with various diamines. For example, 1,4,10,13-tetraoxa-7,16-diazacyclooctadecane provides the cryptand **103**, a quite remarkable [m.n.n](1,3,5) cyclophane!

Although molecules of this size and high degree of functionality are clearly still phanes in the sense that they are bridged aromatic systems, further discussion of these macropolycycles is beyond the scope of this chapter. In fact, because the chemical behavior of these molecules is dominated by the donor centers, they are more properly subsumed under the rubrum of crown ethers.[54]

Returning to simpler systems, there is spectroscopic evidence that the dithiadiazaphane **104** can be prepared from 1-amino-3,7-dialkyl-4-methyl-naphthalenes,[55] whereas the recently synthesized **105** is just waiting to be transformed into a tris-bridged molecule.[56]

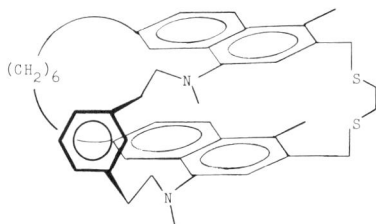

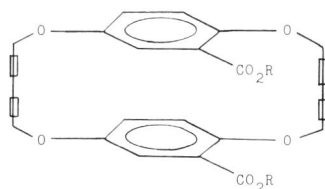

104   105

3. MOLECULAR BRIDGES CONTAINING AROMATIC RINGS

Multibridged phanes in which aromatic systems serve as complete or partial bridging units are an actively studied group of compounds because a number of these systems possess interesting stereochemical (especially conformational) properties. To prepare these phanes, hydrocarbons again in the majority of cases, the sulfone pyrolysis and the Wittig reaction have been employed most often. By the former procedure such interesting molecules as the helical double-clamped **106** (in which two benzene rings are clamped by one *m*-terphenyl and two ethano bridges)[57] and the triply bridged [2.2.2]biphenylophane **107** (in which formally one C_2 bridge of **27** has been replaced by a 1,2-diphenylethane unit)[58] have been synthesized.

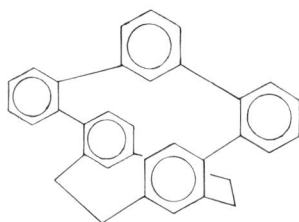

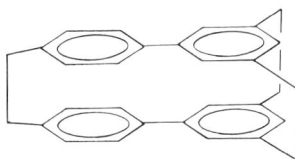

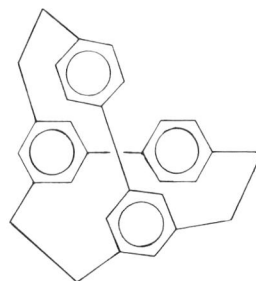

106   107

108

The trissulfide precursor of **107** could also be desulfurized via alkylation, Stevens rearrangement, and elimination (*vide supra*), providing the bridged triene derivative of **107**.[58] It is interesting that the biphenylophane **108,** with its crossed biphenyl units, could not be prepared from the appropriate trissulfide, presumably because there was too much ring strain in the intended target molecule.[58]

If the "replacement principle" connecting **27** and **107** is carried two steps farther, one obtains the tris-bridged hydrocarbon **109**, which has also been obtained by sulfone pyrolysis[59]; the topologically comparable hexasulfides **110** as well as the triene of **109** are also known.[59]

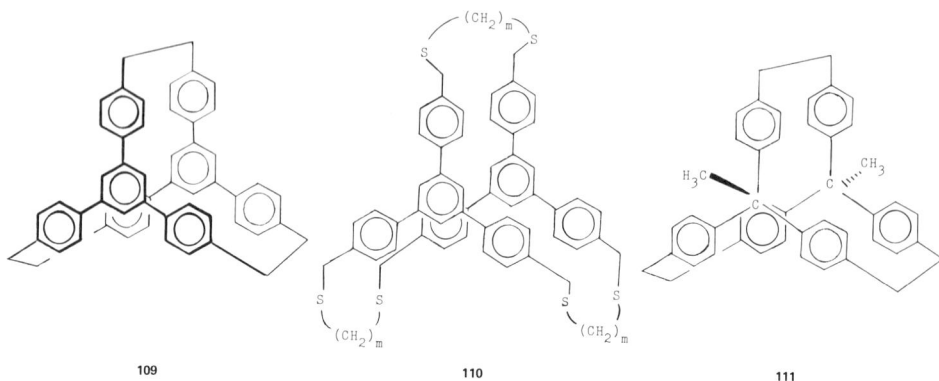

In cyclophanes such as **27, 109,** and **110** the central benzene rings with their 1,3,5 substitution pattern may be regarded as a kind of "enlarged" bridgehead. The question may be asked whether one can prepare phanes possessing a real bridgehead, that is, an sp^3-hybridized carbon atom in this position. As the following examples illustrate, this is indeed the case, and again sulfone pyrolysis is the preparative method of choice. Several triphenylmethanophanes, including **111** and its trisolefin, have been synthesized,[59] as have the monomethyl derivative, the parent hydrocarbon,[60] and several bridgehead halides and alcohols.[60] Hybrid types between **109** and **111** are also known,[61] one-half of the molecule being a triphenylmethane and the other a 1,3,5-triphenylbenzene derivative.

Of particular importance are phanes of this general type with bridgeheads carrying functional groups that can be converted to radicals, carbenium ions, or carbanions. In these species the interaction between the reactive centers across the phane system can be studied by chemical and spectroscopic methods. An initial success in this direction has been the oxidation of the triple-clamped triphenylamine **112** to the radical-cation **113**.[62] Obviously, the preparation of a phane with, for example, radical centers at both bridgeheads and the study of their interaction are the aim of this and related work.[60,61] The tris-bridged cyclophanes **114** (R = CH_3, OCH_3, Cl), composed of a mesitylene and an *m*-substituted triphenylmethane unit, respectively, have also been obtained by sulfone pyrolysis.[63] Whereas the ether could be converted to the bridgehead chloride, the latter could not be reduced to the hoped for radical.

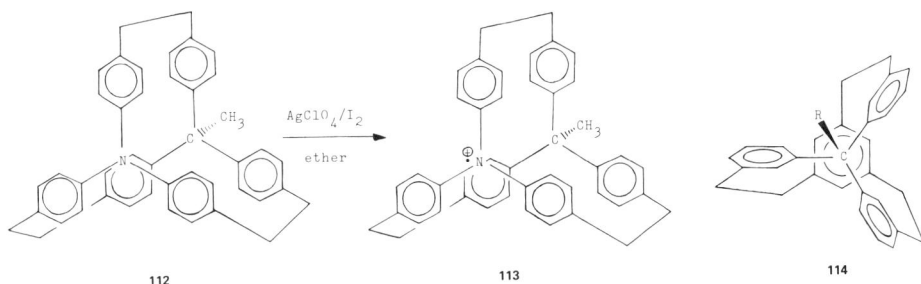

112   113   114

A phane containing two tetraphenylethylene building blocks held to-gether by four 2-thiapropano bridges has been synthesized, but all at-tempts to desulfurize it to the parent hydrocarbon or its tetraolefin deriva-tive have been unsuccessful.[64]

Turning now to multiply arene-bridged phanes prepared by the Wittig reaction, the general principle of this phane synthesis should be men-tioned first. As found by Wennerström and co-workers,[65] the [2⁴]cy-clophanetetraenes **117** are obtained via a one-step Wittig process between the bisaldehyde **115** and the bisylid **116.** More than 20 different cy-clophanes have been prepared by this approach, yields ranging from a few to up to 15%.

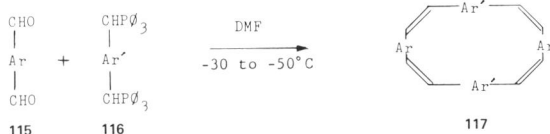

115   116   117

If this method is applied to 1,3,5-benzenetricarbaldehyde **(118)** and the ylid prepared from the bistriphenylphosphonium salt of 1,4-bis(bromo-methyl)benzene **(119),** the triply bridged cyclophane **120** is produced in 1.7% yield.[66] In order to make the naming of this cage compound, which again bears an obvious resemblance to [2₃](1,3,5)cyclophane **(27),** less cumbersome, the authors proposed that this class of phanes be called *bicyclophanes* and that, in particular, **120** be designated [2₆](1,4)₃(1,3,5)₂bicyclophanehexaene. On hydrogenation of **120** over pal-ladium on charcoal only the olefinic double bonds are reduced, and the bicyclophane **121** is produced in quantitative yield (for a detailed discus-

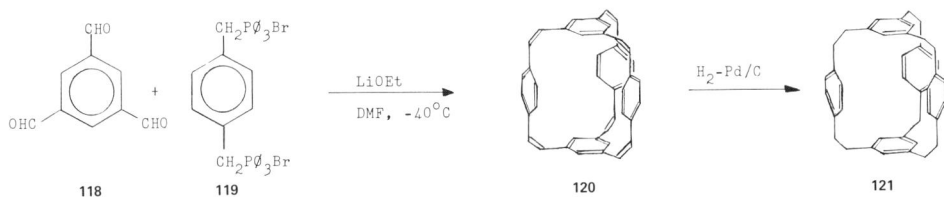

118   119   120   121

sion of the hydrogenation of the $[2_n]$cyclophanes see Section IV,C,2). The preparation of similar cyclophanes by metal-catalyzed cyclotrimerization of p-bis(3-butynyl)benzenes has already been referred to[41,42] (Section III,C,1). The $[2_6]$bicyclophanes exist in different conformations the dynamic properties of which have been investigated extensively by nmr spectroscopy.[67]

Multibridged cyclophanes in which a heteroaromatic ring system constitutes, or is part of, the molecular bridges seem to be unknown. In fact, whereas **122,** after lithiation, can be oxidatively dimerized to the [2.2]paracyclophane derivative **123,**[68] in which the ethano bridges of **1** have been replaced by bispyrazolyl units, **124** fails to undergo the analogous coupling and provides **125** rather than the anticipated tris-bridged phane.[69]

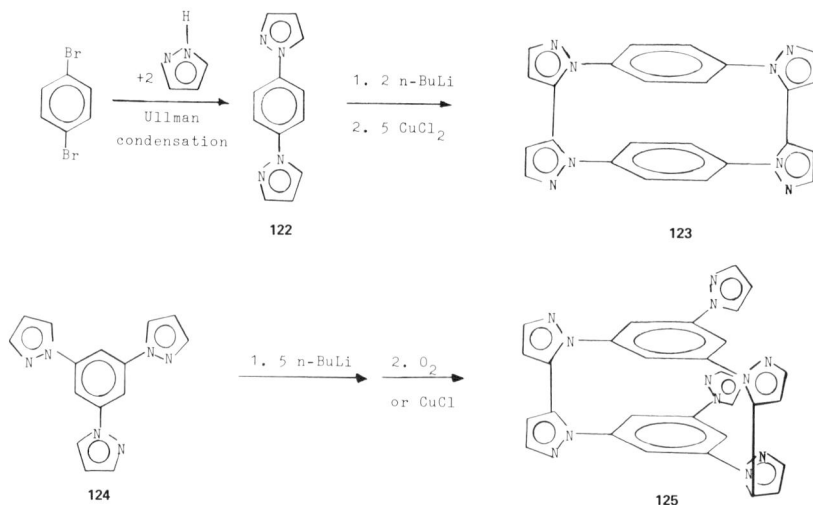

4. MOLECULAR BRIDGES CONTAINING ALIPHATIC RINGS

Cyclophanes bridged by three or more aliphatic ring systems are almost as rare as heteroaromatic bridged phanes; there is only one example in the literature so far. Its preparation involves an interesting photoaddition processs. (The only other photochemical reactions currently used in multi-bridged phane synthesis are the desulfurization and the generation of carbene intermediates by irradiation of salts of tosylhydrazones mentioned in Section III,A). Irradiation of (E,E,E)-1,3,5-tristyrylbenzene **(126)** in benzene at 317 nm affords a high-melting dimer **(127)** the ultimate structure proof of which rests on X-ray analysis.[70] The only other small-ring bridged [2.2]paracyclophane is [1,2;9,10]bismethano[2.2]paracy-clophane, that is, a hydrocarbon carrying two cyclopropane rings instead of the two C_2 units of **1**.[71]

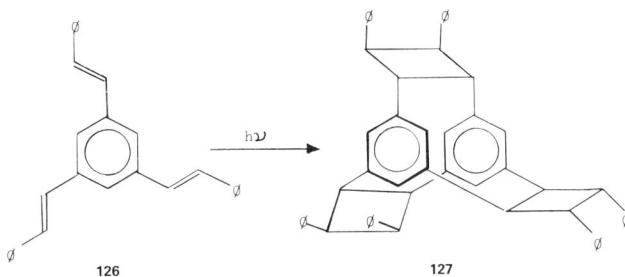

126   127

5. MULTIBRIDGED HETEROAROMATIC PHANES

As the material presented so far illustrates, the vast majority of known multibridged phanes are composed of relatively simple aromatic nuclei and more or less highly functionalized molecular bridges. Of course, the aromatic units may also be varied, and the following examples illustrate that the introduction of heteroaromatic rings could open up a new field of cyclophane chemistry. The results in this area are rather heterogeneous because until now no systematic studies have been reported. Depending on the target molecule, the synthetic approaches may either resemble those developed for the all-carbon systems (Section III,A) or be completely different. This is illustrated, on the one hand, by 4,13-diaza-[2$_4$](1,2,4,5)cyclophane **(128)** and its 4,16 isomer **129** (both heterocycles

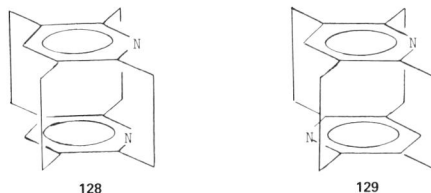

128   129

having been prepared by Boekelheide exploiting his benzocyclobutene–*o*-xylylene route[72]) and, on the other hand, by various synthetic schemes leading to "strati-bisporphyrins,"[73] several "cap" and "homologous cap" porphyrins,[74] and "picket pocket" porphyrins.[75] The topics of cyclophane porphyrins,[76] bis(chlorophyll)cyclophanes,[77] "crowned" porphyrins,[78] and "face-to-face" porphyrin dimers[79] are actively investigated areas of phane chemistry because these complex phanes (*a*) can serve as model compounds for multimetal redox enzymes as encountered in photosynthesis, (*b*) can mimic the active site of heme proteins,[80] (*c*) can be used as binuclear multielectron redox catalysts, and (*d*) possess interesting spectroscopic properties.[81] As in the case of the crown ether phanes (see Section III,C,2) a detailed presentation of the synthetic pathways leading to these novel compounds exceeds the scope of this chapter. The synthesis of the "strati-bisporphyrin" **133**[73] will suffice to illustrate the complexity of the structures and the preparative methods employed.

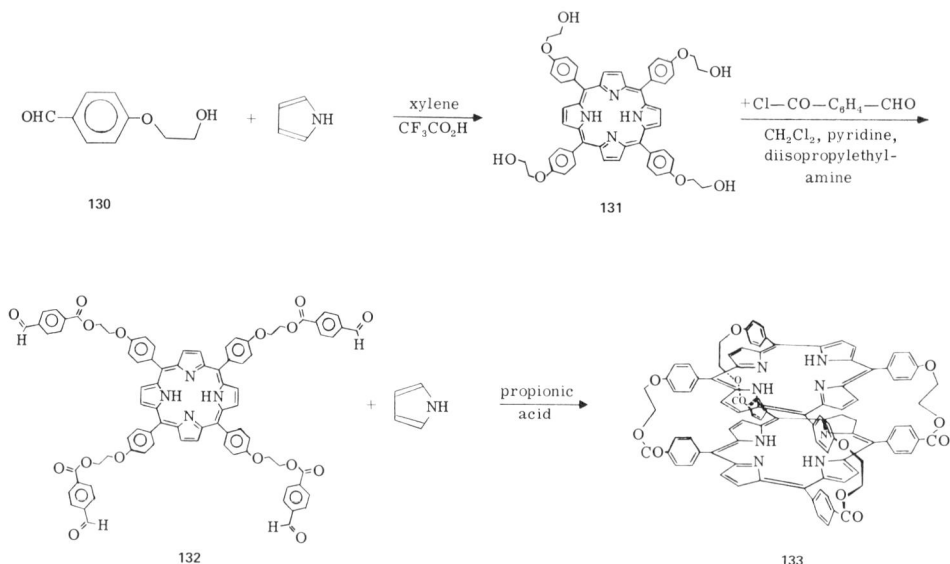

Tetra-*meso*-[*p*-(2-hydroxyethoxy)phenyl]porphyrin **(131)** was prepared by the reaction of *p*-2-hydroxyethoxybenzaldehyde **(130)** with pyrrole (14%) and treated with the acid chloride of *p*-carboxybenzaldehyde to yield the tetraaldehyde **132** (40%). If the latter is added with pyrrole to a refluxing mixture of propionic acid and ethylbenzene, the desired cyclophane **133** is produced in yields of up to 8%.[73]

6. MISCELLANEOUS MULTIBRIDGED PHANES

In formula **6** a multibridged phane is shown consisting of one benzene ring spanned by three independent bridges. Whereas phanes of this type have evidently not yet been prepared, several doubly bridged benzene rings (i.e., [*m*][*n*]paracyclophanes) are known.[82–82b]

To synthesize [8][8]paracyclophane **(139)**, the first representative of this novel class of cyclophanes, 10-methyl[8]paracyclophane **(134)** was converted by established procedures[83] to the Hofmann base **135**, which was subsequently "cross-bred" with the known quaternary salt **136**.[83] The resulting benzene–furan paracyclophane hybrid **137** was hydrolyzed to the ketone **138**, and this [8][8]paracyclophane was reduced via its bis-thioketal by the Mozingo reaction to the desired **139**. A higher homolog of this hydrocarbon, [8][10]paracyclophane, has been obtained analogously using an appropriate [10]paracyclophane derivative as starting material.[82]

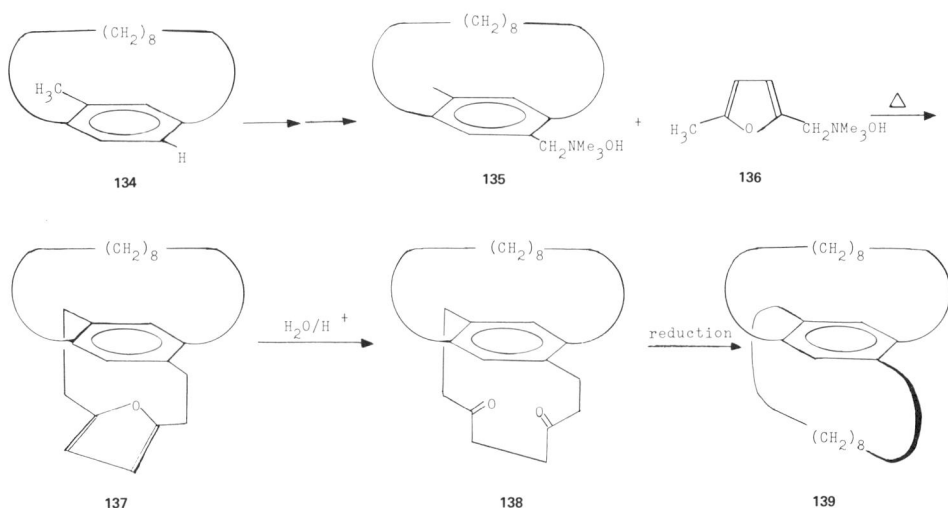

Compound **139** possesses D_2 symmetry and hence is chiral. It has indeed been partially resolved by column chromatography on optically active poly(triphenylmethyl metacrylate).[82a,84]

To conclude this section on the preparation of multibridged phanes, the synthesis of a completely different compound, [4](1,1′)[4](3,3′)[4]-(5,5′)[3](4,4′)ferrocenophane **(145)**, is discussed.[85] [4][4]-α-Oxo[3]ferro-cenophane **(140)**, itself a multibridged molecule,[86] served as substrate; **140** was first ring-enlarged with diazomethane to **141**, which was subsequently reduced to the derivative **142**. Formylation followed by the Reformatsky reaction and catalytic reduction led to **143**, which was transformed to the ketone **144** by standard methods. Another reduction with lithium alumi-num hydride–aluminum chloride completed the synthesis. The tetra-bridged metallophane **145** belongs to a small, but obviously very interest-ing group of new compounds in which an inorganic element is wrapped into an organic "sheath." Other examples of such encapsulated metal atoms are described in Section IV,D,2.

140   141   142

143 (R=CH$_2$CH$_2$CO$_2$Et)   144   145

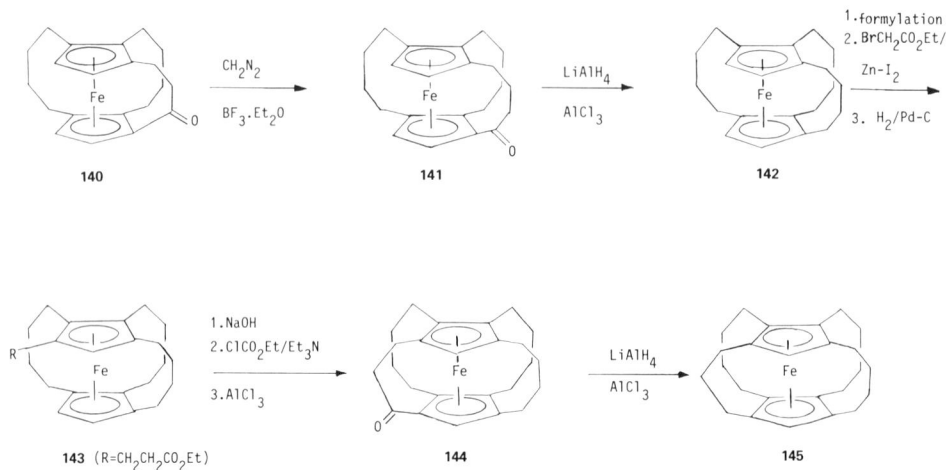

IV. CHEMICAL BEHAVIOR OF MULTIBRIDGED CYCLOPHANES

A. Introductory Remarks: Possible Reaction Pathways

The chemistry of the multibridged cyclophanes at present is more or less a chemistry of the [2$_n$]cyclophanes of general structure **2,** with the ethano bridges in parallel alignment. Most of the other multibridged molecules described in Section III,C have been synthesized for specific purposes (such as the study of stereochemical, conformational, or spectroscopic properties) and not for the detailed investigation of their general chemical behavior. If the appropriate studies have been carried out at all, they have in the majority of cases not been reported in the literature.

Formally, all cyclophanes are composed of two "building units": the bridges and the aromatic nucleus/nuclei. In principle, one can therefore distinguish between two types of reactions: those that take place at or with the molecular bridges and those involving the aromatic subunits, regardless of whether they are retained in the reaction under discussion. This clear separation of centers of reactivity is convenient for organizing the experimental data, although some processes (such as the Lewis acid-catalyzed isomerizations described in Section IV,B,2) involve both parts of the molecules. On the following pages important results on the chemical behavior of the multiclamped [2$_n$]cyclophanes **2** ($n \geq 3$) are discussed.

The chemistry of the parent hydrocarbon **1** is mentioned only in passing (and wherever it is important for comparison) because the spectrum of reactions of this molecule and its simple derivatives not only has been the subject of several reviews,[8,21] but is presented to some extent in other chapters of this volume.

B. Reactions of the Ethano Bridges of [2$_n$]Cyclophanes

So far three types of reactions that involve the C_2 bridges have been described: their cleavage, shift (migration to another bridgehead), and functionalization. The cleavage process may take place between the aromatic nucleus and the bridge or between the methylene groups of the latter. Because this "symmetric" cleavage is the simplest possible reaction of the [2$_n$]cyclophanes in the sense that it initially does not require the presence of other reagents, it is dealt with first.

1. Radical Cleavage of the Ethano Bridges

Homolytic cleavage of the C_2 bridges has been initiated thermally and photochemically. For the most straightforward case, pyrolysis of **1** at temperatures above 200°C in the presence of hydrogen donors such as 1,4-di(isopropyl)benzene or thiophenol leads to 4,4'-dimethylbibenzyl (**147**, 21 and 74% yield, respectively). Evidently, the diradical **146** serves as a precursor that may also be intercepted by methyl maleate or methyl fumarate, yielding the ring-enlarged phane **148** (200°C, 40 hr, 60% yield).[15,26,87] It is interesting that both the yield of **148** and its isomeric composition are practically independent of the configuration of the trapping agent, supporting the diradicaloid nature of the intermediate.

The next higher homolog (**15**) shows similar behavior, although the reaction conditions are slightly more vigorous than those for the homolysis of **1**.[14,16a] Pyrolysis in thiophenol (220°C, 73 hr) or 1,4-di(isopropyl)benzene (270°C, 40 hr) provides only 2,9-dimethyl-dibenzo[*ae*]cyclooctadiene (**150**, 88 and 63% yield, respectively), and trapping experiments with ethyl maleate (220°C, 43 hr, 42%) and fumarate (220°C, 72 hr, 51%) or methyl maleate (250°C, 40 hr, 74%) lead to mixtures of isomeric [4.2.2]cyclophane diesters **151**. That the particular ethano bridge indicated in the pyrolysis scheme is broken has been attributed to the greater strain release during the formation of the diradical **149** than for alternate pathways.[16a]

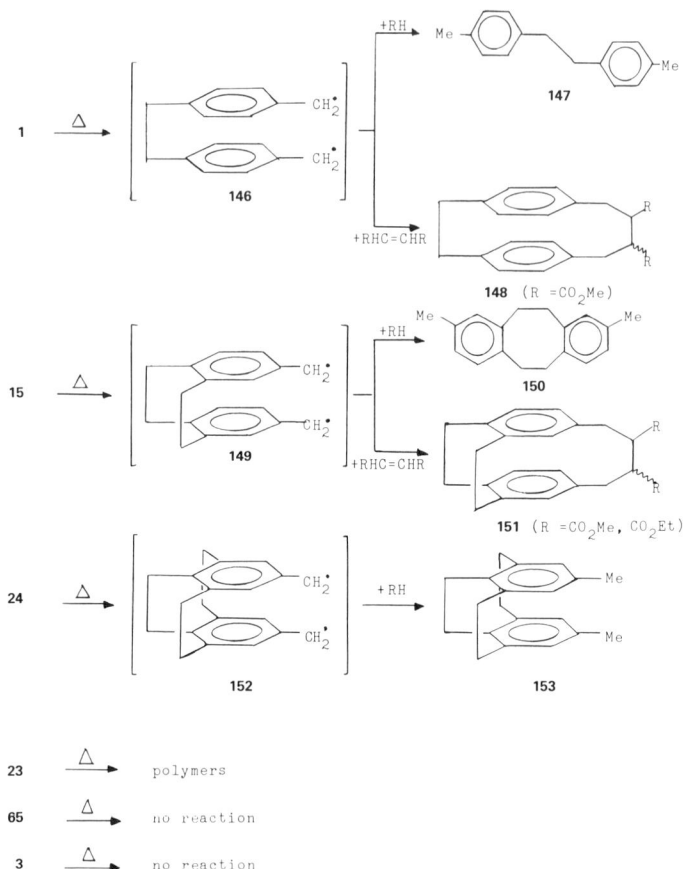

At first glance the [2₄]cyclophane **24** does not show any novel behavior. Extended heating at temperatures above 275°C in 1,4-di(isopropyl)benzene affords the [2₃](1,2,3)cyclophane **153**; that is, the isolated ethano bridge is destroyed.[37] However, a more thorough investigation at temperatures ranging from 240 to 300°C and reaction times between 62 and 73 hr have shown that the situation is more complex.[31] For example, after 68 hr at 270 to 275°C the pyrolyzate contains 34% of **24**, 28% of **160**, 21% of **159**, and only traces of **153**. A slight increase in temperature (275–280°C, 68 hr) leads to 38% of **153**, 46% of **160**, and only traces of **159**. Whereas **24** is stable at 240°C (65 hr), only **161** (20%) is formed from it when the temperature is raised to 300°C (73 hr). To rationalize these results it is proposed that the diradical **152** reacts with 1,4-di(isopropyl)benzene in a stepwise manner, the monoradical **154/155** being produced first. This can either be trapped by a second molecule of the hydrogen donor (formation of **153**) or

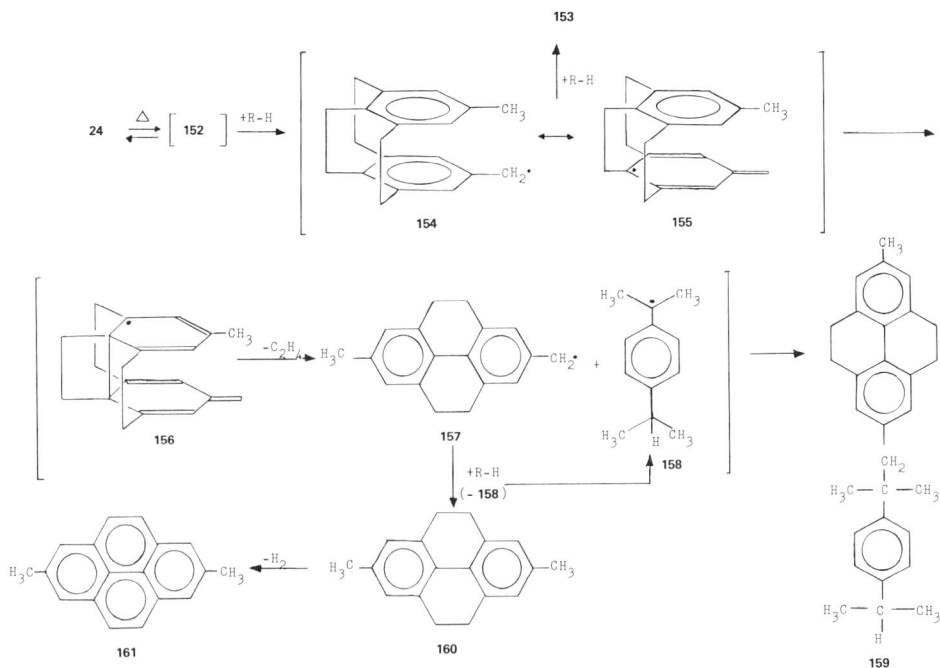

form the isomeric radical **156** by intraannular attack. The later zero-bridged species may expel ethylene, and the resulting **157** can either combine with the benzylic radical **158** to yield **159** in a termination step or abstract another hydrogen atom to provide **160**. Aromatization of this tetrahydropyrene derivative finally leads to the most stable product hydrocarbon 4,9-dimethylpyrene **(161).** It appears that the latter series of steps is favored by higher pyrolysis temperatures.[31]

Compound **23** is considerably more thermally stable than **24.** After 65 hr at 290°C, or even 12 hr at 400°C, this hydrocarbon is recovered unchanged from the pyrolysis–trapping experiment. Only at about 470°C (12 hr) does **23** begin to decompose to polymeric products, possibly via *p*- or *o*-quino-dimethanoid intermediates.[37] The pinnacle of thermal resistence is reached with $[2_5](1,2,3,4,5)$cyclophane **(65)**[88] and superphane **(3),**[89] neither of which is reductively cleaved or ring enlarged in the presence of the appropriate reagents after extended heating at 350°C.

Both **146** and **152** have been generated and observed directly by irradiation of **1** and **24,** respectively, in a glassy matrix of 2-methyltetrahydrofuran at 77K. The uv, visible, and fluorescence spectra of both species measured under these conditions constitute the first spectroscopic proof for the formation of diradicals in a photocycloreversion.[90]

Irradiation of **1** in acetone at room temperature with a medium-pressure mercury lamp leads to *p*-ethylbibenzyl exclusively; that is, under these conditions a rupture between aromatic nucleus and methylene group is taking place.[91]

2. IONIC CLEAVAGE OF THE ETHANO BRIDGES

The experimental results on the ionic cleavage of the ethano bridges in [2$_n$]cyclophanes are rather scattered; in particular, there are no precise studies concerning relative reactivities. The Lewis acid-catalyzed isomerizations are among the best studied processes. The parent molecule **1,** on treatment with aluminum chloride–hydrogen chloride in methylene chloride at 0°C, rearranges to a mixture of hydrocarbons from which [2.2]metaparacyclophane (**164**) can be separated in 44% yield.[40,92] The driving force for this reaction, which very likely takes place via the σ complexes **162** and **163**, is presumably provided by the reduction in strain energy (E_s of **1,** 134 kJ/mol; E_s of **164,** 100 kJ/mol[93]).

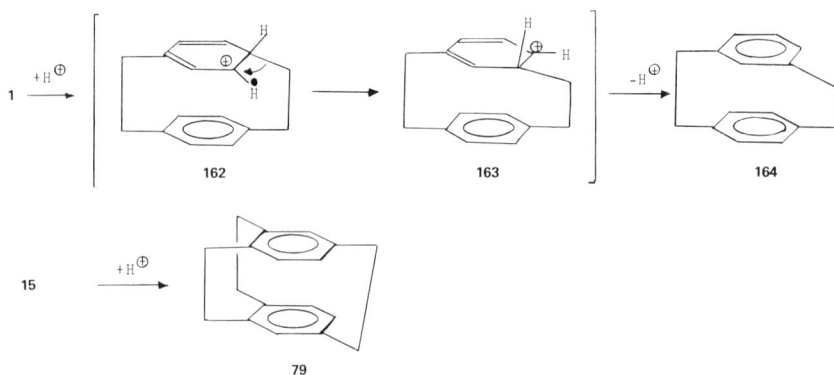

Among the higher bridged [2$_n$]cyclophanes, only **15** has so far been isomerized analogously. After 30 min at −10°C (CH$_2$Cl$_2$, AlCl$_3$–HCl) it rearranged to the "skew" hydrocarbon [2$_3$](1,2,4)(1,2,5)cyclophane (**79,** 44%),[37] a compound previously obtained by sulfone pyrolysis, although in poorer yield (see Section III,B).

The tetra-bridged molecules **23** and **24** do not even isomerize under more severe conditions (3 hr, 25°C). Although the intense red color that appears when hydrochloric acid is added to the reaction mixture signals that a σ complex has been generated, a Wagner–Meerwein rearrangement does not take place.[37] The first step of the ipso substitution of **24,** discussed in Section IV,D,1, is related to these isomerizations.

An anionic cleavage of the C_2 bridges occurs when superphane (3) is reduced with lithium–ethylamine in *n*-propylamine, hydrocarbons 170 and 171 being formed in 57 and 7% yield, respectively.[89] A reasonable mechanism for this interesting reaction begins with the radical-anion 165, which corresponds to the intermediates commonly encountered in the course of the Birch reduction (see Section IV,C,2). The hexamethylbenzene dimer 171 is formed from 170 by a final repetition of the reaction involving intermediates corresponding to 165–169.

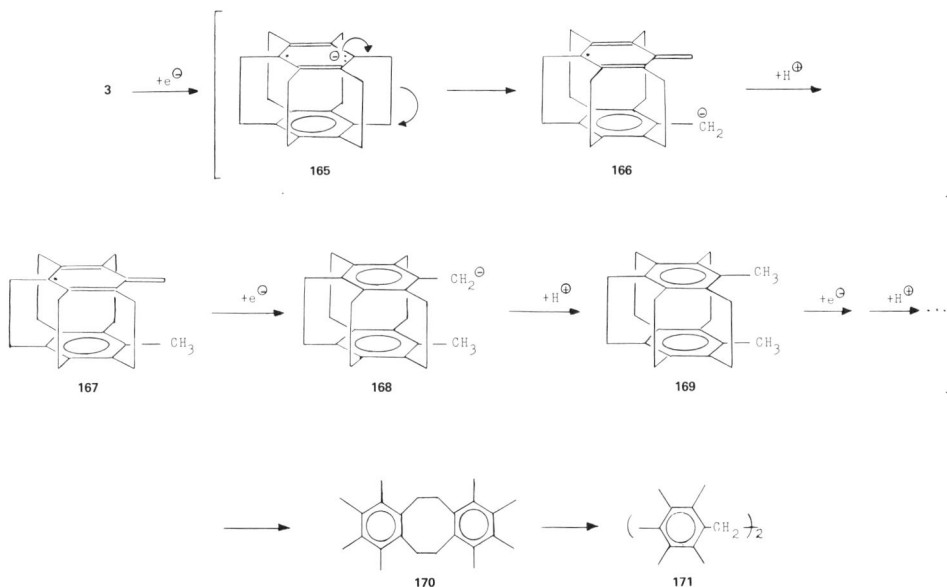

The only report of a heterolytic destruction of a C_2 bridge in one of the [2_n]cyclophanes concerns [2.2]paracyclophane (1) itself, which on irradiation with light of 253.7 nm wavelength in alcoholic solvents is converted to the ether 173 via the zwitterion 172.[91]

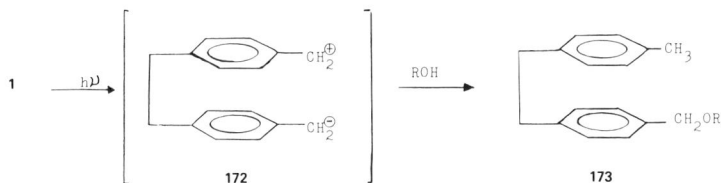

An unusual ring enlargement of unknown mechanism is observed when 3 is treated for 1 hr with zinc in concentrated sulfuric acid. This reaction seems to be characteristic of superphane, because none of the other multibridged cyclophanes investigated yield products comparable to the thio ether **90** when subjected to these reaction conditions.[89]

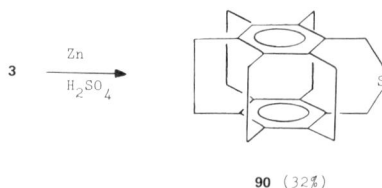

90 (32%)

3. FUNCTIONALIZATION OF THE ETHANO BRIDGES

Although bridge-functionalized [2$_n$]cyclophanes, such as olefins, ketones, or epoxides, are of preparative, mechanistic, and spectroscopic interest, very few of these attractive molecules have been synthesized so far. For example, the monoolefin **174,** prepared from **3** by routine methods, might be regarded as the first step toward the hexaene **175.** The low

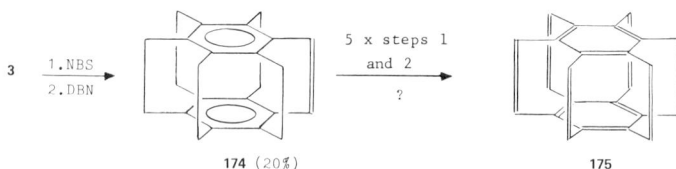

174 (20%)   **175**

yield of only 20%, however, is certainly no invitation to proceed further along this reaction path.[89] Hydrocarbon **175** is an electronically very unusual molecule because it is formally composed of two orthogonal π systems containing 12 electrons each, a central "core," and a perpendicular "sheath."

The synthesis of the triene **177** was accomplished by a threefold Hofmann elimination of the trissulfonium salt **176** with n-butyllithium in tetrahydrofuran.[22]

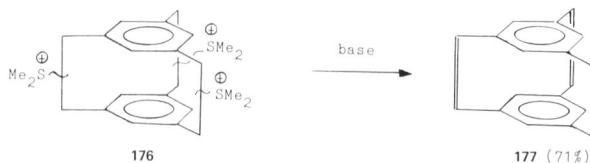

176   **177** (71%)

C. Reactions of the Benzene Rings of [2$_n$]Cyclophanes

The "bent and battered" structure of the benzene rings in [2.2]paracyclophane (1) is one of the most celebrated features of this molecule.[8a] Comparable deformations have meanwhile been observed for most of the multibridged phanes, as discussed, *inter alia,* in Chapter 3.[94]

In a deformed benzene ring the overlap between the p orbitals cannot be as effective as in benzene proper, and one might therefore expect decreased aromatic character for hydrocarbons 2. In contrast to the typical regenerative behavior (electrophilic aromatic substitution) of classical noncondensed arenes, deformed aromatics could hence show an increased propensity for addition. From the very beginning, therefore, the behavior of multibridged cyclophanes toward typical addition reagents has been of great interest, and, as will be seen on the following pages, some [2$_n$]cyclophanes do indeed participate in addition reactions as readily as polyolefinic hydrocarbons. However, electrophilic aromatic substitutions may also be performed with these multibridged molecules, which hence possess some kind of zwitter reactivity.

1. DIELS-ALDER ADDITIONS

Benzene and its simple alkyl and halogen derivatives normally do not play the role of the diene component in [2 + 4] cycloadditions. They are, in fact, routinely used as solvents in these additions because of their low Diels–Alder reactivity. Only under special conditions, such as high reaction temperatures, the presence of very reactive dienophiles, and especially the addition of Lewis acid catalysts, is this inertness overcome.[95] Still, the "superdienophile" 4-phenyl-1,2,4-triazoline-3,5-dione (178) does not add to benzene or any of the poly(methylbenzenes) at room temperature even after several weeks.[96] However, if these molecules are incorporated into a [2$_n$]cyclophane system, that is, [2.2]paracyclophane (1) as a formal dimer of p-xylene, superphane (3) as a dimer of hexamethylbenzene, etc., then a drastic increase in the rate of addition is noted for certain phanes. As the data in Table I show, 1 and 178, for example, react to form a 1 : 2 cycloadduct after 6 days at room temperature. (The positions of the new carbon–carbon bonds are marked with arrows in the table.)

The incorporation of additional bridges (see 15 and especially 22)

TABLE I

Diels–Alder Additions of [2$_n$]Cyclophanes in Benzene[a]

Dienophile	1	15	22	24	23	3
4-Phenyl-1,2,4-triazoline-3,5-dione (178)	20°C, 138 hr, 99% (1:2 adduct)	20°C, 3 hr, 79% (1:1 adduct); 20°C, 24 hr, 65% (1:2 adduct)	20°C, sec, 100% (1:2 adduct)	20°C, 24 hr, nr	20°C, 24 hr, nr	—
Tetracyanoethylene (179)	100°C, 15 hr, π complex; 165 hr (in toluene), nr	20°C, 20 hr, 59% (1:1 adduct) + 26% π complex	20°C, sec, 100% (1:1 adduct)[24,89]	20°C, 20 hr, π complex	100°C, 15 hr, π complex	—
Maleic anhydride (180)	180°C (in toluene), 37 hr, nr	100°C, 10 hr, 53% (1:1 adduct)	—	100°C, 16 hr, nr	100°C, 17 hr, nr	—
Diacyanoacetylene (181)	120°C, 32% (1:1 adduct); 170°C, 72% (1:2 adduct)[97]	20°C, 14 days, 78% (1:1 adduct); 81°C, 24 hr, 30% (1:2 adduct)	60°C, 6 hr, 93% (1:1 adduct); 70°C, 2 days, 76% (1:2 adduct)[24]	—	—	Prolonged heating, nr[89]
Perfluoro-2-butyne (182)	—	—	100°C, 12 hr, 66% (1:2 adduct)[24]	—	—	—
Dimethyl acetylene dicarboxylate (183)	nr at higher temperatures[14]	170°C, 1 hr, 61% (1:1 adduct)[14]	—	—	170°C, 1 hr, nr	—
Dimethyl maleate (184)	165°C, 13 hr, nr	100°C, 10 hr, nr	—	—	—	—

[a] nr, no reaction.

causes an extreme rate increase; in the presence of $[2_4](1,2,4,5)$cy-
clophane (22) the intense red color of 178 vanishes in a few seconds. In its
cycloaddition capacity this monocyclic arene can therefore compete with
acyclic polyolefins![96] Comparable trends are noted for the additions of 1,
15, and 22 with tetracyanoethylene (179) and dicyanoacetylene (181),
respectively, as well as maleic anhydride (180), perfluoro-2-butyne (182),
and dimethyl acetylenedicarboxylate (183), although the experiments

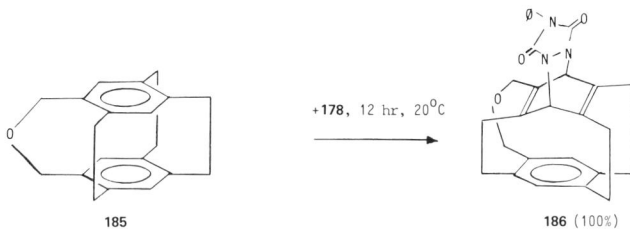

185   186 (100%)

with these dienophiles are not as comprehensive as those discussed earlier. Only the comparatively unreactive dimethyl maleate (184) does not
add to 1 and 15. That a large substrate strain does not suffice to induce
Diels–Alder addition for a multibridged phane, however, is demonstrated
by the behavior of hydrocarbons 24,[96] 23,[96] and 3,[89] none of which reacts
with the various dienophiles listed in Table I. From these experiments one
would have predicted that the $[2_5]$cyclophane 65 would also be a very
poor Diels–Alder partner. Surprisingly, however, 65 reacts with both
dicyanoacetylene (181, 60°C, 3 days, 44%) and perfluoro-2-butyne (182,
100°C, 7 days, 100%), 1 : 1 adducts being produced in both instances.[88]

It appears that a subtle balance between the ring strain of the substrate
and the cycloadducts controls the outcome of these [2 + 4] additions. As
seen in Table I, 22 initially forms a 1 : 1 adduct with dicyanoacetylene
(181). This intermediate is reactive enough either to add a second equivalent of 181 or to form a "mixed" 2 : 1 adduct with tetracyanoethylene
(179) within a brief period.[24] If, however, the ring-enlarged, and hence
probably less strained, ether 185 is treated with 178, the addition stops at
the 1 : 1 stage (186).[31] In contrast to 15, its skew isomer 79 does not
participate in Diels–Alder additions.[39]

An interesting cascade of addition reactions was observed for
superphane (3) when attempts were made to overcome its inertness in
[2 + 4] cycloadditions with Lewis acid catalysis. When 181 and 3 were
combined at room temperature in methylene chloride for 3 days in the
presence of aluminum chloride, the polycycle 188, the complex structure
of which was determined by X-ray structural analysis, was produced
(40%).[89]

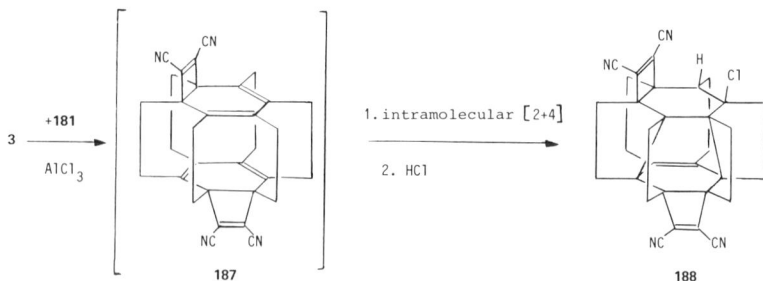

To explain the formation of **188** it has been proposed that **3** is first transformed into **187** by double [2 + 2] cycloaddition, that this 2 : 1- intermediate is stabilized by an intramolecular [2 + 4] cycloaddition, and that the resulting tetraene is intercepted by hydrochloric acid (formed during the reaction in an ionic process) to yield the isolated product.[89]

In another [2 + 4] cycloaddition singlet oxygen was added to **22**. The primary adduct, the endoperoxide **189**, is a very useful starting material

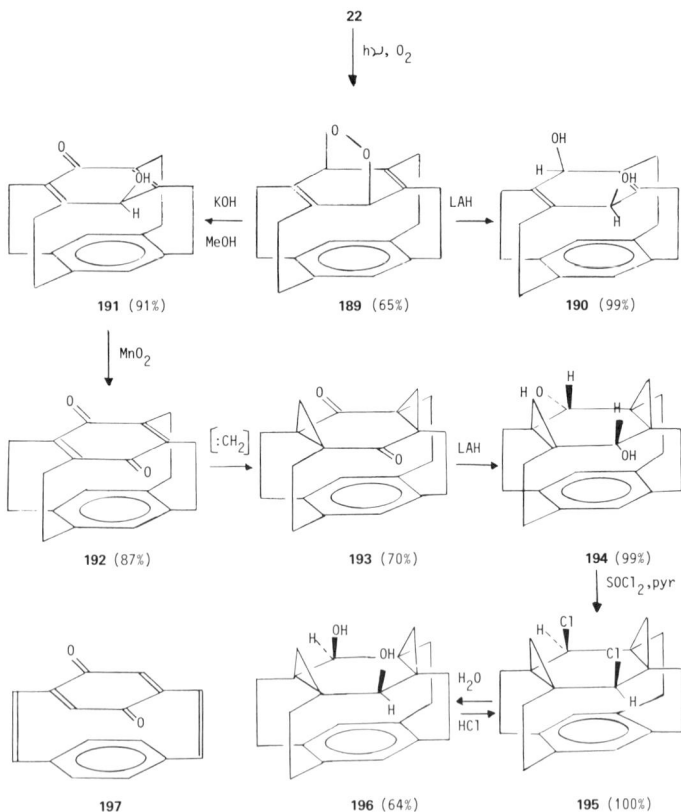

for a host of other oxygenated multibridged phanes.[24] As shown in the scheme, **189** can be reduced to the diol **190** in quantitative yield or isomerized to the dienone **191**. The latter provides the quinone **192** on oxidation with activated manganese dioxide (cf. the preparation of the four-fold-bridged bisquinone **36** in Section III,A). Cyclopropanation of **193** with dimethyloxosulfonium methylide leads to **193,** which via the *endo*-diol **194** was converted to the dichloride **195**. Unfortunately, this derivative could not be ring-expanded to a homotropilidene. On treatment with water, **195** hydrolyzed to the *exo*-diol **196,** which could be reconverted to **195** on acid treatment, no *endo*-dichloride being formed.[24]

The unsaturated [2.2]cyclophanequinone **197** has been synthesized using an approach analogous to the oxidation of **22** to **192**.[98]

Another Diels–Alder adduct of a multibridged cyclophane that has been employed for additional transformations is the 1 : 1 adduct **198** prepared from dicyanoacetylene **(181)** and **22** (see Table I). Irradiation of **198** in tetrahydrofuran at room temperature with a low-pressure mercury lamp converts it to a single yellow compound (44%), the spectroscopic

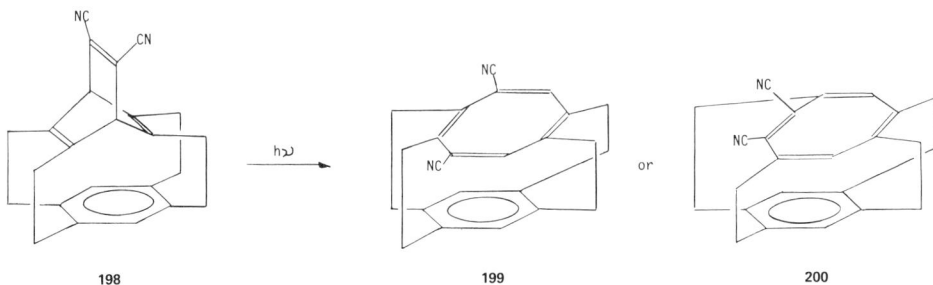

198   199   or   200

properties of which indicate that a multibridged cyclophane containing a cyclooctatetraene unit has been produced.[24] Mechanistic considerations indicate that either **199** or **200** can be formed, the former structure being favored from molecular model considerations.[24]

The parent system of the $[2_n]$cyclooctatetraenophanes, [2.2](11,14)cyclooctatetraenylparacyclophane, has been prepared from the cycloadduct of **1** and **181**.[99]

2. HYDROGENATIONS

Two types of hydrogenation, catalytic hydrogenation and Birch reduction, have been carried out with $[2_n]$cyclophanes, and both show a reactivity pattern comparable to that noted in the Diels–Alder addition (see Section IV,C,1). It may hence be concluded that these three addition processes are controlled by similar parameters, among which the buildup of strain in the σ framework of the adducts seems to be particularly important. This is illustrated by qualitatively comparing the rates of hydrogenation of [2.2]paracyclophane (1) and the triply bridged hydrocarbon 15. The parent compound 1 is reduced under relatively mild conditions to a diene that possesses either structure 201 or structure 202 (the

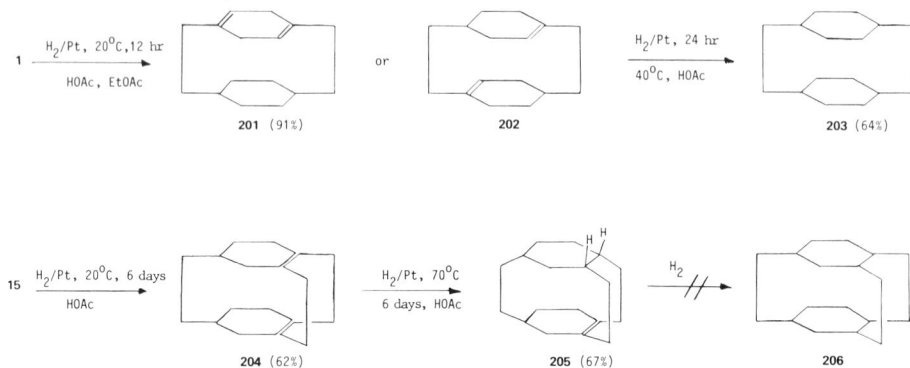

1 $\xrightarrow[\text{HOAc, EtOAc}]{\text{H}_2/\text{Pt, 20°C,12 hr}}$

201 (91%)   or   202

$\xrightarrow[\text{40°C, HOAc}]{\text{H}_2/\text{Pt, 24 hr}}$ 203 (64%)

15 $\xrightarrow[\text{HOAc}]{\text{H}_2/\text{Pt, 20°C, 6 days}}$ 204 (62%)

$\xrightarrow[\text{6 days, HOAc}]{\text{H}_2/\text{Pt, 70°C}}$ 205 (67%)

$\xrightarrow{\text{H}_2 \,/\!/}$ 206

former being more likely on spectroscopic grounds) and to the fully saturated perhydro[2.2]paracyclophane (203) under more severe conditions.[100] Introduction of the third bridge makes the hydrogenation more difficult, but a diene (204) can nevertheless be prepared from 15.[16a]

The addition of a fifth equivalent of hydrogen requires much more drastic conditions (6 days, 70°C[16a] or 5 days, room temperature, 5 atm H_2[35a]), and the resulting monoene 205 is resistant to the final reduction. Inspection of molecular models indicates that the intended product 206 is a very strained hydrocarbon, and evidently the energy gained by hydrogenating 205 does not suffice to overcome this high strain barrier. From these results one would conclude that the higher bridged [2_n]cyclophanes should show an even lesser tendency to combine with hydrogen, and this is indeed the case; Neither 24[31] nor 65[88] or superphane (3)[89] is converted to olefinic products of any type when exposed to hydrogen in the presence of various catalysts.

Qualitative indications of the ease of reduction of the benzene rings can also be obtained from Birch reduction experiments, which so far have been carried out with 1, 27, 22, 24, 65, and 3. This reduction should take place particularly readily for cyclophanes having boat-shaped rings, because their transformation into 1,4-cyclohexadiene units having a boat configuration should be accompanied by a substantial strain reduction.

Under the conditions given in the reaction scheme, hydrocarbon 1, in addition to providing a small amount of the dihydro compound 207, affords mainly the tetrahydro derivative 208, the "crossed" arrangement of its double bonds having been inferred from subsequent reactions and the

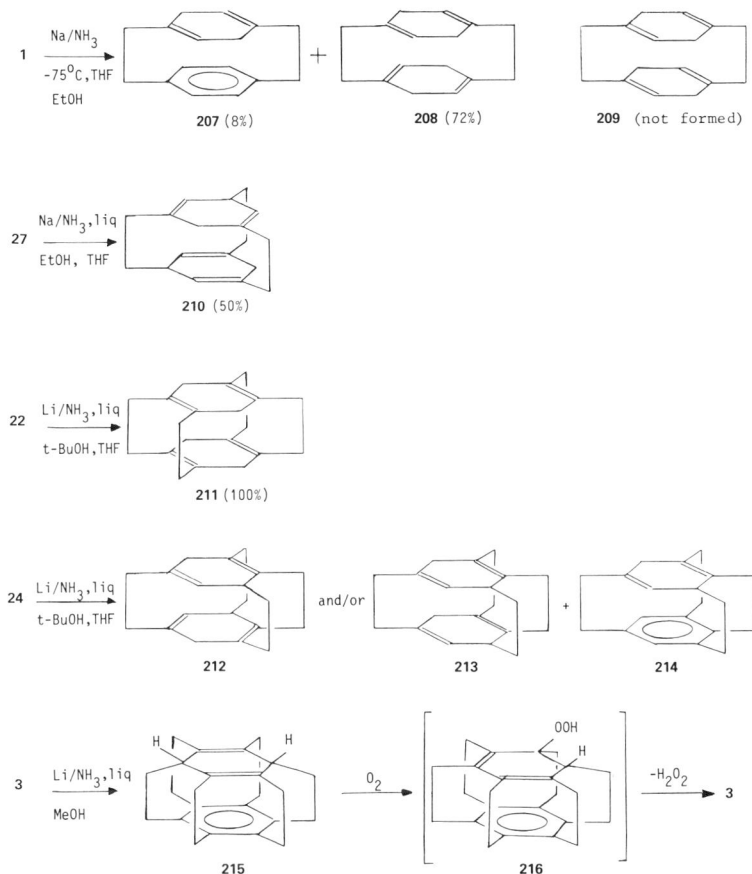

spectroscopic properties of the products formed thereby.[101a-c] Heating **208** to 150 to 160°C in a closed system leads to **207**.[101] The known[102] strong dependence of the product composition in Birch reductions on seemingly small variations of the reaction conditions can also be observed for **1**. For example, the addition of a solution of **1** in tetrahydrofuran to a refluxing solution–suspension of sodium in liquid ammonia, followed by the addition of ethanol, yields 4,4'-dimethylbibenzyl (**147,** 94%),[101a] whereas the slow addition of **1** in tetrahydrofurane–ethanol to a solution of sodium in boiling ammonia leads to **208** quantitatively.[101b] As indicated by molecular models, the six-membered-ring sp^3 carbon atoms of **208** point toward the "outside"; that is, the cyclohexadiene rings are boat-shaped. An ethano bridge anchored in these positions very likely would have to be stretched considerably, causing a severe strain energy increase. It is hence not surprising that in the Birch reduction of **22** the energetically more favorable tetrahydro[2.2](1,2,4,5)cyclophane **211,** in which all bridges are bonded to sp^2-hybridized ring carbon atoms,[24] is formed exclusively and quantitatively. The observation that **27** is readily reduced to **210**[22] does not contradict this conclusion, because a meta arrangement of bridges, as in **27** and the tetraene **210**, tolerates strain more easily. It was therefore even more interesting to study the Birch reduction of **24**, formally a hybrid of a [2.2]para- and a [2.2]metacyclophane. Reaction with lithium in liquid ammonia–*tert*-butanol yields a mixture of **212** and/or **213** and **214**.[9] A distinction between the two tetrahydro compounds based on spectral evidence alone is difficult, as in the analogous case of **1**. Steric arguments favor **212** since sp^3 hybridization of both bridgehead carbon atoms of one C_2 bridge (see **213**) probably cannot be realized. When the reaction mixture was investigated immediately after the reduction was terminated, it was found that **212/213** was practically the only primary product. Hydrocarbon **214** was present only in traces, and substrate **24** was consumed completely. However, after purification of the products by column chromatography, 6% of **214** and the same amount of **24** was isolated, along with 83% of **212/213**. The highly strained tetraolefin hence began to rearomatize during work-up at room temperature (it was stable for several days only at $-20°C$). The dihydro derivative **214** is somewhat more stable than **212/213**; however, purification by either recrystallization or sublimation again fails, with **24** being reformed. Superphane **(3)** seemingly fits into this series because it yields the Birch product **215** (although in only 10% yield). This compound, however, rearomatizes to **3** at room temperature in a short time.[89] Mechanistically, this dehydrogenation could proceed as a thermal, symmetry-allowed process, as known for other, simpler 1,4-cyclohexadiene derivatives,[103] or alternatively as an ene reaction via **216**, initiated by atmospheric oxygen.[89] It is interesting that hydrocarbon **65** behaves unusually in this reaction as well: It survives the Birch reduction unchanged.[88]

The bridge-cleaving hydrogenation of **3** has been discussed in Section IV,B,2. A similar reduction of **27** to a dimethyl[2.2]metacyclophane has also been accomplished.[22] Among the complex phanes, the catalytic reduction of **120** to **121** has already been mentioned (Section III,C,3). All attempts to prepare the [4]cryptand **218** from **96** (see Section III,C,2) by catalytic hydrogenation were unsuccessful, **217** being the only partially saturated product isolated.[52]

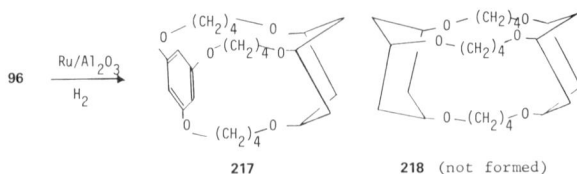

217   **218** (not formed)

3. OTHER ADDITION REACTIONS

Considering the great ease with which some $[2_n]$cyclophanes participate in such typical polyene reactions as the Diels–Alder addition (see Section IV,C,1), it is apparent that these hydrocarbons should be submitted to other classical olefin reactions as well.

Indeed, the tetraester **219** reacts at room temperature with diazomethane–cuprous chloride to give a complex product mixture, from which the

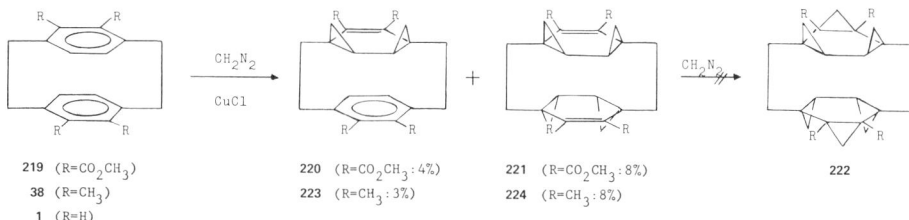

219 (R=CO$_2$CH$_3$)   220 (R=CO$_2$CH$_3$:4%)   221 (R=CO$_2$CH$_3$:8%)   222
38 (R=CH$_3$)   223 (R=CH$_3$:3%)   224 (R=CH$_3$:8%)
1 (R=H)

methylenation products **220** and **221** can be separated by column chromatography.[104] However, another carbene addition to yield a derivative of the so far unknown all-*cis*-trishomobenzene, namely **222,** does not take place.[105] The electron-attracting effect of the methoxycarbonyl groups is not responsible for this inertness, because the tetramethyl[2.2]paracyclophane **38** also yields only bis **(223)** and tetrakis adducts **(224),** respectively.[105] [2.2]Paracyclophane **(1)** can also be cyclopropanated under these conditions, providing the norcaradiene **225** and the ring-enlarged phane **226,** both in about 10% yield.[106,106a] It is surprising that [2.2]paracyclophane-1,9-diene is attacked by carbene at one of the benzene rings preferentially, hydrocarbon **227** being isolated as the monomethylenation product.[106a] Normally, olefinic double bonds are cyclopropanated more easily than aromatic ones.[107]

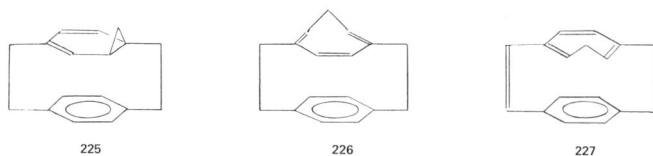

225   226   227

The norcaradiene derivative **228** is the most likely primary product in the reaction of **22** with ethyl diazoacetate. Under the reaction conditions, however, **228** isomerizes spontaneously to the seven-membered-ring system **229.**[24]

Such tropilidenes as **226, 227,** and **229** are interesting substrates for the preparation of phane systems in which an uncharged aromatic nucleus such as the benzene ring is connected to a tropylium ion by C$_2$ bridges and

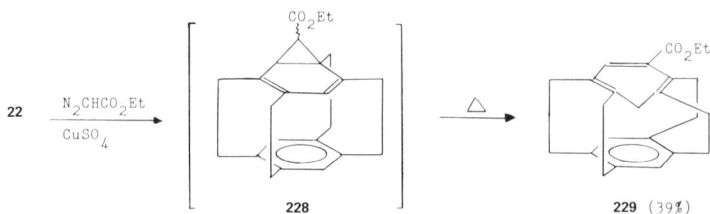

$$22 \xrightarrow[\text{CuSO}_4]{\text{N}_2\text{CHCO}_2\text{Et}} \quad [\ 228\ ] \quad \xrightarrow{\triangle} \quad 229\ (39\%)$$

is fixed in a stereochemically unambiguous position. The parent molecule of this class of compounds, [2.2](1,4)tropylioparacyclophane, has indeed been prepared from **226** by hydride abstraction.[106] A synthesis of [2₄](1,2,4,5)-7′-methyltropyliocyclophane from **229** has also been reported.[24]

The application to **3** of the techniques developed during the synthesis of **229** resulted in the "circular process" shown here.[89] As in the case of **22**,

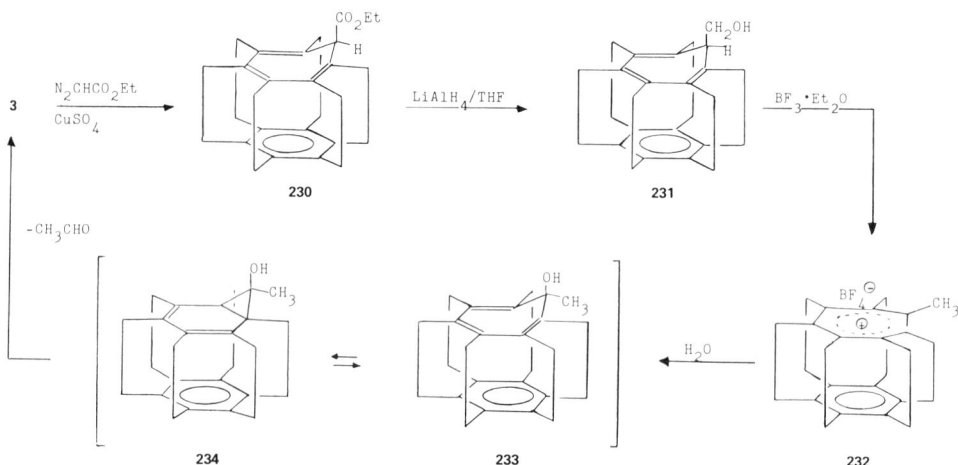

the addition of ethyl diazoacetate first leads to a tropilidene ester **(230)**. Reduction with lithium aluminum hydride provides the alcohol **231**, which, on treatment with boron trifluoride etherate, isomerizes to the tropyliophane **232**. When this salt is either recrystallized from methanol–water or kept standing in the open air, it is reconverted to **3**! This remarkable decomposition could occur via the carbinol **233** and its valence isomer **234**, with the latter splitting off acetaldehyde to yield superphane **(3)**.[89]

The occurrence of ionic additions to the benzene rings of the multibridged cyclophanes strongly hints at a reduced aromatic character for these molecules. Indeed, there are several reports describing phanes that

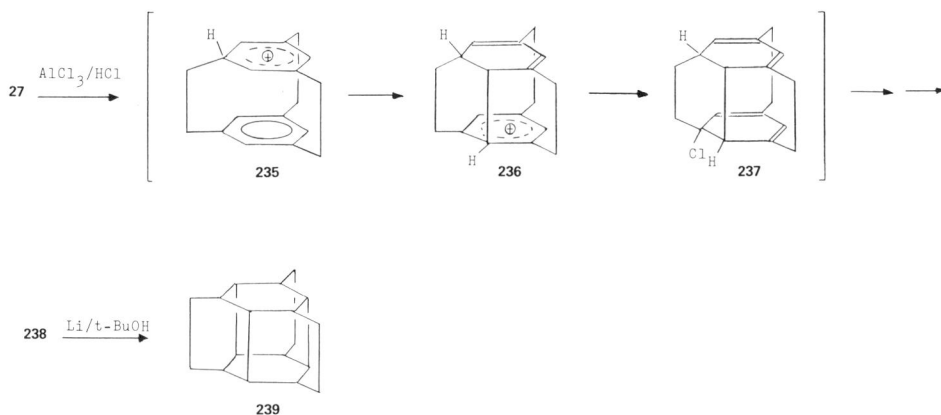

react most readily with polar reagents. [2₃](1,3,5)Cyclophane **(27),** for example, is transformed into a complex product mixture of the composition $C_{18}H_{18} \cdot H_xCl_{6-x}$ ($x < 6$, **238**) when treated with hydrochloric acid–aluminum chloride in dichloromethane for 10 min at 0°C. Although the separation and identification of the different chlorides failed, **238** could be dehalogenated to a homogeneous hydrocarbon, $C_{18}H_{24}$, which, according to spectral data, possesses the cage structure **239**.[22] A reasonable precursor for **238** is the HCl adduct **237,** formed from **27** via the σ complexes **235** and **236**. Monochloride **237** is well suited for a multiple repetition of this ionic addition process.

In another ionic reaction, **3** was alkylated with Meerwein's reagent to the cation **240,** which on treatment with sodium borohydride provided **241**. When this hydrocarbon was recrystallized from methanol, it formed the epoxide **242** or a positional isomer thereof.[89]

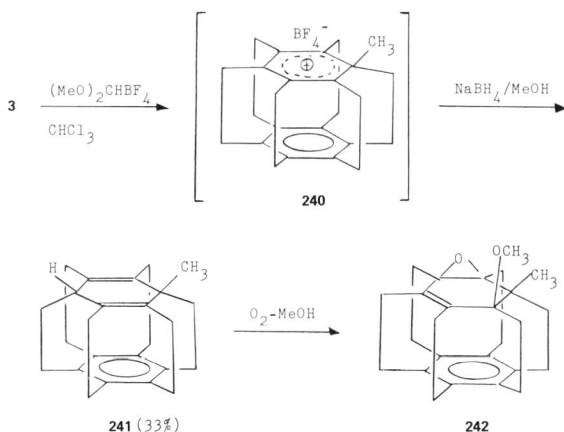

D. Reactions with Retention of the Aromatic Rings

The most frequently used method for the preparation of simple derivatives of [2.2]paracyclophane (1) is electrophilic aromatic substitution. Bromination, Friedel–Crafts acylation, nitration, and other methods have been reported[8,15,21] and are carried out under the standard conditions for this type of substitution. It is therefore reasonable to assume that in the majority of cases the standard reaction mechanism involving σ complexes is also involved. Because precise mechanistic studies are rare, however, there is *a priori* no reason to exclude substitutions that take place by an addition–elimination sequence.

1. ELECTROPHILIC AROMATIC SUBSTITUTION

Bromination of $[2_3](1,2,4)$cyclophane (15) provides the expected monobromide (243) as well as traces of a dibromide (244).[16a] Compound 15 is

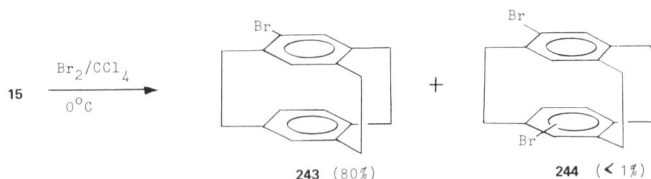

243 (80%)   244 (< 1%)

more reactive than 1, and the fourfold-bridged phane 24 does not compete with 15 either. Under these conditions 24 does not react with bromine, and stronger reaction conditions (addition of iodine, increase in reaction time) are initially of no consequence. Only when the reaction times become very long (16 hr) does the bromine solution decolorize, although no monosubstituted derivative could be isolated from the resulting multiproduct mixture.[9]

The Friedel–Crafts acylation of 15 and its positional isomer 27 proceeds as expected.[16a,22] Both methyl ketones 245 and 246 are stable toward fur-

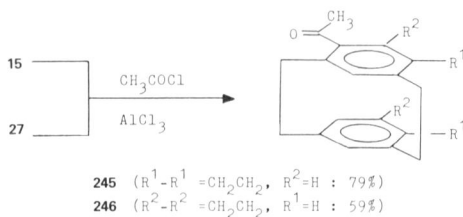

245 (R^1-R^1 =CH_2CH_2, R^2=H : 79%)
246 (R^2-R^2 =CH_2CH_2, R^1=H : 59%)

ther acylation. [2₄](1,2,3,5)Cyclophane **(24),** however, affords ketone **250** in small yields only, the major products being the two (1,2,3)cyclophanes **251** and **252** (16% of **24** recovered).[9] The formation of the two methyl

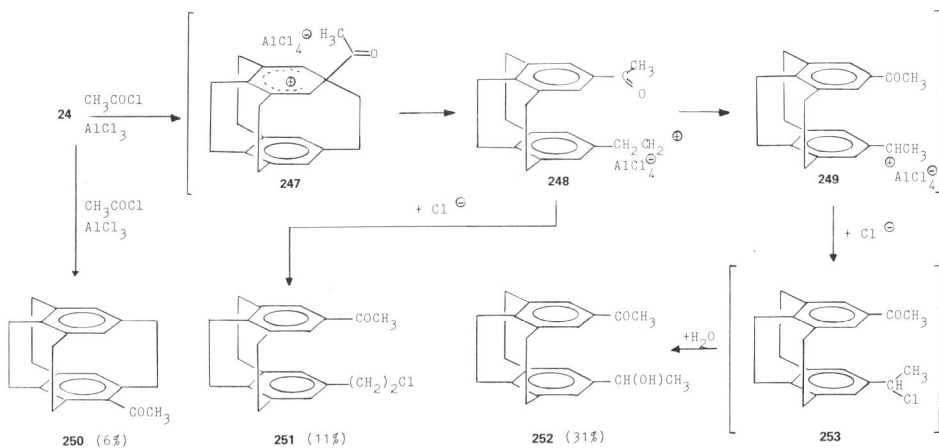

ketones requires the breaking of an ethano bridge between the aromatic nucleus and a methylene group, and the mechanism accounting most readily for this observation involves an ipso substitution of **24.**[109] The σ complex **247,** which loses part of its strain by isomerization to the primary carbenium ion **248,** may be an initial intermediate. Ion **248** is then either intercepted by chloride ion (formation of **251**) or stabilized by hydrogen migration to form the secondary cation **249.** The combination of this species with chloride leads to the secondary and benzylic halogen derivative **253,** which is hydrolyzed to the alcohol **252** isolated on work-up.

Formylation according to Rieche[29] converts **24** in nearly quantitative yield (94%) to the aldehyde **254,** whereas nitration (fuming nitric acid, glacial acetic acid, 70°C, 2 min) yields only traces of the nitro compound

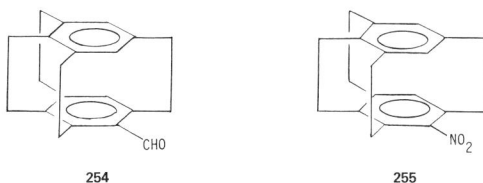

255.[9] Both the bromination (bromine–iron) and the Rieche formylation transform [2₅](1,2,3,4,5)cyclophane **(65)** in excellent yields to the expected monosubstitution products.[89]

2. Preparation of Metal Complexes

Provided that the molecular bridge in a cyclophane cannot move unre-
strictedly around the aromatic nucleus (or equivalently that the latter
cannot rotate freely "under" the bridge), the two faces of the aromatic
system are nonequivalent. In principle, therefore, an [*n*]cyclophane can
form two types of metal complexes, with its aromatic subsystem function-
ing as a donor: The metal or a metal-atom-containing fragment can be
bound either to the "inside" of the aromatic ring, as in **256,** or to the

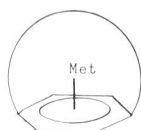

256

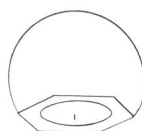

257

"outside," as in **257.** The same two possibilities exist for [2.2]paracy-
clophanes **(258** and **259).** However, because of the bidentate nature of
binuclear phanes, such 2:1 complexes as **260** and even such polymeric
structures as **261** could also be formed. The latter molecules, in which the

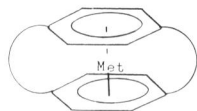

258

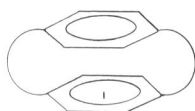

259

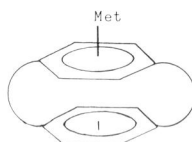

260

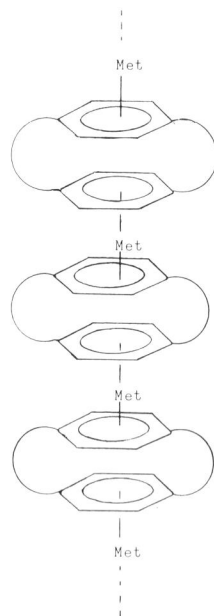

261

organic and inorganic "monomers" alternate in a columnar structure, are of considerable current interest because they might show unusual electronic properties (even conduct electricity) if, for example, the built-in metal atoms are in different oxidation states. So far, however, no polymer like **261** has been synthesized.

Although the number of known metal complexes of phanes is rather small, considerable progress has been achieved since 1960, when Cram prepared the first tricarbonylchromium complex of [2.2]paracyclophane **(262)**.[110] The method used, reaction of the phane with hexacarbonylchromium in dry diglyme at 130 to 150°C in an inert atmosphere, is rather

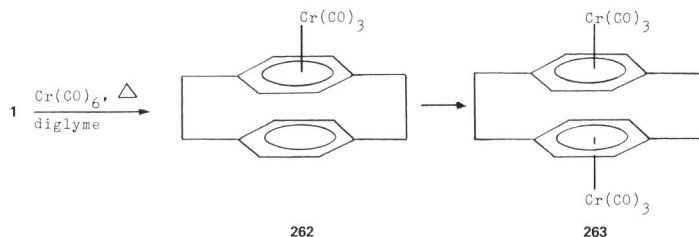

general, having been applied to the synthesis of several derivatives of **262**.[111,111a] The 2:1 complex **263** could not be obtained by the original authors,[110] but later Misumi and co-workers showed that small yields (8%) can be obtained by employing an excess of the hexacarbonyl.[112] These workers also prepared a tricarbonylchromium complex of [8]paracyclophane (external arrangement as in **259**) as well as several mono and bis complexes of multilayered cyclophanes.[112] Only one multibridged cyclophane, [2₃](1,2,4)cyclophane **(15),** has been treated with hexacarbonylchromium, and again the 1:1 product was formed (40%).[111a]

Using another technique, the reaction of **1** with chromium atoms, Elschenbroich was successful in synthesizing (η^{12}-[2.2]paracyclophane)chromium(0) **(264)** and bis(η^6-[2.2]paracyclophane)chromium(0) **(265)**.[113]

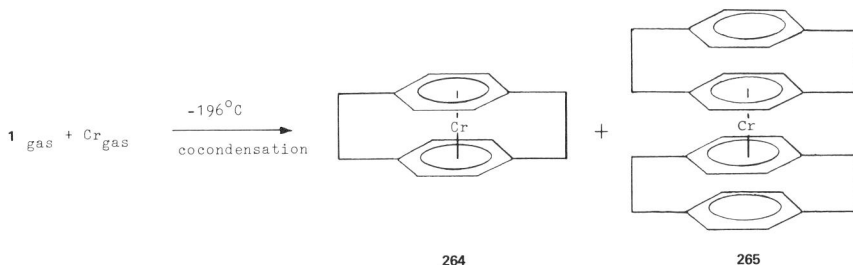

These two compounds are the first representatives of the group of molecules symbolized by the general structure **258** as well as a section of the columnar structure **261**. The same technique has also been applied to the synthesis of (η^{12}-[3.3]paracyclophane)chromium(0)[114] and of (η^{12}-[2_3](1,2,4)cyclophane)chromium(0).[115]

Because neither the metal carbonyl nor the metal atom technique seems to be suitable for preparing oligomers and/or polymers of cyclophane transition metal complexes, Boekelheide and co-workers developed a new method, which involves the reaction of arene–ruthenium complexes with phanes in the presence of silver salts.[116] For example, when bis(p-cymene)dichlorodi-μ-chlorodiruthenium(II) (**266**: arene = p-cymene), silver tetrafluoroborate, and acetone are stirred at room temperature, and when **1** in trifluoroacetic acid is added to this solution after the precipitated silver chloride has been removed and the resulting mixture heated under reflux, the two multilayered complexes **267** and **268** (arene = p-cymene) are formed, the latter being the preferred product in the presence of a large excess of **266**.[116]

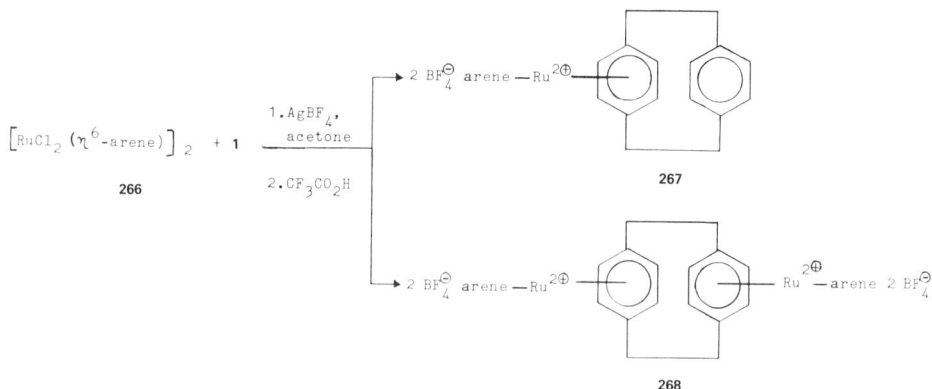

Essentially the same approach has been applied to the preparation of ruthenium complexes **269** and **270** from [2_5](1,2,3,4,5)cyclophane (**65**). The derivative **271** was isolated in 68% yield when (η^6-p-xylene)(η^5-cyclopentadienyl)iron(II) hexafluorophosphate and **65** were irradiated in dichloromethane.[88]

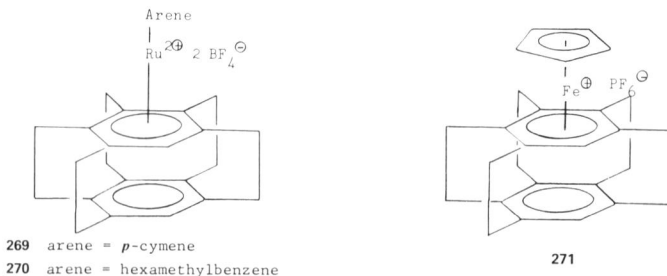

269   arene = p-cymene
270   arene = hexamethylbenzene

271

V. CONCLUDING REMARKS: OUTLOOK

Essential for all work in this and, of course, all other areas of chemistry are efficient synthetic schemes that provide enough material for chemical investigation. Although many of the problems in the synthesis of multibridged cyclophanes have been solved during recent years, it must be pointed out that some of the reaction sequences, brilliant as they may be, have not yet been sufficiently optimized to yield useful (i.e., multigram) amounts of the cyclophane substrates. Sizable quantities of these compounds are necessary for further mechanistic and chemical studies. From the mechanistic viewpoint, precise kinetic data for selected reactions would certainly be desirable. Furthermore, attempts should be made to study postulated reaction intermediates (the various σ complexes) directly. Physical data, such as heats of hydrogenation or combustion, would make arguments involving molecular strain less qualitative. A number of reactions should be studied in greater detail; among these are photo processes of all kinds, nucleophilic substitution reactions of metal complexes of the multibridged cyclophanes, and radical reactions.

Although it is to be expected that the synthesis of novel multibridged cyclophane hydrocarbons will be described in the future, the hexaene **175** being a particularly attractive target molecule, it seems safe to predict that the main emphasis of future (and ongoing) research in this area of phane chemistry will be on complex, multibridged, functionalized, and heteroaromatic systems as well as on metal complexes of all kinds. It is hoped that the few examples reported in Section IV,D,2 are only the beginning.

REFERENCES

1. Y. Sekine, M. Brown, and V. Boekelheide, *J. Am. Chem. Soc.* **101**, 3126 (1979).
2. F. Vögtle and P. Neumann, *Angew. Chem.* **84**, 75 (1972); *Angew. Chem., Int. Ed. Engl.* **11**, 73 (1972).
3. For a review of benzocyclobutenes and related compounds, see R. P. Thummel, *Acc. Chem. Res.* **13**, 70 (1980).
4. W. Nutakul, R. P. Thummel, and A. D. Taggart, *J. Am. Chem. Soc.* **101**, 770 (1979).
5. H. Hopf and J. Kleinschroth, *Angew. Chem.* **94**, 485 (1982); *Angew. Chem., Int. Ed. Engl.*, **21**, 469 (1982).
6. A. Greenberg and J. F. Liebman, "Strained Organic Molecules," p. 176. Academic Press, New York, 1978.
7. L. Trabert and H. Hopf, unpublished; cf. L. Trabert and H. Hopf, *Liebigs Ann. Chem.* p. 1786 (1980).
8. B. H. Smith, "Bridged Aromatic Compounds." Academic Press, New York, 1964.
8a. D. J. Cram and J. M. Cram, *Acc. Chem. Res.* **4**, 204 (1971).

8b. F. Vögtle and P. Neumann, *Tetrahedron* **26**, 5847 (1970).

9. J. Kleinschroth, Ph.D. Dissertation, University of Würzburg (1980).

10. M. Nakazaki, Y. Yamamoto, and Y. Miura, *J. Org. Chem.* **43**, 1041 (1978).

11. V. Boekelheide, *Acc. Chem. Res.* **13**, 65 (1980).

12. H. Iwamura, M. Katoh, and H. Kihara, *Tetrahedron Lett.* p. 1757 (1980).

13. W. F. Gorham, *J. Polym. Sci., Part A-1* **4**, 3027 (1966); cf. W. F. Gorham, German Patent, 1,085,763 (1960).

14. E. A. Truesdale and D. J. Cram, *J. Org. Chem.* **45**, 3974 (1980). For a preliminary account of this work, see E. A. Truesdale and D. J. Cram, *J. Am. Chem. Soc.* **95**, 5825 (1973).

15. D. J. Cram, R. B. Hornby, E. A. Truesdale, H. J. Reich, M. H. Delton, and J. M. Cram, *Tetrahedron* **30**, 1757 (1974).

16. H. Hopf and F. T. Lenich, *Chem. Ber.* **107**, 1891 (1974); I. Böhm, H. Herrmann, H. Hopf, and K. Menke, *ibid.* **111**, 523 (1978); H. Hopf and A. E. Murad, *ibid.* **113**, 2358 (1980).

16a. H. Hopf, K. Menke, and S. Trampe, *Chem. Ber.* **110**, 371 (1977).

17. J. Kleinschroth, unpublished.

18. J. Kleinschroth and H. Hopf, *Angew. Chem.* **91**, 336 (1979); *Angew. Chem., Int. Ed. Engl.* **18**, 329 (1979).

19. H. Hopf, W. Gilb, and K. Menke, *Angew. Chem.* **89**, 177 (1977); *Angew. Chem., Int. Ed. Engl.* **16**, 191 (1977).

20. H. Hopf and J. Kleinschroth, *Tetrahedron Lett.* p. 969 (1978).

21. F. Vögtle and P. Neumann, *Synthesis* p. 85 (1973); F. Vögtle and P. Neumann, *Top. Curr. Chem.* **48**, 67 (1974); F. Vögtle and G. Hohner, *ibid.* **74**, 1 (1978); S. Misumi and T. Otsubo, *Acc. Chem. Res.* **11**, 251 (1978); F. Vögtle and L. Rossa, *Angew. Chem.* **91**, 534 (1979); *Angew. Chem., Int. Ed. Engl.* **18**, 514 (1979).

22. V. Boekelheide and R. A. Hollins, *J. Am. Chem. Soc.* **95**, 3201 (1973); cf. *idem., ibid.* **92**, 3512 (1970).

23. N. Nakazaki, K. Yamamoto, and Y. Miura, *Chem. Commun.* p. 206 (1977).

24. R. Gray and V. Boekelheide, *J. Am. Chem. Soc.* **101**, 2128 (1979). For a preliminary account of this work, see R. Gray and V. Boekelheide, *Angew. Chem.* **87**, 138 (1975); *Angew. Chem., Int. Ed. Engl.* **14**, 107 (1975). *Note added in proof:* As shown by X-ray crystallography, structure **199** is correct (A. W. Hanson, personal communication to V. Boekelheide).

25. V. Boekelheide, I. D. Reingold, and M. Tuttle, *J. Chem. Soc., Chem. Commun.* p. 406 (1973); J. Bruhin and W. Jenny, *Tetrahedron Lett.* p. 1215 (1973).

26. H. J. Reich and D. J. Cram, *J. Am. Chem. Soc.* **91**, 3505, 3517, 3527 (1969).

27. H. A. Staab and V. M. Schwendemann, *Liebigs Ann. Chem.* p. 1258 (1979); cf. *Angew. Chem.* **90**, 805 (1978); *Angew. Chem., Int. Ed. Engl.* **17**, 756 (1978) for a preliminary account of this work.

28. S. H. El-tamany and H. Hopf, *Tetrahedron Lett.* p. 4901 (1980); see also A. E. Murad, Ph.D. Dissertation, University of Würzburg (1979).

29. A. Rieche, H. Gross, and E. Höft, *Chem. Ber.* **93**, 88 (1967).

30. H. Hopf, *Nachr. Chem. Tech. Lab.* **28**, 311 (1980). *Note added in proof:* By modification of the reaction conditions, this approach to superphane **3** has recently also been accomplished (H. Hopf and S. H. El-tamany, *Chem. Ber.* **116**, 1682 (1983).

31. J. Kleinschroth and C. Mlynek, unpublished results.

32. M. P. Cava and A. A. Deana, *J. Am. Chem. Soc.* **81**, 4266 (1959); L. A. Errede, *ibid.* **83**, 949 (1961).

32a. F. R. Jensen, W. E. Coleman, and A. J. Berlin, *Tetrahedron Lett.* p. 15 (1962).

33. H. Hart and R. W. Fish, *J. Am. Chem. Soc.* **82,** 749 (1960); H. Hart, J. A. Hartlage, R. W. Fish, and R. R. Refos, *J. Org. Chem.* **31,** 2244 (1966); R. Gray, L. G. Harruff, J. Krymowski, J. Peterson, and V. Boekelheide, *J. Am. Chem. Soc.* **100,** 2892 (1978); P. Schiess and M. Heitzmann, *Helv. Chim. Acta* **61,** 844 (1978); P. Schiess, M. Heitzmann, S. Rutschmann, and R. Stäheli, *Tetrahedron Lett.* p. 4569 (1978).

34. V. Boekelheide and G. D. Ewing, *Tetrahedron Lett.* p. 4245 (1978).

35. G. D. Ewing and V. Boekelheide, *J. Chem. Soc., Chem. Commun.* p. 207 (1979).

35a. W. G. L. Aalbersberg and K. P. C. Vollhardt, *Tetrahedron Lett.* p. 1939 (1979).

36. B. Neuschwander and V. Boekelheide, *Isr. J. Chem.* **20,** 288 (1980).

37. H. Hopf, J. Kleinschroth, and A. E. Mourad, *Isr. J. Chem.* **20,** 291 (1980).

38. P. F. T. Schirch and V. Boekelheide, *J. Am. Chem. Soc.* **101,** 3125 (1979).

39. The lithium aluminum hydride reduction of the pseudo-geminal dibromide leads to [2$_3$](1,2,4)cyclophane **(15),** and the thermal decomposition of the bistosylhydrazone of the pseudo-geminal dialdehyde under basic conditions provides the monoolefin of **15** (K. Broschinski, C. Mlynek, J. Kleinschroth, and H. Hopf, unpublished).

40. M. H. Delton, R. E. Gilman, and D. J. Cram, *J. Am. Chem. Soc.* **93,** 2329 (1971). For a preparation of [2.2]metaparacyclophanes by the dithiacyclophane–sulfur extrusion route, see S. A. Sherrod, R. C. da Costa, R. A. Barnes, and V. Boekelheide, *ibid.* **96,** 1565 (1974), and references quoted.

41. A. J. Hubert and J. Dale, *J. Chem. Soc.* p. 3160 (1965).

42. A. J. Hubert, *J. Chem. Soc. C* p. 6 (1967); A. J. Hubert and M. Hubert, *Tetrahedron Lett.* p. 5779 (1966); A. J. Hubert, *J. Chem. Soc. C* pp. 11, 13 (1967).

43. R. D. Stephens, *J. Org. Chem.* **38,** 2260 (1973).

44. J. A. Gladysz, J. G. Fulcher, S. J. Lee, and A. B. Bocarsley, *Tetrahedron Lett.* p. 3421 (1977).

45. J. Kleinschroth and C. Mlynek, unpublished observation.

46. F. Vögtle, *Justus Liebigs Ann. Chem.* **735,** 193 (1970).

47. A. W. Hanson and E. W. Macaulay, *Acta Crystallogr., Sect. B* **B28,** 1255 (1972).

48. F. Vögtle and R. G. Lichtenthaler, *Tetrahedron Lett.* p. 1905 (1972).

49. F. Vögtle and R. G. Lichtenthaler, *Chem. Ber.* **106,** 1319 (1973).

50. F. Vögtle and P. Neumann, *J. Chem. Soc., Chem. Commun.* p. 1464 (1970).

51. K. Böckmann and F. Vögtle, *Chem. Ber.* **114,** 1965 (1981).

52. W. D. Curtis, J. F. Stoddart, and G. H. Jones, *J. Chem. Soc., Perkin Trans. 1* p. 785 (1977).

53. N. Wester and F. Vögtle, *Chem. Ber.* **112,** 3723 (1979).

54. Another group of "crown ether phanes" consists of the so-called catapinands prepared by F. Vögtle and N. Wester, *Chem. Ber.* **113,** 1487 (1980).

55. D. S. Kemp, M. E. Garst, R. W. Harper, D. D. Cox, D. Carlson, and S. Denmark, *J. Org. Chem.* **44,** 4469 (1979).

56. E. T. Jarvi and H. W. Whitlock, Jr., *J. Am. Chem. Soc.* **102,** 657 (1980).

57. E. Hammerschmidt and F. Vögtle, *Chem. Ber.* **113,** 3550 (1980).

58. F. Vögtle and G. Steinhagen, *Chem. Ber.* **111,** 205 (1978).

59. G. Hohner and F. Vögtle, *Chem. Ber.* **110,** 3052 (1977). This publication reports an interesting experiment in which an attempt was made to combine hexakis(bromomethyl)benzene with the corresponding hexathiol. Rather than yielding the desired hexathia[3$_6$](1,2,3,4,5,6)cyclophane, the reaction led to a monomeric trithiabenzene derivative formed by ring closure between vicinal methylenethiol groups (P. Neumann, Ph.D. Dissertation, University of Heidelberg, 1973). Similar experiments leading to tetrathia[3$_4$]phanes have evidently been more successful (B. Klieser and F. Vögtle, unpublished; private communication of Professor Vögtle).

60. S. Korbach and F. Vögtle, unpublished results; private communication of Professor Vögtle.
61. J. Winkel and F. Vögtle, unpublished results; private communication of Professor Vögtle.
62. J. Winkel and F. Vögtle, *Tetrahedron Lett.* p. 1561 (1979).
63. M. Nakazaki, K. Yamamoto, and T. Toya, *J. Org. Chem.* **45**, 2553 (1980); **46**, 1611 (1981).
64. F. Vögtle and N. Wester, *Liebigs Ann. Chem.* p. 545 (1978).
65. B. Thulin, O. Wennerström, and H.-E. Högberg, *Acta Chem. Scand., Ser. B* **B29**, 138 (1975); B. Thulin and O. Wennerström, *ibid.* **B30**, 369, 688 (1976).
66. H.-E. Högberg, B. Thulin, and O. Wennerström, *Tetrahedron Lett.* p. 931 (1977).
67. T. Olsson, D. Tanner, B. Thulin, and O. Wennerström, *Tetrahedron* **37**, 3485 (1981); cf. *idem., ibid.* p. 3491.
68. T. Kauffmann and H. Lexy, *Angew. Chem.* **90**, 804 (1978); *Angew. Chem., Int. Ed. Engl.* **17**, 755 (1978).
69. H. Lexy, Ph.D. Dissertation, University of Münster (1978), reported in T. Kauffmann, *Angew. Chem.* **91**, 1 (1979); *Angew. Chem., Int. Ed. Engl.* **18**, 1 (1979).
70. J. Juriew, T. Skorochodowa, J. Merkuschew, W. Winter, and H. Meier, *Angew. Chem.* **93**, 285 (1981); *Angew. Chem., Int. Ed. Engl.* **20**, 269 (1981).
71. E. A. Truesdale and R. S. Hutton, *J. Am. Chem. Soc.* **101**, 6475 (1979).
72. H. C. Kang and V. Boekelheide, *Angew. Chem.* **93**, 587 (1981); *Angew. Chem., Int. Ed. Engl.* **20**, 571 (1981).
73. N. E. Kagan, D. Manzerall, and R. B. Merrifield, *J. Am. Chem. Soc.* **99**, 5484 (1977).
74. J. E. Baldwin and J. F. DeBernardis, *J. Org. Chem.* **42**, 3986 (1977), and references cited; J. R. Budge, P. E. Ellis, R. D. Jones, J. E. Linard, T. Szymanski, F. Basolo, J. E. Baldwin, and R. L. Dyer, *J. Am. Chem. Soc.* **101**, 4762 (1979); P. E. Ellis, J. E. Linard, T. Szymanski, R. D. Jones, J. R. Budge, and F. Basolo, *ibid.* **102**, 1889 (1980).
75. J. P. Collmann, J. I. Braumann, T. J. Collins, B. Iverson, and J. L. Sessler, *J. Am. Chem. Soc.* **103**, 2450 (1981).
76. T. G. Traylor, D. Campbell, and S. Tsuchiya, *J. Am. Chem. Soc.* **101**, 4748 (1979); T. G. Traylor, M. J. Mitchell, S. Tsuchiya, D. H. Campbell, D. V. Stynes, and N. Koga, *ibid.* **103**, 5234 (1981), cf. A. R. Battersby, S. G. Hartley, and M. D. Turnbull, *Tetrahedron Lett.* p. 3169 (1978).
77. M. R. Wasielewski, W. A. Svec, and B. T. Cope, *J. Am. Chem. Soc.* **100**, 1961 (1978).
78. C. K. Chang, *J. Am. Chem. Soc.* **99**, 2819 (1977).
79. J. P. Collman, A. O. Chong, G. B. Jameson, R. T. Oakley, E. Rose, E. R. Schmitton, and J. A. Ibers, *J. Am. Chem. Soc.* **103**, 516 (1981).
80. For a recent review, see J. P. Collman, *Acc. Chem. Res.* **10**, 265 (1977); cf. J. W. Buchler, *Angew. Chem.* **90**, 425 (1978); *Angew. Chem., Int. Ed. Engl.* **17**, 407 (1978).
81. J. C. Waterton and J. K. M. Sanders, *J. Am. Chem. Soc.* **100**, 1295 (1978).
82. M. Nakazaki, K. Yamamoto, and S. Tanaka, *Tetrahedron Lett.* p. 341 (1971).
82a. M. Nakazaki, K. Yamamoto, and M. Ito, *J. Chem. Soc., Chem. Commun.* p. 433 (1972).
82b. K. Yamamoto and M. Nakazaki, *Chem. Lett.* p. 1051 (1974); M. Nakazaki, K. Yamamoto, and S. Tanaka, *J. Org. Chem.* **41**, 4081 (1976); M. Nakazaki, K. Yamamoto, M. Ito, and S. Tanaka, *ibid.* **42**, 3468 (1977); M. Nakazaki, K. Yamamoto, S. Tanaka, and H. Kametani, *ibid.* p. 287.
83. D. J. Cram, C. S. Montgomery, and G. R. Knox, *J. Am. Chem. Soc.* **88**, 515 (1966).
84. H. Yuki, Y. Okamoto, and I. Okamoto, *J. Am. Chem. Soc.* **102**, 6356 (1980).

85. M. Hisatome, Y. Kawaziri, K. Yamakawa, and Y. Iitaka, *Tetrahedron Lett.* p. 1777 (1979).
86. M. Hisatome, N. Watanabe, T. Sakamoto, and K. Yamakawa, *J. Organomet. Chem.* **125,** 79 (1977). *Note added in proof:* A fully bridged ferrocenophane, [4₄.3]-(1,2,3,4,5)ferrocenophane, has recently been prepared: M. Hisatome, Y. Kawaziri, K. Yamakawa, Y. Hareda, and Y. Iitaka, *Tetrahedron Lett.,* 1713 (1982).
87. D. J. Cram and H. J. Reich, *J. Am. Chem. Soc.* **89,** 3078 (1967).
88. V. Boekelheide and P. F. T. Schirch, *J. Am. Chem. Soc.* **103,** 6873 (1981).
89. V. Boekelheide and Y. Sekine, *J. Am. Chem. Soc.* **103,** 1777 (1981).
90. G. Kaupp, E. Teufel, and H. Hopf, *Angew. Chem.* **91,** 232 (1979); *Angew. Chem., Int. Ed. Engl.* **18,** 215 (1979).
91. R. C. Helgeson and D. J. Cram, *J. Am. Chem. Soc.* **88,** 509 (1966).
92. D. J. Cram, R. C. Helgeson, D. Lock, and L. A. Singer, *J. Am. Chem. Soc.* **88,** 1324 (1966).
93. H. J. Lindner, *Tetrahedron* **32,** 753 (1976), and private communication; cf. also K. Nishiyama, M. Sakiyama, S. Saki, H. Horita, T. Otsubo, and S. Misumi, *Tetrahedron Lett.* p. 3739 (1977).
94. Besides the references mentioned in Chapter 3, see H. Irngartinger, J. Hekeler, and B. M. Lang, *Chem. Ber.* **116,** 527 (1983) for the x-ray analysis of **23** and **24,** respectively.
95. For a recent review, see T. Wagner-Jauregg, *Synthesis* pp. 165, 769 (1980).
96. H. Hopf, J. Kleinschroth, and A. E. Mourad, *Angew. Chem.* **92,** 388 (1980); *Angew. Chem., Int. Ed. Engl.* **19,** 389 (1980).
97. E. Ciganek, *Tetrahedron Lett.* p. 3321 (1967).
98. A. de Meijere, I. Erden, P. Gölitz, and R. Näder, *Angew. Chem.* **93,** 605 (1981); *Angew. Chem., Int. Ed. Engl.* **20,** 583 (1981).
99. J. E. Garbe, V. Boekelheide, E. Heilbronner, and Y. Zhong-zhi, *Abstr. Pap., 183rd Am. Chem. Soc. Natl. Meet.* No. 88 (1982).
100. D. J. Cram and N. L. Allinger, *J. Am. Chem. Soc.* **77,** 6289 (1955); cf. D. J. Cram, *Rec. Chem. Prog.* **20,** 71 (1959).
101. W. Jenny and J. Reiner, *Chimia* **24,** 69 (1970).
101a. J. L. Marshall and T. K. Folsom, *Tetrahedron Lett.* p. 757 (1971).
101b. J. L. Marshall and B.-H. Song, *J. Org. Chem.* **39,** 1342 (1974).
101c. J. L. Marshall and B.-H. Song, *J. Org. Chem.* **40,** 1942 (1975); J. L. Marshall and L. Hall, *Tetrahedron* **37,** 1271 (1981).
102. H. O. House, "Modern Synthetic Reactions," 2nd ed., p. 145. Benjamin, Menlo Park, California, 1972.
103. H. M. Frey and R. J. Ellis, *J. Chem. Soc. A* p. 552 (1966); cf. S. W. Benson and R. Shaw, *Trans. Faraday Soc.* **63,** 985 (1967).
104. H. Hopf and K. Menke, *Angew. Chem.* **88,** 152 (1976); *Angew. Chem., Int. Ed. Engl.* **15,** 165 (1976).
105. D. Wullbrandt, Ph.D. Dissertation, University of Braunschweig (1982).
106. H. Horita, T. Otsubo, Y. Sakata, and S. Misumi, *Tetrahedron Lett.* p. 3899 (1976).
106a. R. Näder and A. de Meijere, *Angew. Chem.* **88,** 153 (1976); *Angew. Chem., Int. Ed. Engl.* **15,** 166 (1976).
107. D. Wendisch, *in* "Houben-Weyl Methoden der organischen Chemie" (E. Mueller, ed.), Vol. IV, Part 3. Thieme, Stuttgart, 1971.
108. J. G. O'Connor and P. M. Keehn, *J. Am. Chem. Soc.* **98,** 8446 (1976).
109. R. B. Moodie and K. Schofield, *Acc. Chem. Res.* **9,** 287 (1976).
110. D. J. Cram and D. I. Wilkinson, *J. Am. Chem. Soc.* **82,** 5721 (1960).

111. E. Langer and H. Lehner, *Tetrahedron* **29,** 375 (1973); F. Cristiani, D. DeFilippo, P. Deplano, F. Devillanova, A. Diaz, E. F. Trogu, and G. Verani, *Inorg. Chim. Acta* **12,** 119 (1975).

111a. A. E. Mourad and H. Hopf, *Tetrahedron Lett.* p. 1209 (1979).

112. H. Ohno, H. Horita, T. Otsubo, Y. Sakata, and S. Misumi, *Tetrahedron Lett.* p. 265 (1977).

113. C. Elschenbroich, R. Möckel, and U. Zenneck, *Angew. Chem.* **90,** 560 (1978); *Angew. Chem., Int. Ed. Engl.* **17,** 531 (1978).

114. A. R. Koray, M. L. Ziegler, N. E. Blank, and M. W. Haenel, *Tetrahedron Lett.* p. 2465 (1979); cf. R. Benn, N. E. Blank, M. W. Haenel, J. Klein, A. R. Koray, K. Weidenhammer, and M. L. Ziegler, *Angew. Chem.* **92,** 45 (1980); *Angew. Chem., Int. Ed. Engl.* **19,** 44 (1980).

115. C. Elschenbroich and U. Zenneck, unpublished observations; private communication of Professor Elschenbroich.

116. E. D. Laganis, R. G. Finke, and V. Boekelheide, *Tetrahedron Lett.* p. 4405 (1980); cf. E. D. Laganis, R. G. Finke, and V. Boekelheide, *Proc. Natl. Acad. Sci. U.S.A.* **78,** 2657 (1981).

CHAPTER **10**

Multilayered Cyclophanes

SOICHI MISUMI

The Institute of Scientific and Industrial Research
Osaka University
Osaka, Japan

I. INTRODUCTION

A number of [m.n]paracyclophanes have been studied extensively by Cram and many groups from the viewpoint of transannular electronic interaction, and many interesting chemical and physical properties have been reported.[1,1a] Additional information related to the electronic interaction of stacked aromatic systems can be obtained through the study of multilayered compounds, in which more than two layers of aromatic rings are stacked face-to-face with each other.

Several types of multiply layered compounds have been reported: (1) multilayered cyclophanes,[1a] (2) 1,8,9-triphenylanthracene,[2] (3) [14]helicene,[3] and (4) triple-decker nickelocene,[4] tris(cyclooctatetraene)dititanium dianion,[5] and related triple- and quadruple-layered compounds.[6] This chapter deals with multilayered systems that have been studied systematically in a series of layered cyclophane compounds.

II. MULTILAYERED PARACYCLOPHANES

A. Multilayered [2.2]Paracyclophanes

1. SYNTHESIS

Of various procedures developed for the preparation of the [2.2]paracyclophane framework, the method of Hofmann elimination[7] of quaternary ammonium hydroxides was considered to be appropriate for the synthesis of multilayered compounds because the intermediates required for the

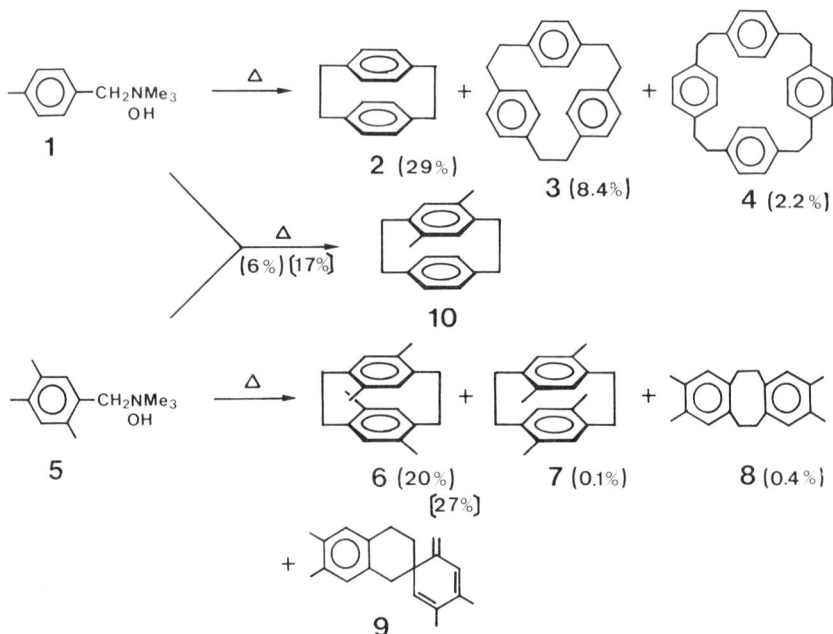

SCHEME 1. Numbers in brackets indicate yield by the modified method.

method are relatively accessible. Compounds **20** and **21** (isolated as a mixture) were the first multilayered cyclophanes prepared by this method.[8]

The yields of the Hofmann elimination product were generally low in the early experiments. Thus, cross-breeding pyrolysis of two ammonium bases in a 1 : 1 ratio afforded a low yield of the desired cyclophane, as seen in the preparation of 4,7-dimethyl[2.2]paracyclophane **(10)** in Scheme 1.[9] However, the yields of cross-breeding products **(10, 13, 14,** etc.) were increased up to 15 to 38% by the use of greater than a catalytic amount of phenothiazine, an excess of the more accessible ammonium base, a pear-shaped, three-necked flask, and the least amount of solvent (xylene) possible (the improved yields are shown in brackets in Schemes 1 and 2.)[10]

Similarly, triple- and quadruple-layered [2.2]paracyclophanes were prepared from di- and tetramethyl[2.2]paracyclophanes (**6** and **10**), as shown in Scheme 2.[9,11,12] Cross-coupling of **11** with **1** gave the highest yield of

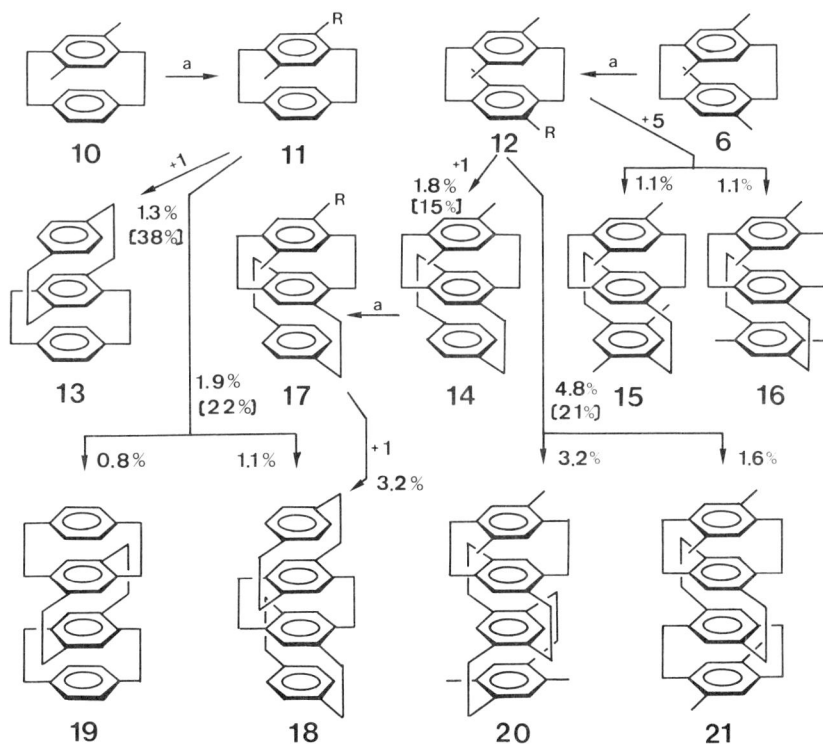

SCHEME 2. Numbers in brackets indicate yield by the modified method [R = CH₂N-Me₃OH; a: (1) NBS, (2) NMe₃, (3) OH⁻].

triple-layered compound **13**. Tetramethyl triple-layered compounds **15** and **16** are easily assigned by comparing the chemical shifts of their aromatic protons (see Table III). In contrast to the large difference in the yields of tetramethyl[2.2]paracyclophanes **6** and **7**, the formation of **15** and **16** in a 1 : 1 ratio indicates that the steric repulsion between methyl and benzylmethylene groups eclipsed with each other (designated as the pseudo-geminal position) in **16** is significantly less than the steric repulsion between the two pseudo-geminal methyl groups in **7**.

Isomeric quadruple-layered compounds **18** (D_2 symmetry) and **19** (C_{2h} symmetry), obtained by pyrolysis of **11**, show similar uv, nmr, and ir spectra and identical mass spectra and elemental analyses. To determine their structures, an alternative, stepwise synthesis of **18** was carried out by the cross-breeding pyrolysis of ammonium base **17**, derived from **6**, with **1**.[9,12] The D_2 symmetric isomer **18** thus obtained proved to be identical in all respects with the readily soluble isomer of the mixture of **18** and **19**. Consequently, the sparingly soluble isomer should be assigned to the C_{2h} symmetric structure **19**. The structures of the isomeric tetramethyl quadruple-layered compounds **20** and **21** were assigned by comparing their spectra with those of **18** and **19**, as well as by an X-ray crystallographic analysis[13] of the less soluble isomer (Fig. 1, C_{2h} symmetric **21**).

Pyrolysis of an equimolar mixture of **17** and **11** gave a mixture of quintuple-layered cyclophanes **22** and **23**, along with quadruple-layered compounds (Scheme 3). Attempts to separate the mixture into its component isomers were unsuccessful. However, the dimethyl derivative **25** of D_2 symmetric framework was obtained by cross-coupling pyrolysis of **24** and **1**.[9,14] A 2 : 1 mixture of the sextuple-layered compounds was obtained from **17** itself, and these were separated by column chromatography on alumina using benzene for elution.[9,15] By analogy with the quadruple-layered cyclophanes, the structure of the major, readily soluble isomer is assigned to **26** of D_2 symmetry, whereas the minor, less soluble isomer has structure **27** of C_{2h} symmetry. (See note added in proof.)

2. STRUCTURE AND STRAIN ENERGY

The first X-ray crystallographic analysis of the multilayered cyclophanes was achieved with the C_{2h} symmetric tetramethyl quadruple-layered compound **21**.[13] As shown in Fig. 1 intramolecular nonbonded distances between adjacent benzene rings are in the range 2.74–3.13 Å, indicating face-to-face fixation of four benzene rings within van der Waals contact distance (3.4 Å). Similar nonbonded distances (2.74–3.19 Å) are observed in the crystal structure (Fig. 2) of the 12-bromo triple-layered [2.2]paracyclophane **32**.[16]

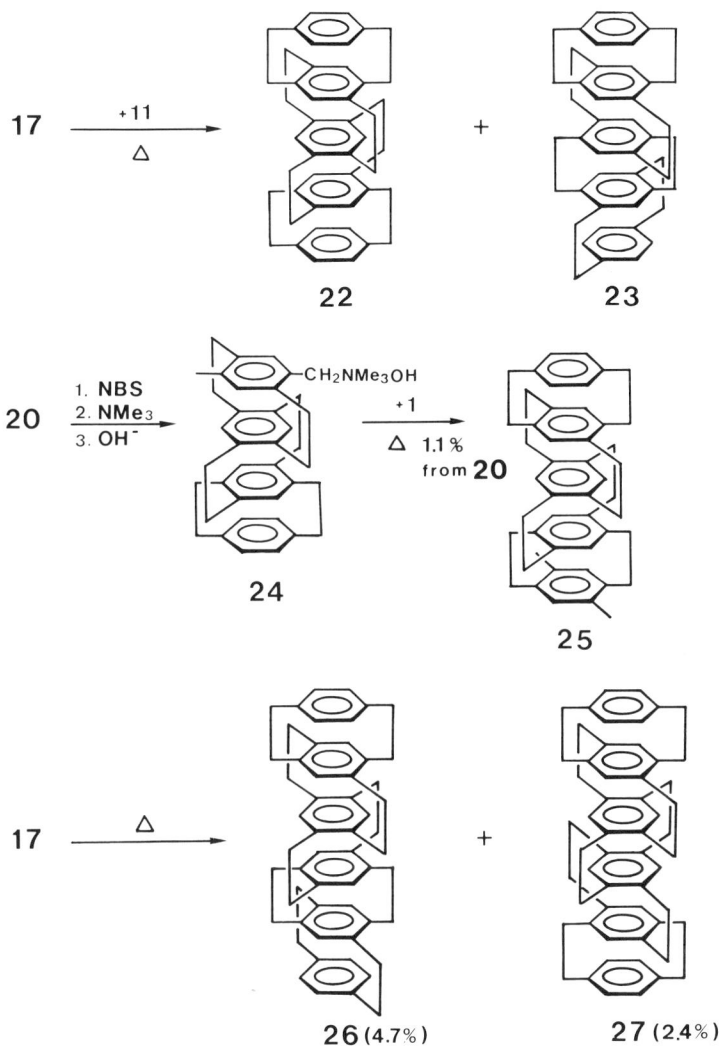

17 $\xrightarrow[\Delta]{+11}$

22   +   23

20 $\xrightarrow[\text{3. OH}^-]{\substack{\text{1. NBS} \\ \text{2. NMe}_3}}$ $-CH_2NMe_3OH$ $\xrightarrow[\substack{\Delta \quad 1.1\% \\ \text{from } \mathbf{20}}]{+1}$

24

25

17 $\xrightarrow{\Delta}$ 26 (4.7%)   +   27 (2.4%)

SCHEME 3

All the outer benzene rings of the triple- and quadruple-layered compounds are bent into a boat form with bending angles similar to those of
the double-layered compound **2**,[17] as seen in Table I. The inner benzene
rings, however, are extraordinarily distorted into a twist form. (The twist
angles γ for **2**, **32**, and **21** are given in Table I.) The sp^3–sp^3 single bond of
the ethylene bridges connecting the two inner benzene rings in **21** is 1.550
Å long and slightly shorter than all the sp^3–sp^3 bonds (~1.58 Å) of the

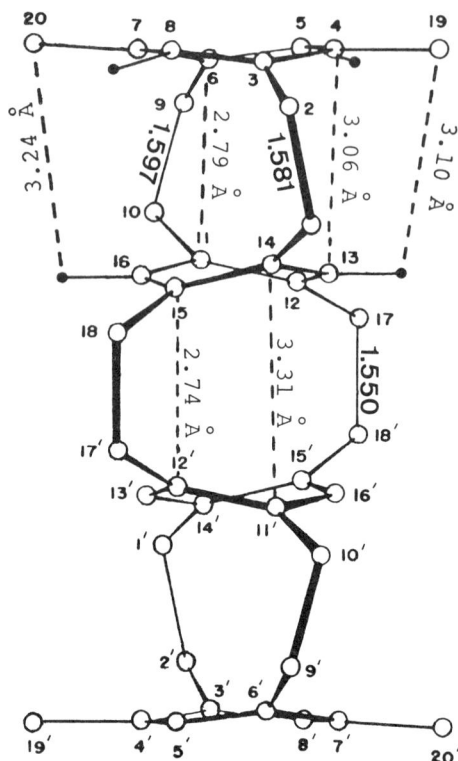

Fig. 1. Molecular structure of tetramethyl quadruple-layered [2.2]paracyclophane centro-symmetric isomer **21**.

ethylene bridges in **2** and **34** and the others in **21**. Compared with the symmetric rotational vibration (~6.4°) of the two benzene rings (around the axis centered in, and perpendicular to, the rings) in the crystal of **2,** rotation of the inner and outer benzene rings of **21** and **32** is restricted to one side (with angles of ~8 and 6.8°, respectively) owing to the repulsion between the two ortho methylene groups.

Such highly strained molecular structures as those of the triple- and quadruple-layered [2.2]paracyclophanes are suggestive of higher strain energy than that of [2.2]paracyclophane. In fact, the strain energy of the nonsubstituted triple-layered [2.2]paracyclophane **13** is 245.6 ± 9.8 kJ/mol (59.0 kcal/mol),[18] which is almost twice the corrected value (122.6 ± 5.4 kJ/mol, 29.6 kcal/mol) of [2.2]paracyclophane (**2**) (cf. Boyd's[19] value of 31.5 kcal/mol and Westrum's[20] value of 31.4 kcal/mol). Because crystallographic analysis shows a similar bending angle of the boat form in the

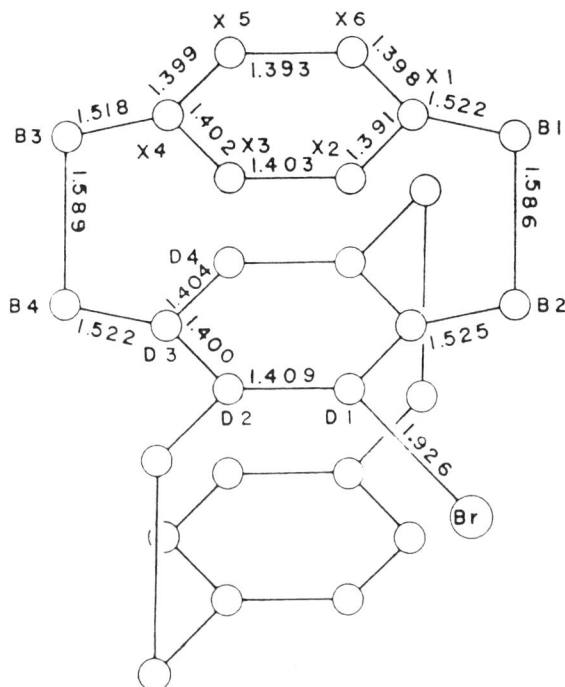

Fig. 2. Molecular structure of bromo triple-layered [2.2]paracyclophane **32.**

benzene rings of the double- and triple-layered paracyclophanes, it may
be concluded that the twisted inner benzene ring in the triple-layered
compound **13** is more strained than the boat form of the outer benzene
ring, and the contribution from the inner twisted benzene ring to the total
strain energy is approximately twice that from the outer boat-form benzene ring.

3. ELECTRONIC SPECTRA

a. Absorption Spectra

The absorption spectra of nonsubstituted, multilayered [2.2]paracy-
clophanes are shown in Fig. 3.[9] The absorption curves are almost independent of the methyl substitution and the configurational stereo-
isomerism. The figure demonstrates increasingly strong bathochromic
and hyperchromic shifts as the number of layers increases successively.
In particular, these shifts are prominent as the layer is multiplied from

TABLE I
Bending Angles of Benzene Rings in Multilayered Paracyclophanes

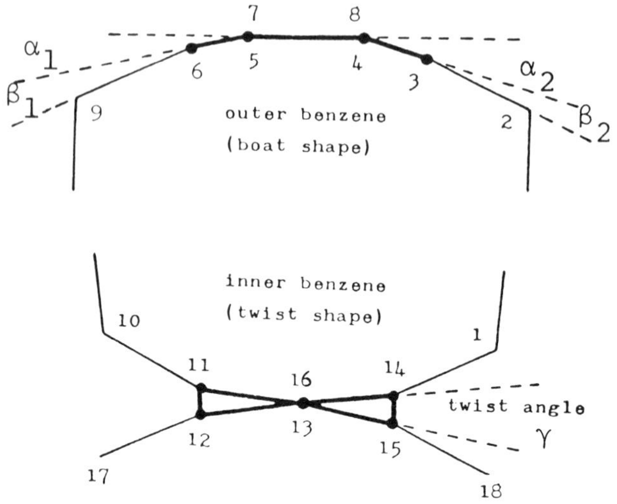
7
8
α1
6
5
4
3
α2
β1
9
outer benzene
(boat shape)
2
β2
inner benzene
(twist shape)
10
1
11
16
14
twist angle
γ
12
13
15
17
18

Compound	Outer Benzene (deg)		Inner benzene (deg)
	α_1, α_2	β_1, β_2	γ
2	12.6	11.2	—
32	12.1	10.9	13.6
	12.3	11.5	
21	12.5	10.8	13.4
	12.9	11.8	

single to double and double to triple, while the absorption curves become markedly structureless. These spectral features are explained mainly by transannular π–π electronic interaction or transannular delocalization, because the absorption band shift is little affected by the second important factor, distortion of the benzene ring originally observed in [*n*]paracyclophanes.[21] A theoretical study of the electronic spectra of double- to quadruple-layered [2.2]paracyclophanes was achieved semiempirically by considering the configuration interaction between the ground, locally excited, and charge-transfeı configurations, and the spectra were interpreted with band assignments that were more reliable than those in previous studies.[22]

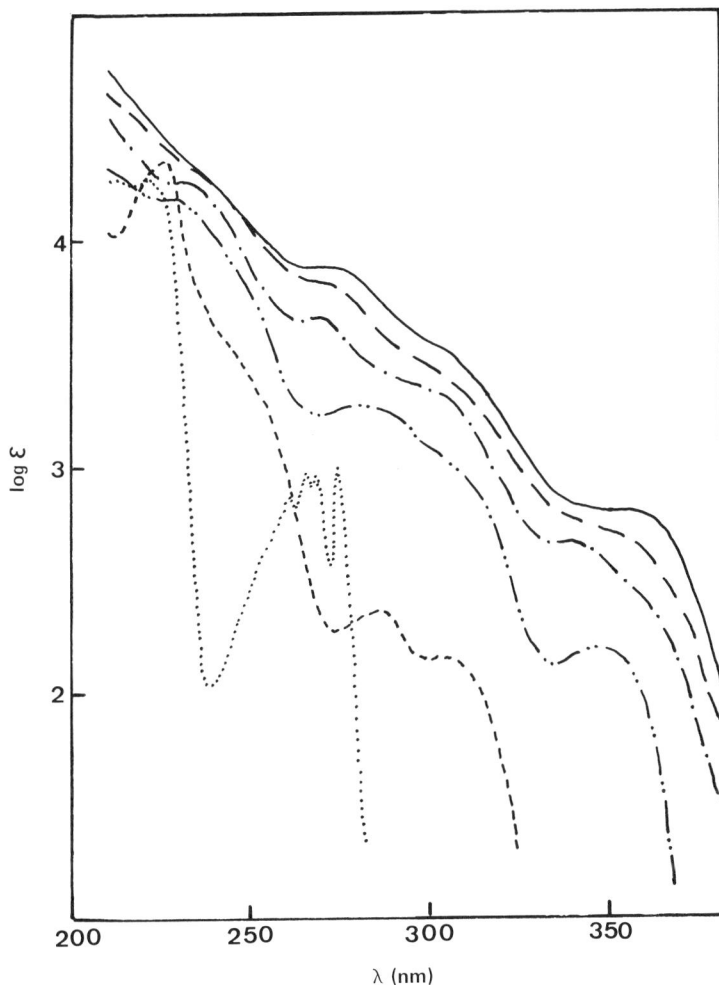

Fig. 3. Electronic spectra of multilayered [2.2]paracyclophanes in cyclohexane: ·····, p,p′-dimethylbibenzyl; ----, **2**; -·-·, **13**; —·—, **18**; ———, **22**; ———, **26.**

b. Emission Spectra

Additional information concerning the transannular interaction of multilayered cyclophanes is derived from the study of their emission spectra. Both fluorescence and phosphorescence spectra show bathochromic shifts with successive increases in the number of layers, just as in the case of the absorption spectra.[9] In contrast to the ambiguous positions of the longest-wavelength bands in the absorption spectra, the fluorescence maxima are clearly distinguished. No further red shift of the maximum is observed between the maxima of the quintuple- and sextuple-layered compounds, indicating no further increase in the transannular interaction for more than the quintuple-layered system.

Lifetimes of phosphorescence[9] tend to become progressively shorter and to converge in the same manner as the band shift in the absorption and fluorescence spectra, suggesting a lowering of the energy gap between ground and triplet states, which thereby increases the probability of radiationless deactivation.

4. Charge-Transfer Complexes with Tetracyanoethylene and Trinitrobenzene

In a series of π complexes of various methylated benzenes with tetracyanoethylene (TCNE), there is a good correlation between the charge-transfer (CT) band position and the association constant K of the complexes.[23] Increased transannular interaction was reported to be reflected in the π basicity of the tetramethyl quadruple-layered paracyclophane (as a mixture of 20 and 21) compared with 2.[8] Transannular electron release to the complexed ring from the remaining benzene rings in the donor is to be expected. Table II and Fig. 4 indeed show that, with an increase in the number of layers, the long-wavelength maxima of the TCNE complexes shift to longer wavelength, and hence the donor character of the cyclophanes increases progressively. It is noteworthy that, even in the case of the quintuple-layered compound, such an electron release is still effective for stabilization of the complex 28 (Fig. 4).

Although the CT spectrum of the sextuple-layered compound–TCNE complex could not be measured owing to the very low solubility of the complex, the use of 1,3,5-trinitrobenzene (TNB) makes it possible to measure the CT spectrum, as shown in Table II. This table also shows identical spectra for the quintuple- and sextuple-layered compounds, indicating no difference in their transannular electron-releasing capacity.

5. Nuclear Magnetic Resonance Spectra

a. ¹H-nmr Spectra

The aromatic protons of a series of [m.n]paracyclophanes appear at higher field, beyond the usually accepted range of alkylbenzenes, with decreasing number of methylene groups in the bridges, m and n.[24] This upfield shift is ascribed to the anisotropic effect of one benzene ring on the

TABLE II

Absorption Maxima of CT Complexes of Multilayered Paracyclophanes with TCNE and 1,3,5-Trinitrobenzene and Equilibrium Constants

	TCNE complex	1,3,5-TNB complex	
Compound	λ_{max}, nm (CH_2Cl_2)	λ_{max}, nm ($CHCl_3$)	K_x
p-Xylene	420, 465[b]	312	2.08
p,p'-DMB[a]	420, 465[b]	312	—
Durene	480	—	—
2	521	362 (365)[c]	—
10	555	—	—
6	584 (580)[c]	405	10.2 (9.6)[c]
7	595	—	—
13	530,[b] 630	405	6.84
14	430,[b] 650	430	15.9
15	430,[b] 655	—	—
16	430,[b] 643	—	—
18	530, 710	400, 450[b]	7.46
20	720 (690)[c]	430	15.7 (21.6)[c]
22	540, 750	400, 470[b]	—
26	—	400, 470[b]	11.1

[a] DMB, dimethylbibenzyl.
[b] Shoulder.
[c] Longone and Chow.[8]

aromatic protons of the other ring and to the rehybridization of the benzene carbon atoms due to ring deformation. It is expected that the aromatic protons of multilayered [2.2]paracyclophanes are also affected by the anisotropy of the more remote benzene rings (in addition to the effect of the nearest facing benzene ring) and that these cyclophanes are quite suitable as a model for the study of magnetic anisotropy of benzene rings.

The chemical shift of the aromatic protons (ArH) of multilayered [2.2]paracyclophanes (**2, 6, 7, 10, 13–16, 18–22, 25–27** in Fig. 5) are summarized in Table III.[25] The signals are assigned to the corresponding protons mainly with the aid of integration. However, the aromatic protons of some multilayered compounds cannot be assigned simply by their integrated intensities, for example, **14** (H_b and H_c), **20, 21,** and **25** (H_b–H_e). For these compounds the solvent effect is useful for assigning the aromatic protons. As seen in Table III the aromatic protons (ArH) of alkylbenzenes, double-layered compounds, and outer benzene rings of multilayered compounds reveal a large difference ($\Delta\delta$ 0.10–0.16 ppm) in chemical shift between carbon tetrachloride and deuteriochloroform, whereas the inner ArH protons show only a small difference ($\Delta\delta$ 0.03–

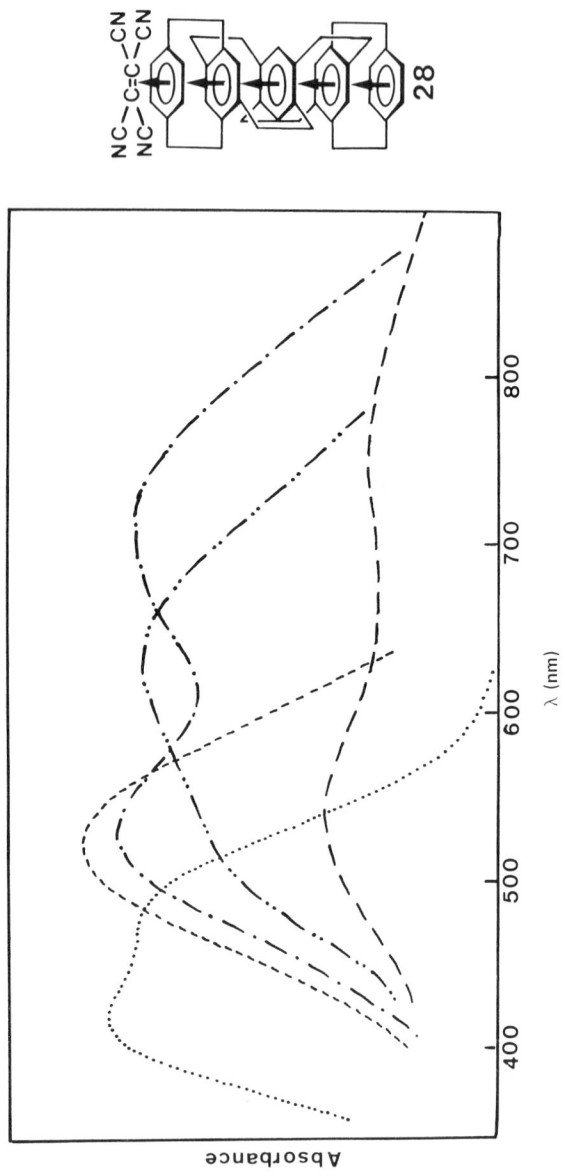

Fig. 4. Charge-transfer specta of multilayered [2.2]paracyclophane–TCNE complexes in CH$_2$Cl$_2$: ······, p,p'-dimethylbibenzyl; ----, **2**; — · —, **13**; — · · —, **18**; ———, **22**.

TABLE III

Chemical Shifts of Aromatic Protons of Multilayered [2.2]Paracyclophanes[a]

| Compound | ArH | Solvent | | $\Delta\delta$[b] | Calculated value |
		CCl$_4$	CDCl$_3$		
p-Xylene	—	6.95	7.06	0.11	—
Durene	—	6.74	6.90	0.16	—
2	—	6.35	6.47	0.12	6.26
10	H$_a$	6.68	6.78	0.10	6.62
	H$_b$	6.26	6.36	0.10	6.26
	H$_c$	5.88	6.00	0.12	5.84
6	—	6.25	6.36	0.11	6.20
7	—	6.07	6.19	0.12	6.05
13	H$_a$	6.08	6.19	0.11	6.04
	H$_b$	5.35	5.40	0.05	5.36
14	H$_a$	6.12	6.23	0.11	6.04
	H$_b$	5.70 ⎫	5.75	0.05	5.72
	H$_c$	5.64 ⎭		0.11	5.62
15	H$_a$	5.67	5.79	0.12	5.62
	H$_b$	6.06	6.12	0.06	6.08
16	H$_a$ or H$_c$	—	5.83 ⎧	0.16	—
	H$_a$ or H$_c$	5.67	5.79 ⎨	0.12	—
	H$_b$	—	5.72 ⎩	0.05	—
18	H$_a$	5.98	6.09	0.11	5.94
	H$_b$	5.12	5.15	0.03	5.14
19	H$_a$	5.98	6.12	0.14	—
	H$_b$	5.10	5.13	0.03	—
20	H$_a$	5.57	5.69	0.12	5.52
	H$_b$	5.47	5.52	0.05	5.50
21	H$_a$	5.55	5.68	0.13	—
	H$_b$	5.48	5.52	0.04	—
22	H$_a$	5.96	6.06	0.11	5.91
	H$_b$	5.03	5.07	0.04	5.04
	H$_c$	4.88	4.92	0.04	4.92
25	H$_a$	5.96	6.09	0.13	5.91
	H$_b$	5.04	5.09	0.05	5.04
	H$_c$	4.92	4.95	0.03	4.92
	H$_d$	5.36	5.40	0.04	5.40
	H$_e$	5.53	5.65	0.12	5.49
26	H$_a$	5.94	6.05	0.11	5.89
	H$_b$	5.00	5.04	0.04	5.01
	H$_c$	4.80	4.83	0.03	4.82
27	H$_a$	—	6.05	—	—
	H$_b$	—	5.05	—	—
	H$_c$	—	4.81	—	—

[a] Values are expressed as parts per million.
[b] $\Delta\delta = \delta(CCl_4) - \delta(CDCl_3)$.

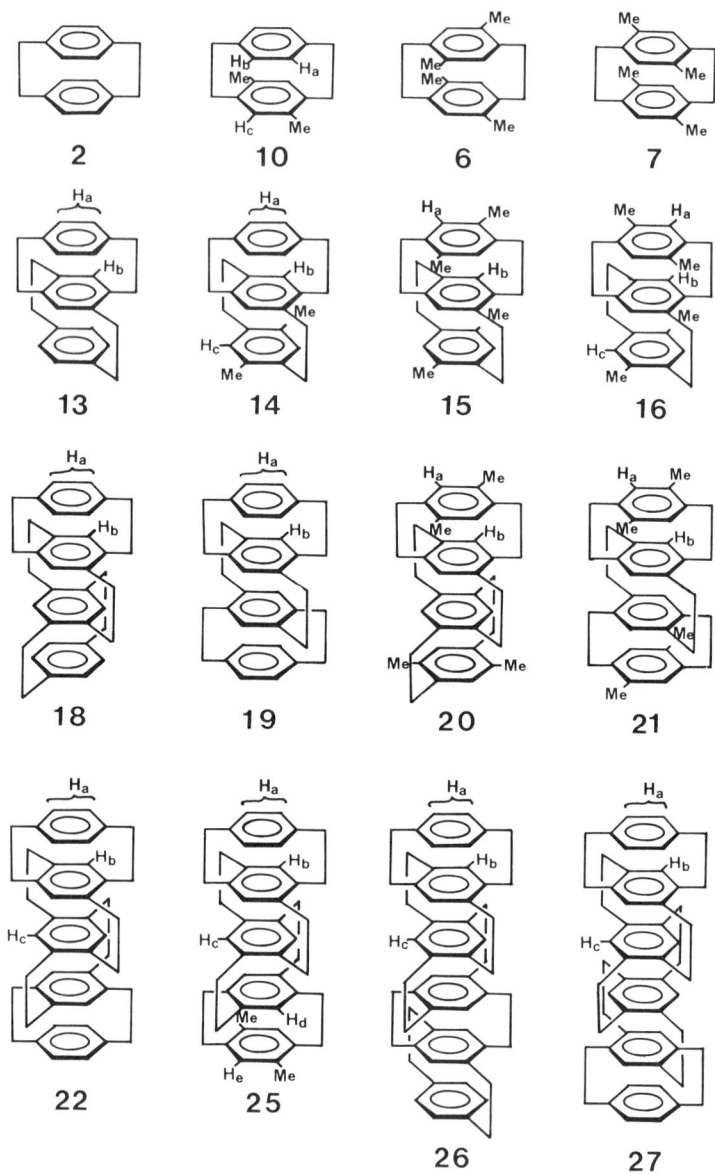

Fig. 5. Multilayered [2.2]paracyclophanes and their methyl derivatives.

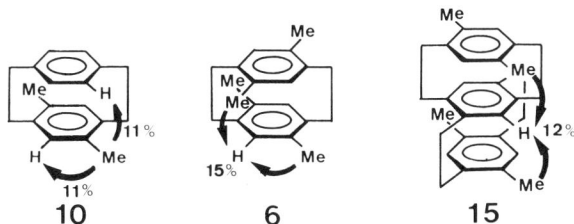

Fig. 6. Nuclear Overhauser effects of layered paracyclophanes in deuteriochloroform.

0.06 ppm). On the basis of such a solvent effect, ArH protons of **14, 20,** and **21** are easily assigned, and the ArH protons of dimethyl quintuple-layered compound **25** are also assigned by the combined use of the solvent effect and comparison with its parent compound **22.**

i. Nuclear Overhauser Effect, Steric Compression Effect, and En-hanced Ortho Substitution Shift. The ArH protons (H_a and H_b) of **10** appear as an A_2B_2 pattern due to a transannular proximity effect of the methyl group in the opposite benzene ring (Fig. 6). Since the nuclear Overhauser effect (NOE) has been used successfully to assign such protons, the aromatic protons of some layered compounds were assigned by applying the effect, as shown in Fig. 6. The observed NOE value for the inner ArH of **15** indicates efficient relaxation by the two methyl groups, which closely sandwich the proton.

Hydrogen atoms that are significantly compressed generally exhibit downfield shifts in their nmr spectra.[26] Downfield shifts of ArH protons pseudo-geminal to the substituent were observed in the spectra of 4-substituted [2.2]paracyclophane.[27] Similar steric compression effects by a transannularly located methyl group have been observed in the presently considered multilayered compounds.[25] For example, obvious downfield shifts are recognized by comparing the chemical shifts of H_a in **10** and ArH of **6** with that of H_b in **10**. Similar deshielding of compressed protons is observed for H_b of **14, 16, 20,** and **21** and H_d of **25** (Table III). An average value for the compression effect of a methyl group, σ_{CE} (pseudo-geminal shift), is -0.36 ppm, which is nearly equal to the value for 4-methyl[2.2]paracyclophane. Of greater interest is a downfield shift (-0.71 ppm) of H_b in the triple-layered compound **15.** Such a value, nearly twice the average methyl pseudo-geminal shift, indicates that compound **15** presents a novel example of steric compression from both sides and an additivity of the compression shift. A chemical shift difference (0.19 ppm) for the methyl protons of **6** and **7** can also be ascribed to a compression effect between the two methyl groups.

TABLE IV

Shielding Effect of Additional Benzene Rings

Benzene ring	Shielding effect (ppm)
σ_1 (neighboring ring)	0.69
σ_2 (next ring but one)	0.22
σ_3 (next ring but two)	0.10
σ_4 (next ring but three)	0.03
σ_5 (next ring but four)	0.02

The introduction of a substituent into the benzene ring of [2.2]paracyclophane causes a large chemical shift of the ortho aromatic proton compared with that observed in the case of the corresponding noncyclophane compound.[27] Such an enhanced ortho shift σ_{OE} is also observed in methylated multilayered cyclophanes. Thus, the protons ortho to a methyl group appear at higher field by an average of 0.42 ppm than the corresponding proton of nonsubstituted, parent cyclophanes, for example, outer protons in each pair of **15** versus **13, 20** versus **18,** and **25** versus **22.** A value of 0.21 ppm, obtained by subtracting the difference (0.21 ppm) in chemical shifts of ArH protons of durene and *p*-xylene from the above average value, is the enhanced ortho substituent effect of a methyl group in the layered [2.2]paracyclophane system.

ii. Magnetic Anisotropy. Table III shows remarkable upfield shifts of all the aromatic protons as the number of layers increases. A certain regularity among chemical shifts of the aromatic protons is expected because each benzene is stacked at a regular interval in this series of multilayered cyclophanes, as described in Section II,A,2.[13] By examining a number of examples of spectral data, the shielding effects of additionally stacked benzene rings are empirically estimated, as shown in Table IV. It can be seen in the table that the chemical shift of a given aromatic proton is calculated by subtracting the shielding effect of the additional benzene ring from the chemical shift of aromatic protons of standard alkylbenzenes such as *p*-xylene (6.95 ppm) or durene (6.74 ppm). For example, a calculated value for H_a of the quintuple-layered compound **22** is 5.91 ppm ($6.95 - \sigma_1 - \sigma_2 - \sigma_3 - \sigma_4$), that for H_b is 5.04 ppm ($6.74 - 2\sigma_1 - \sigma_2 - \sigma_3$), and that for H_c is 4.92 ppm ($6.74 - 2\sigma_1 - 2\sigma_2$). This treatment can also be applied to the cases of methyl derivatives by considering the secondary effects of methyl groups (i.e., σ_{CE} and σ_{OE}). The calculated value for H_d of **25,** for instance, is 5.40 ppm ($6.74 - 2\sigma_1 - \sigma_2 - \sigma_3 - \sigma_{CE}$), and that for H_e is 5.49 ppm ($6.74 - \sigma_1 - \sigma_2 - \sigma_3 - \sigma_4 - \sigma_{OE}$). These calculated values are in excellent agreement with the observed

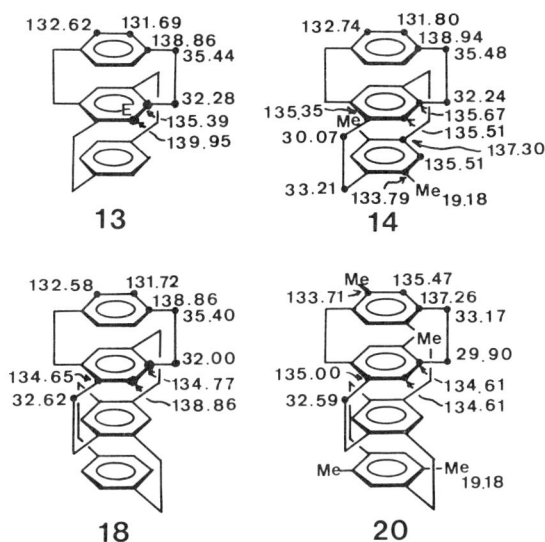

Fig. 7. ^{13}C-nmr data for multilayered [2.2]paracyclophanes.

values, as shown in Table III, supporting this as a reasonable evaluation of each shielding magnitude.

b. ^{13}C-nmr Spectra

The π–π compression has pronounced effects on the ^{13}C-nmr spectra of multilayered [2.2]paracyclophanes (Fig. 7).[28] Thus, aryl carbons of all the outer benzene rings reveal the same extent of downfield shifts as those of double-layered compounds. It is noteworthy that triple-layered compound 13 provides a novel example of the double compression effect. The signal for the aryl carbon C_E of 13 is shifted to lower field (by 7.95 ppm) than the corresponding carbon resonance of a reference compound, 6,9-dimethyl[4.4]paracyclophane. This downfield shift is about twice as large as the difference value (4.25 ppm) between the corresponding carbons of dimethyl[4.4]paracyclophane and 4,7-dimethyl[2.2]paracyclophane (10), indicating an additivity of the steric compression contribution on carbon resonances.

6. Other Physical Properties

a. Optical Properties

A series of optically active multilayered [2.2]paracyclophanes (13, 18, and 22, Schemes 1 and 2) with known absolute configuration were synthe-

sized to study the relationship between optical activity and stacking interaction. The synthesis was carried out by Hofmann elimination pyrolyses
with the double- and triple-layered ammonium bases **11** and **17** of known
absolute configuration.[29] All the optical rotations $[(R)$-$(-)$-**13**, $[\alpha]_D^{28}$
$-256°$ (c 0.67, $CHCl_3$); (R,R)-$(-)$-**18**, $[\alpha]_D^{21}$ $-285°$ (c 0.39, $CHCl_3$);
(R,R,R)-$(-)$-**22**, $[\alpha]_D^{22}$ $-362°$ (c 0.22, $CHCl_3$)] of these R-form compounds
are levorotatory in the visible region and tend to increase with an increase
in the number of layers.[30]

b. esr Spectra

The esr and ENDOR spectroscopy of radical-anions of multilayered
[2.2]paracyclophanes shows the spin distributions in these radical-anions
to be largely polarized by interaction with counterions.[31] It has been
shown by MO calculation that the observed ion-pairing effects can be
explained by the ion-pair models in which the cation is located above the
center of the outermost benzene ring and that the potassium ion migrates
from one side of the radical ion molecule to the other side in a loosely
bound ion pair of the triple-layered radical-anion in a DMF–THF (1:1)
mixed solvent, as in the ion pair of [2.2]paracyclophane anion-radical–
potassium ion in DMF–THF (2:1).[32] (See note added in proof.)

7. CHEMICAL PROPERTIES

The chemistry of double-layered [$m.n$]paracyclophanes ($m, n < 4$) has
been widely investigated by many groups, and a number of intriguing
chemical properties due to transannular electronic interaction and molecular strain have been reported. Multilayered paracyclophanes are expected to be highly reactive species and to reveal interesting chemical
properties from the viewpoint of both electronic interaction and molecular strain, which are described in Sections II,A,2 and II,A,3.

a. Electrophilic Substitution

The enhancement in chemical reactivity due to transannular electronic
interaction has been observed in some studies on electrophilic aromatic
substitutions of double-layered compounds such as [$m.n$]paracy-
clophanes[33] and janusene.[34] However, the triple-layered paracyclophanes
revealed an enhanced reactivity in bromination but gave poor yields in the
Friedel–Crafts reaction and nitration because of their more severe strain.

  i. Bromination. Triple-layered [2.2][2.2]-, [2.2][3.3]-, and [3.3][3.3]-
paracyclophanes[34a] (**29–31**) easily gave monobromides **32–34** as the sole
product by noncatalytic bromination with pyridinium hydrobromide

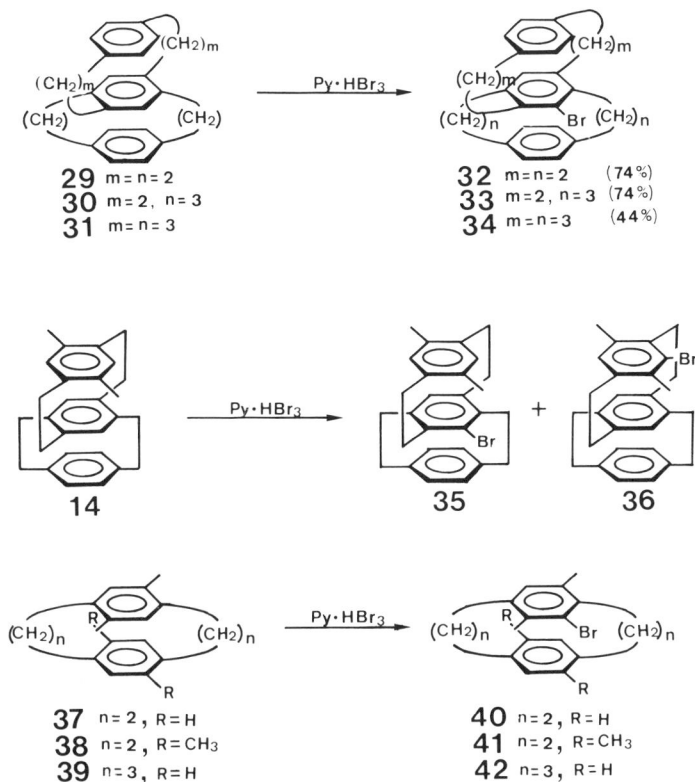

SCHEME 4

perbromide (Scheme 4).[35] These brominations are complete in a shorter time than are those for [n]paracyclophanes and double-layered paracy-clophanes, indicating an acceleration of the reaction due to transannular electronic stabilization in the transition state. No dibromo derivative is formed even if an excess of the reagent is used or the reaction time is prolonged. The bromine substituent first introduced may prevent the second substitution from taking place because the steric requirement and the inductive effect of the bromo group probably inhibit the formation of the second σ complex at the inner benzene.

The bromination of layered paracyclophanes is surely accelerated by the transannular electronic interaction in the transition state. The magni-tude of the acceleration is demonstrated by the competitive reactions summarized in Table V. The reactions are obviously favored by an in-crease in the number of layers regardless of the type of reagent. The reactions of all triple-layered compounds are overwhelmingly faster than

TABLE V

Competitive Bromination of Paracyclophanes

Run	Compounds	Reagent[a]	Product ratio
1	**29 + 30**	A or B	**32/33**, 20 : 1
2	**30 + 31**	A	**33/34**, 3 : 1
3	**14**	B	**35/36**, 1 : 1
4	**29 + 14**	B	**32/35 + 36**, 1.5 : 1
5	**29 + 37**	B	**32/40**, 1 : 0
6	**29 + 38**	B	**32/41**, 1 : 0
7	**30 + 37**	B	**33/40**, 1 : 0
8	**31 + 39**	B	**34/42**, 1 : 0
9	**14 + 38**	B	**35 + 36/41**, 1 : 0

[a] Reagents: A, pyridinium hydrobromide perbromide; B, bromine.

those of the double-layered compounds (runs 5–9). This can be explained mainly by a difference in the stabilization of the σ complexes **43** and **44**

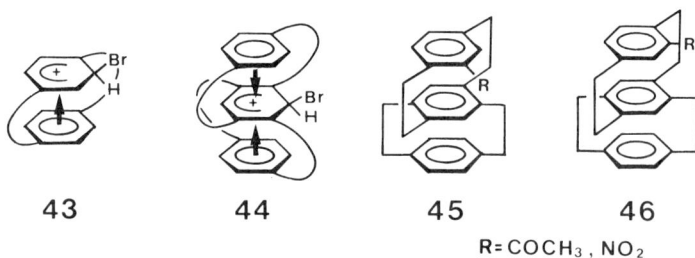

43   **44**   **45**   **46**

R= COCH$_3$, NO$_2$

due to the transannular charge delocalization. The relative rate of reaction of triple-layered compounds **29–31** with bridge chains of different length is 60 : 3 : 1 (runs 1 and 2). This order is in accord with the relative strength of the transannular electronic interaction.

ii. Other Electrophilic Substitutions. Acetylation and nitration of **29** were carried out to give in 44% yield the monoacetyl derivatives **45** (R = COCH$_3$, 23%) and **46** (R = COCH$_3$, 21%) and 8% of the mononitro derivatives **45** and **46** (R = NO$_2$).[10] These electrophilic substitutions on the outer benzene ring are in striking contrast to the bromination of the inner benzene.

b. Skeletal Rearrangement

An interesting skeletal rearrangement of [2.2]paracyclophane to [2.2]metaparacyclophane was reported by Cram and co-workers.[36] In contrast, treatment of its tetramethyl derivative **6** with the same catalyst,

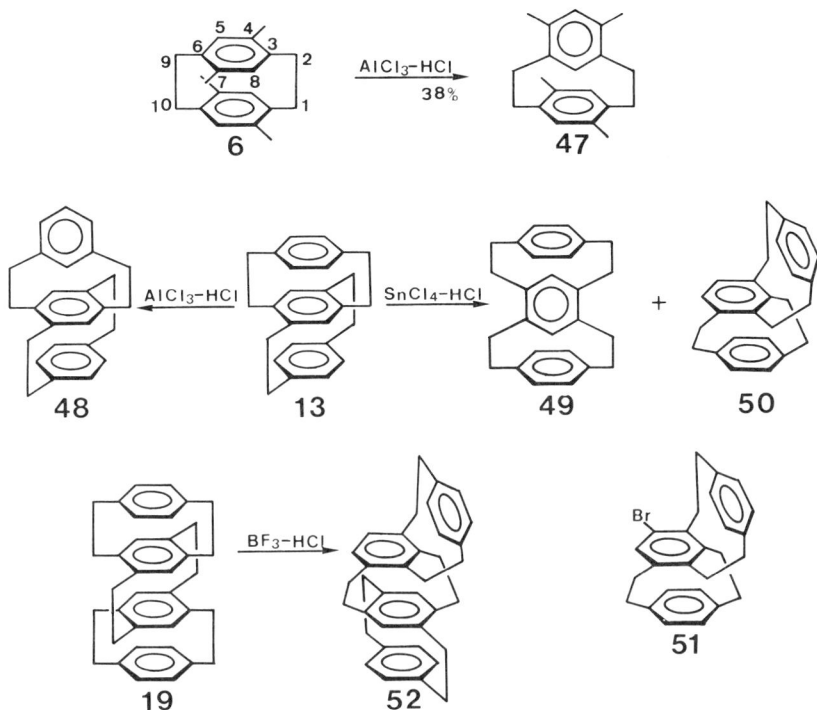

Scheme 5

AlCl$_3$–HCl, gives a 38% yield of **47,** which is derived by an intriguing double rearrangement of the ethylene bridge at C-3 and the methyl group at C-4 (Scheme 5).[37]

Treatment of the more strained triple-layered compound **13** under the same conditions gave only polymeric material. When the reaction was carried out at lower temperature, triple-layered metaparacyclophane **48** was the main product, together with two isomeric metaparacyclophanes (**49** and **50**), as shown in Table VI.[37,37a] The structure of **48** was confirmed by an alternative stepwise synthesis (Section IV,A) as well as by ^{1}H-nmr data.

However, when treated with a large excess of a weak Friedel–Crafts catalyst such as SnCl$_4$–HCl at room temperature, **13** gave an 80% yield of a mixture of **49** and **50** in 1.2 : 1 ratio. Similar results were obtained by the use of TiCl$_4$–HCl, BF$_3$ · Et$_2$O–HCl, ZnCl$_2$–HCl, or I$_2$ as a catalyst. It is of great interest that the formation of **49** and **50** (Table VI) is also achieved by the just described double rearrangement of the two ethylene bridges attached to the inner benzene ring and that the molar ratio, 1.2 : 1, of the

TABLE VI

Rearrangements of 13 with Friedel–Crafts Acids

Acid	Temp (°C)	Time (min)	Yield (%)	Product ratio (%)		
				48	**49**	**50**
AlCl$_3$–HCl	−17	15	51	76	11	10
SnCl$_4$–HCl	29	30	80	—	54	46
TiCl$_4$–HCl	29	30	72	—	54	46
BF$_3$·Et$_2$O–HCl	30	180	79	—	53	47
ZnCl$_2$–HCl	40	180	80	—	56	43
I$_2$	23	180	88	—	55	45

two isomers remains constant even with increased reaction time, higher temperature, and the use of different types of catalysts. Treatment of **6** with the weak proton acid SnCl$_4$–HCl afforded no migration product, indicating that the triple-layered compound is much more reactive than the double-layered one.

The structure of **49** was confirmed by stepwise synthesis (Section IV,A), X-ray crystallographic analysis (see Fig. 12), and nmr data. Compound **50** was assigned to a C_2 symmetric structure by nmr, and its detailed molecular structure was determined by X-ray analysis of the monobromo derivative **51,** which was obtained by bromination of **50** in quantitative yield (Fig. 8).

A similar rearrangement was observed in a treatment of the quadruple-layered compound **19** with BF$_3$·Et$_2$O–HCl at 0°C. The structure of product **52** was determined by comparing its nmr spectrum with that of **50.**[37]

The structure of the protonated species **55,** formed during catalytic rearrangement with a weak Friedel–Crafts acid, was determined by nmr analysis in FSO$_3$H–SO$_2$ClF–CD$_2$Cl$_2$ at −100°C. The preferential protonation at a bridgehead carbon of the inner benzene ring is ascribed to stronger π basicity of the inner benzene ring, pronounced strain relief, and larger stabilization of the intermediate σ complex **55** compared with those of the outer benzene ring.

As shown in Scheme 6, the ethylene bridge attached to the protonated carbon in **55** at first migrates intramolecularly to the adjacent bridgehead carbon C-4 to give cation **56.** Cation **56** is probably a common intermediate for the two rearranged products **49** and **50,** because a constant product ratio in the reaction mixtures was obtained regardless of the type of weak proton acids, reaction time, and temperature. Second, another ethylene bridge at C-4 migrates in two ways, that is, 1,2 shifts to C-3 and C-5 to

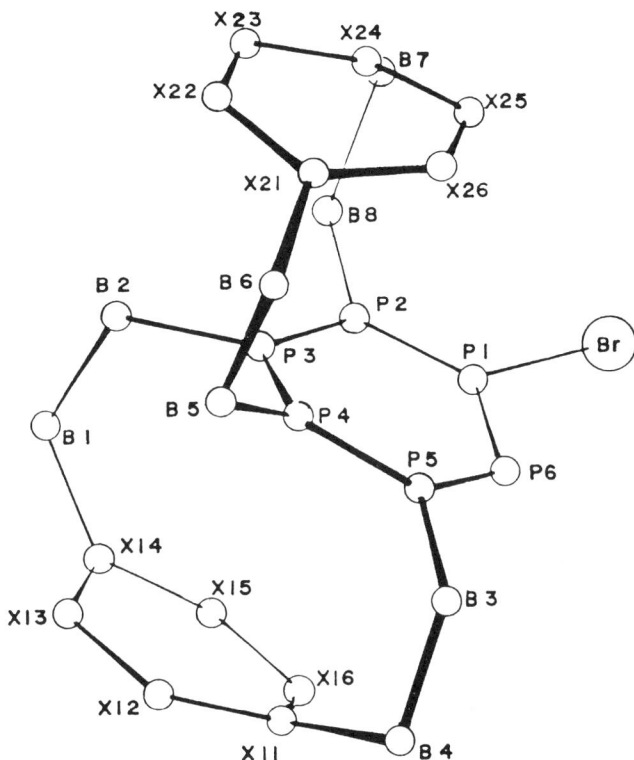

Fig. 8. Profile of bromide **51.**

give cations **57** and **58,** respectively, from which products **49** and **50** are formed by deprotonation. No migration from C-3 to C-8 (**55–59**) was observed due to the severe molecular strain in structure **60.**

The initial protonation with the strong proton acid $AlCl_3$–HCl takes place at a bridgehead carbon of the outer benzene rings with strain relief, followed by bridge migration to **54** and deprotonation to give **48.** The preference for such a different position of initial protonation may be associated with strong acidity and the steric requirement of this proton acid. In other words, these rearrangements may proceed via a thermodynamically controlled process for the weak proton acids and via a kinetically controlled one for the strong acid.

c. Reaction with Carbene[10]

The carbene addition reaction was carried out by introducing diazomethane to a suspension of double- or triple-layered [2.2]paracyclophane

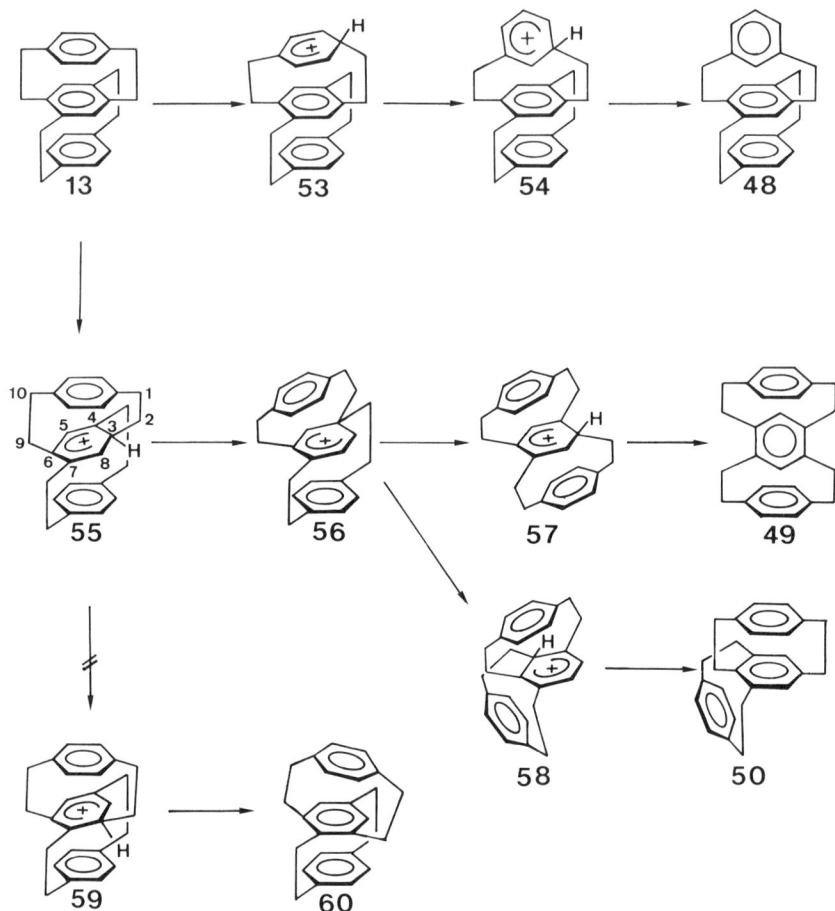

Scheme 6

(2 or 13) and cuprous chloride (in dry dichloromethane at room temperature), according to the method of Näder and deMeijère.[38] The resulting tropylidene cyclophanes (61 and 62, Scheme 7), separated from the isomeric mixture by silica gel column chromatography, were subjected to hydride abstraction with trityl fluoroborate in acetonitrile to give the double- and triple-layered tropylioparacyclophanes 63 and 64, respectively.[39] The stacking interaction of the tropylium ring on the benzene ring is clearly reflected in the upfield shifts of all the aromatic protons in their nmr spectra and the bathochromic shift of the CT bands [λ_{max} (CH$_2$Cl$_2$) 345 nm for 63 and 434 nm for 64] in their electronic spectra.

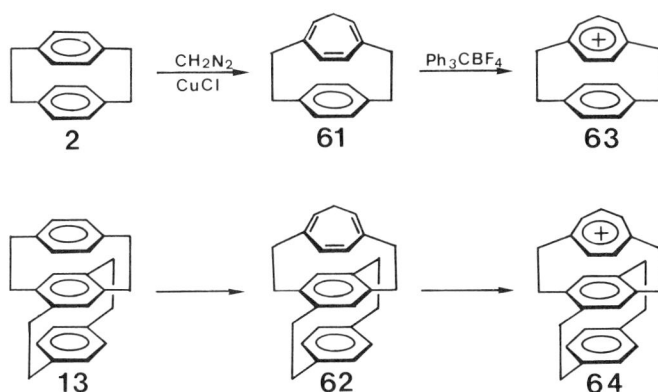

SCHEME 7

d. Tricarbonylchromium Complexes

From the viewpoint of studying the transannular electronic interaction between benzene rings in ligand cyclophanes, tricarbonylchromium complexes of multilayered paracyclophanes **65–70** were prepared[40] and compared with those of the double-layered cyclophanes.[41] After treatment of

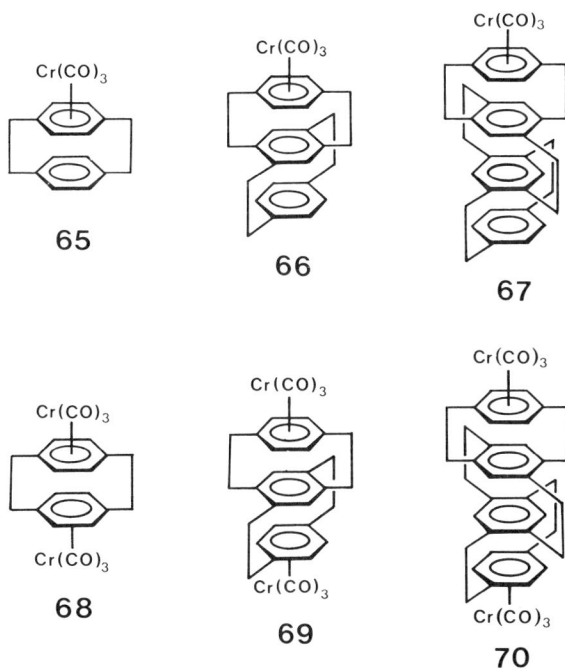

the corresponding cyclophanes with hexacarbonylchromium in purified diglyme at 140 to 150°C for 2 to 3 hr in a nitrogen atmosphere, successive extractions of the reaction mixture with dichloromethane and THF gave mono complexes 65–67 and bis complexes 68–70, respectively. The yields of mono complexes decrease with the increase in the number of layers, whereas the bis/mono ratio increases. These results indicate that the first attached tricarbonylchromium group exerts less transannular electronic interaction on the other outside benzene ring as the number of layers increases. X-ray crystallographic analysis shows that the inter-planar distance between two benzene rings of [2.2]paracyclophane is shorter than usual due to the coordination of tricarbonylchromium group, as seen in the mono complex 65.[42]

The three mono complexes 65–67 show significant bathochromic shifts of their CT bands compared with the p-xylene tricarbonylchromium com-plex, but strangely show no difference in magnitude of the shifts regard-less of the number of layers in the ligand cyclophanes. In contrast, the CT bands of the bis complexes 68–70 shift to shorter-wavelength region with an increase in the number of layers.

B. Triple-Layered [m.m][n.n]Paracyclophanes

A series of triple-layered [m.m][n.n]paracyclophanes (m, n = 2–4) was synthesized to specifically study the transannular electronic interaction effect without the strain effect which is often responsible for various intriguing properties of the multilayered [2.2]paracyclophanes.[43]

Following the thermal desulfurization and photodesulfurization meth-ods applied to the syntheses of [3.3]- and [4.4]paracyclophanes,[34a,44] five new [m.m][n.n]paracyclophanes (71b–71f) were prepared via a way simi-lar to the synthesis of the triple-layered [3.3][3.3]paracyclophane 71c in Scheme 8.

The absorption spectra of these cyclophanes and their TCNE complexes reveal the same order of interaction as seen in the case of the double-layered [m.m]paracyclophanes: [2.2] > [3.3] > [4.4] in the neutral state and [3.3] > [2.2] > [4.4] in the CT state. The five compounds 71a–71e show characteristic excimer fluorescence bands with broadening and large Stokes shift, whereas the [4.4][4.4] compound 71f shows a band at 395 nm, which is assigned to an excited trimer emission.[43]

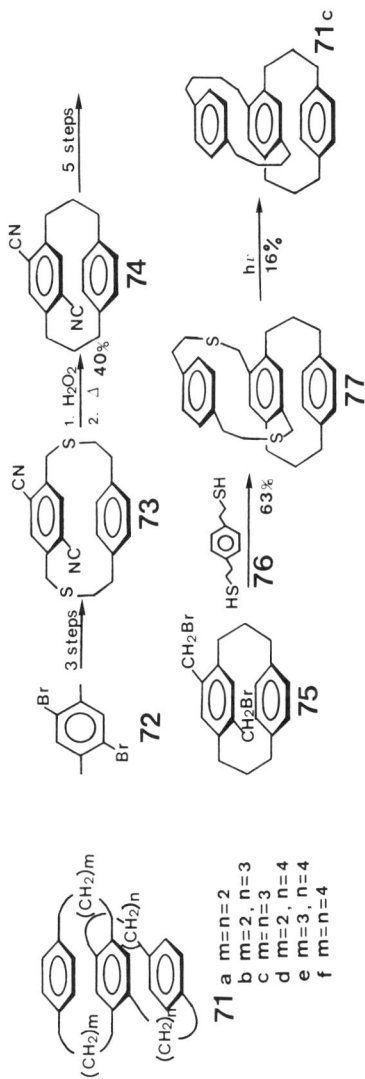

71
a m=n=2
b m=2, n=3
c m=n=3
d m=2, n=4
e m=3, n=4
f m=n=4

SCHEME 8

III. MULTILAYERED [2.2]METACYCLOPHANES

Since initial reports of unique structure[45] and unusual electrophilic substitution,[46] the chemistry of [2.2]metacyclophane has been widely developed by many groups, and various interesting properties, such as transannular reactions[47] and bridged [14]annulene synthesis,[48] have been reported. Accordingly, it is particularly interesting from this viewpoint to prepare and study multilayered metacyclophanes.

A. Synthesis

By applying the Stevens rearrangement[49]–Raney nickel desulfurization (method S) and/or direct photodesulfurization of the disulfide, two conformers of the triple-layered [2.2]metacyclophane and three conformers of the quadruple-layered compound were prepared, as shown in Schemes 9 and 10.[50]

Bromide **78**, a key compound in these synthetic sequences, was obtained in 60% overall yield by the Stevens rearrangement–Raney nickel desulfurization–NBS treatment of 5,7-dimethyl-2,11-dithia[3.3]metacyclophane, which was prepared by coupling 4,6-bis(chloromethyl)-*m*-xylene with 1,3-bis(mercaptomethyl)benzene **(79)** in 70 to 80% yield.[51,52] Irradiation of disulfide **80** in triethyl phosphite at room temperature for 19 hr afforded the u,d isomer **81** and the u,u isomer **82** (where u denotes up, and d down) of the triple-layered [2.2]metacyclophane. An alternative route (S-methylation–Stevens rearrangement–Raney nickel desulfurization: method S) gave the sole product **81** in 80% overall yield. Similarly, the methyl derivatives **86** and **87** were prepared via disulfides **84** and **85,** as seen in Scheme 9, and three conformers of the quadruple-layered compounds **91–93** were synthesized via the disulfides **89** and **90,** as shown in Scheme 10.

Both conformers **81** and **82** of the triple-layered series were also prepared by Raney nickel reduction of the bisdithiane derivative **94**[53] according to the 1,3-dithiane method.[54] The u,u form of the triple-layered metacyclophane **96,**[55] which was a desired key compound for the study of a new [4*n* + 2]annulene system **(97),** was prepared by the Wittig rearrangement of the dithia compound **95** followed by Raney nickel desulfurization in the same manner as for the synthesis of the parent compound **100** (Scheme 11).[56]

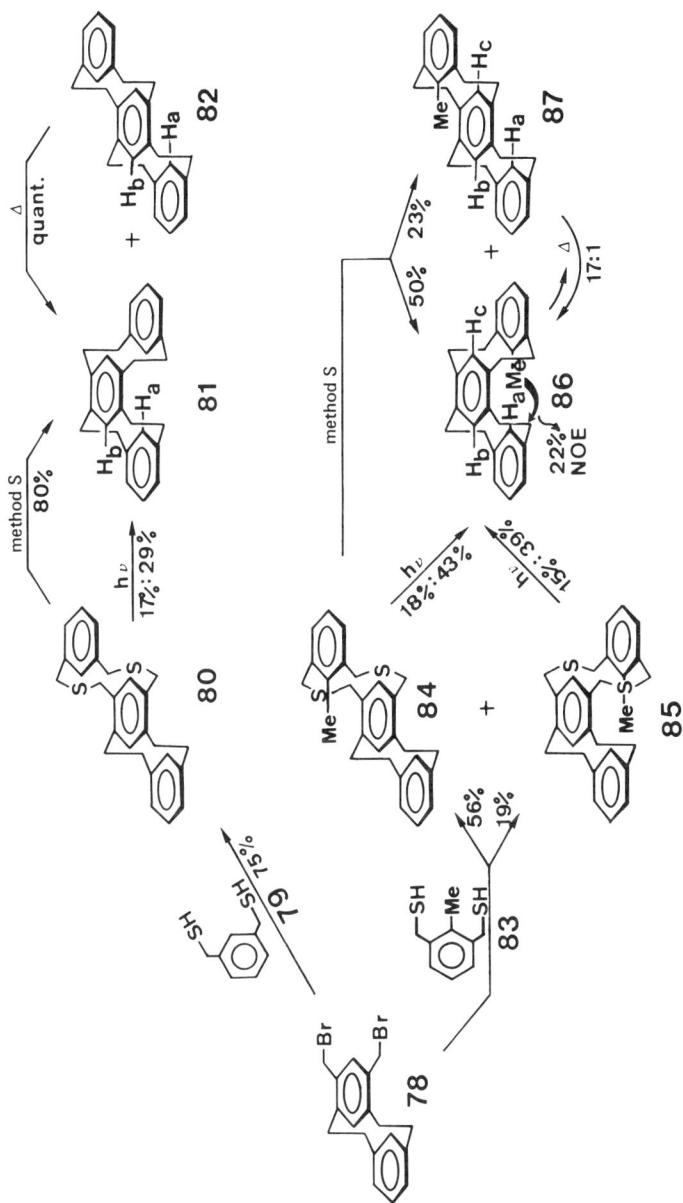

SCHEME 9. Method S: 1. $(CH_3O)_2CHBF_4$; 2. *tert*-BuOK; 3. Raney nickel.

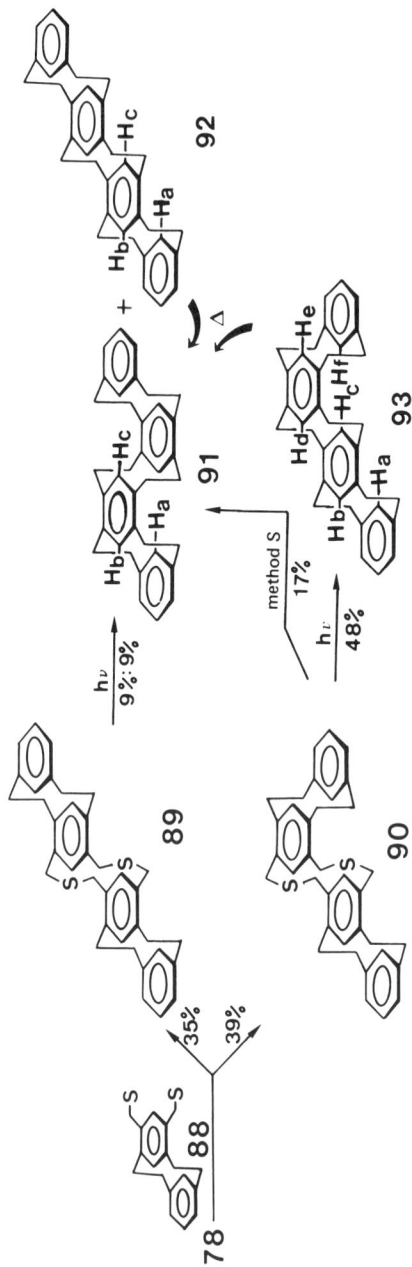

SCHEME 10. Method S: 1. $(CH_3O)_2CHBF_4$; 2. *tert*-BuOK; 3. Raney nickel.

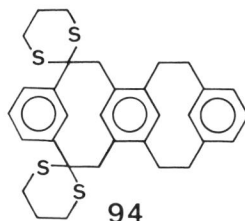

94

B. Physical Properties

1. ^{1}H-NMR SPECTRA[50,51]

One of the most important characteristics of multilayered metacy-clophanes is the chemical shift of the aromatic inner protons that are positioned on adjacent benzene rings and are affected significantly by magnetic anisotropy of the rings, as shown in Table VII. The downfield shift of H_a of **81** compared with that of H_a of **82** is explained by a deshielding effect of the third closely situated benzene ring and allows the assignment of both conformers. A marked downfield shift of H_a of **86** can be accounted for by an additional deshielding effect due to steric compression between H_a and the closely placed methyl group in addition to the effect of magnetic anisotropy of the third benzene ring. Also, a downfield shift of the methyl group signal of **86** compared with that of **87** was interpreted as a sum of the anisotropy and the steric compression effects. In fact, a 22% NOE was observed on irradiation of the methyl protons of **86**, whereas no effect was observed for **87**.

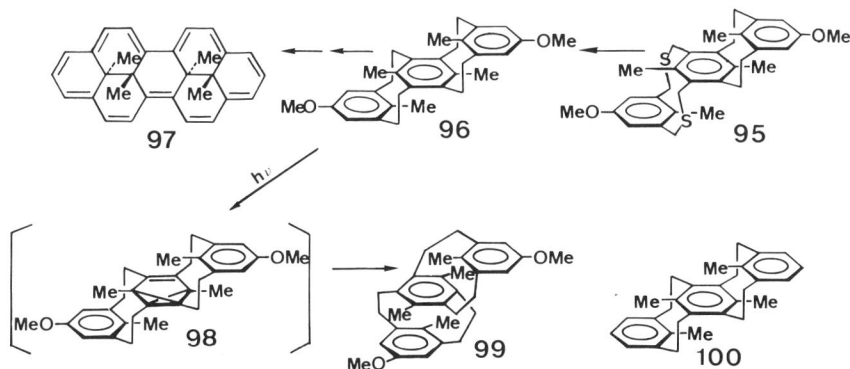

SCHEME 11

TABLE VII

¹H-nmr Data of Aromatic Inner Protons of Multilayered Metacyclophanes[a]

Compound	H_a	H_b	H_c	H_d	H_e	H_f
[2.2]MCP[b]	4.27	—	—	—	—	—
81	5.03	4.12	—	—	—	—
82	4.41	4.04	—	—	—	—
86	5.64	3.67	3.57	1.07 (CH_3)		—
87	4.21	3.85	3.59	0.58 (CH_3)		—
91	5.10	4.20	4.95	—	—	—
92	4.45	4.07	4.21	—	—	—
93	4.48	4.14	4.84	4.35	4.14	5.10

[a] δ (ppm) in CCl_4.
[b] MCP, metacyclophane.

2. MOLECULAR STRUCTURE

The detailed molecular structure of u,d isomer **81** was determined by X-ray analysis, as shown in Fig. 9.[57] The important structural characteristics of **81** are that all three benzene rings are bent into a boat form (with larger bending angles for the inner benzene ring), their mean planes being stacked partially and nearly in parallel.

In contrast, the three benzene rings of methyl derivatives **86** and **87** deviate from the parallel stacking due to a repulsive interaction of the bulky methyl group, as shown in Fig. 10.[58] The marked deviation in u,d isomer **86** was interpreted as a severe steric repulsion between the proton H_a and the methyl group as shown by NOE (Scheme 9).

C. Chemical Properties

1. CONFORMATIONAL AND CONFIGURATIONAL ISOMERIZATION

In marked contrast to the great difficulty in conformational flipping of [2.2]metacyclophane (energy barrier, ~33 kcal/mol),[59] thermal interconversions between the conformers of multilayered metacyclophanes take place quite easily.[50] As shown in Scheme 9, u,u isomer **82** in toluene-d_8 affords a 1 : 1 mixture of **81** and **82** after 3 to 4 min at 100°C and is nearly quantitatively transformed to **81** after 16 to 18 min. Under the same conditions, methyl u,u isomer **87** gives an equilibrium mixture of **86** and **87** in a

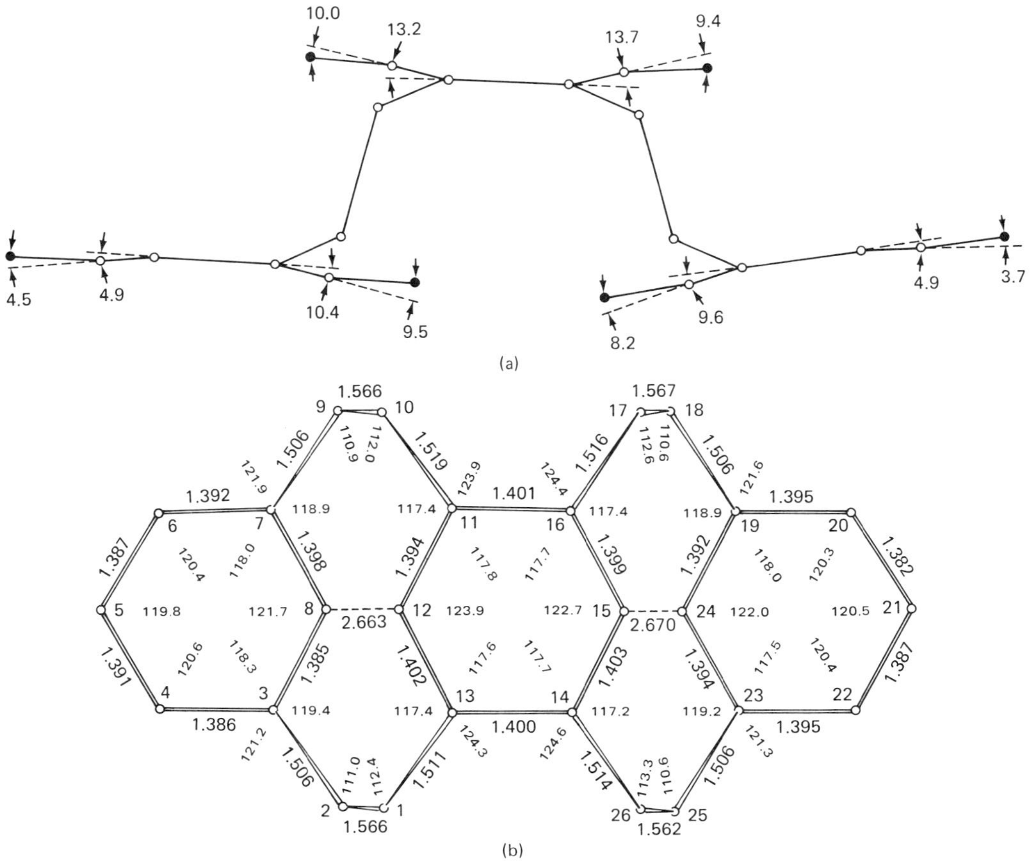

Fig. 9. Molecular projections of u,d isomer **81**: (a) side view and (b) bond lengths and bond angles.

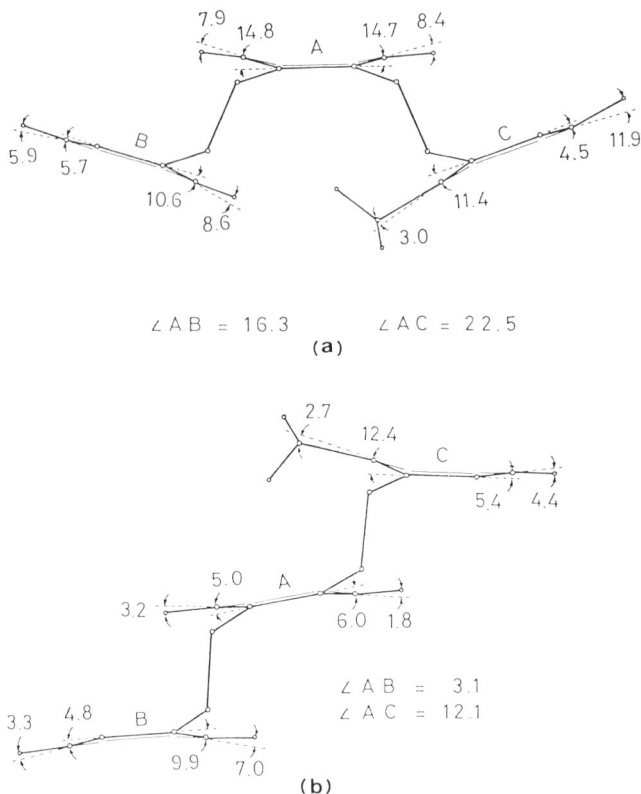

Fig. 10. Side views of 8-methyl triple-layered [2.2]metacyclophane: (a) u,d isomer **86** and (b) u,u isomer **87**.

17:1 ratio, which is also reached by treatment of **86** under the same conditions. This incomplete transformation is explained by a degree of destabilization of the more stable u,u conformer due to the steric repulsion between H_a and the methyl group in **86**. Similarly, isomers **92** and **93** of the quadruple-layered series were thermally isomerized to the most stable isomer **91** under the same conditions.

In contrast, an interesting skeletal photoisomerization of triple-layered metacyclophane **96** was observed. Thus, when irradiated with a medium-pressure mercury lamp, compound **95** gave a triple-layered metaparacyclophane **(99)**, which could be derived via a benzvalene intermediate **(98)**.[55] Similar skeletal photoisomerizations were also observed in the photodesulfurization reaction of disulfides **134** (Scheme 14), **140,** and **141.**[60]

From an nmr study of the thermal isomerization of the triple-layered metacyclophanes, u,d isomer **81** was calculated to be more stable by at least 4 kcal/mol than u,u isomer **82.** Because there appears to be no particular difference in the structures of the two conformers **81** and **82** except boat or chair deformation of the central benzene ring, as can be seen in Figs. 9 and 10, the difference in their stability is ascribed mainly to the manner of deformation of the central benzene ring.

To learn more about the difference in stability, two bending modes of the benzene ring to boat and chair forms were examined theoretically by SCF MO calculations using the MINDO/2 method.[61] The deformation to the boat conformer was predicted to be more facile with less loss of resonance energy than that for the chair conformer. As a result it was concluded that the decrease in resonance energy of the inner benzene ring was mainly responsible for the difference in thermal stability of conformers **81** and **82.**

2. Transannular Reaction: Synthesis of Pyrene-like Polycondensed Aromatic Hydrocarbons and Their Cyclophane Derivatives

It is well known that the electrophilic substitution reactions of [2.2]-metacyclophane give a transannularly dehydrogenated product, tetrahydropyrene **103,** via elimination of a proton and HX from an intermediate cation **(101)**[62] or the corresponding ipso cation **(102)**[63] or via a one-electron transfer mechanism.[64] By the application of such a transannular reaction and subsequent dehydrogenation, a series of condensed aromatic cyclophanes and polycondensed aromatic hydrocarbons was prepared from multilayered [2.2]metacyclophanes, as shown in Schemes 12 and 13.

Both conformers **81** and **82** of triple-layered metacyclophane were treated with 2 molar equivalents of pyridinium hydrobromide perbromide (Py·HBr$_3$) to give a mixture of mono-bridged product **104** and bis-bridged product octahydroperopyrene **(105).**[65] Compound **104** gave a quantitative yield of [2.2]metacyclopyrenophane **(106)** by refluxing with DDQ in benzene or by treatment with NBS in carbon tetrachloride.

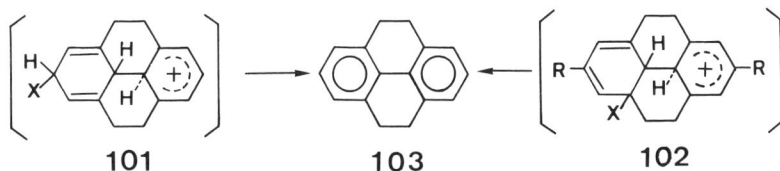

101   **103**   **102**

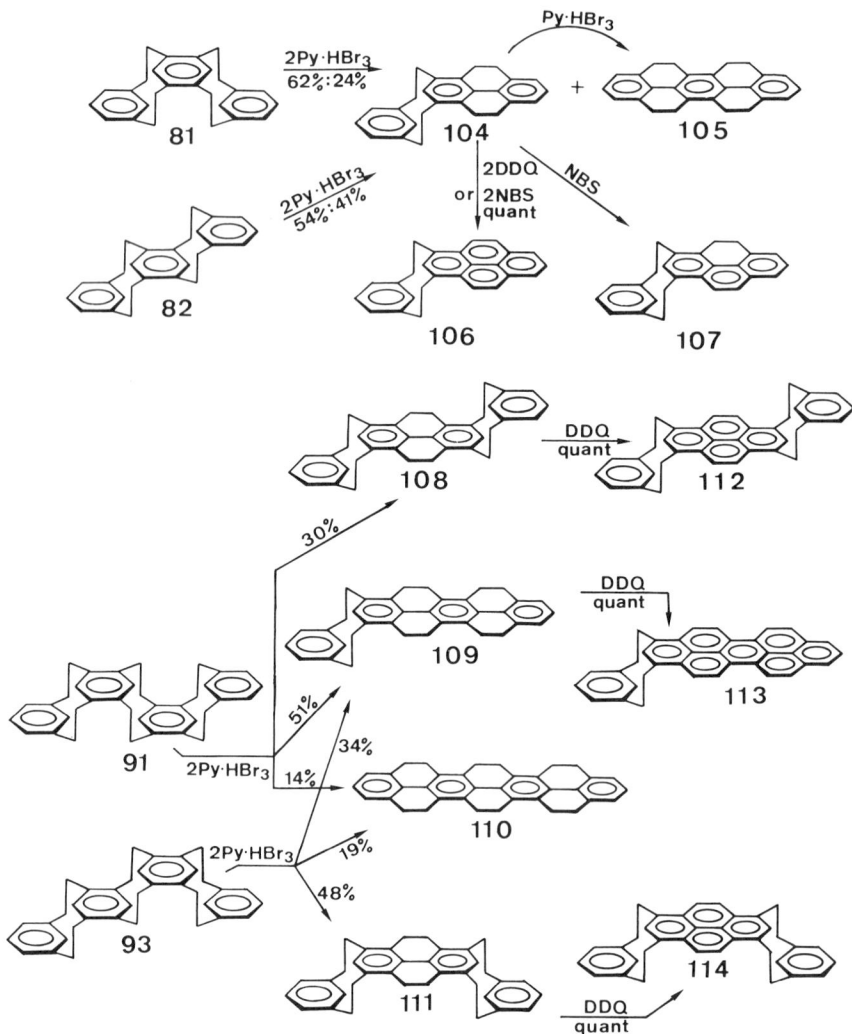

SCHEME 12

Similarly, two conformers (**91** and **93**) afforded the mono-bridged products **108** and **111,** respectively, a bis-bridged compound **(109),** and a tris-bridged compound **(110).**[65] By dehydrogenation with DDQ, triple-layered metacyclopyrenophanes **112** and **114** and metacycloperopyrenophane **113** were quantitatively obtained. The fact that **112** shows no thermal isomerization to **114** is in great contrast to the ready conversion of **82** to **81** and is probably best interpreted in terms of the dispersion of molecular strain

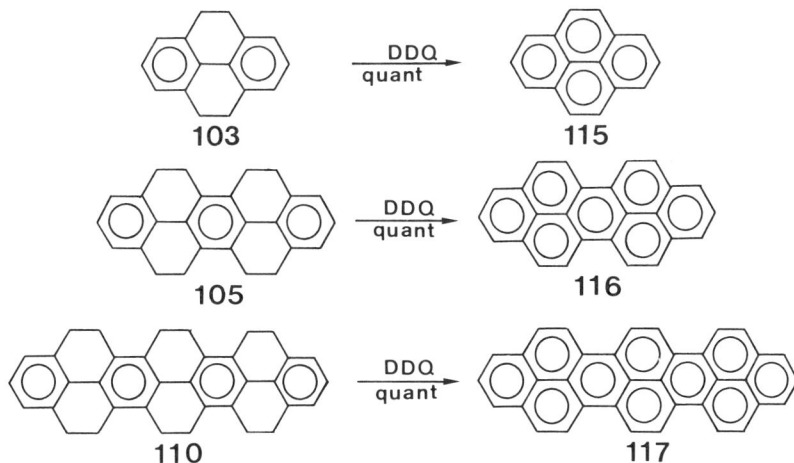

from $\pi-\pi$ repulsion over a larger pyrene ring or nearly the same resonance energy loss in the chair- and boat-deformed pyrene rings of **112** and **114.**

As shown in Scheme 13 transannular reaction products **103, 105,** and **110** were dehydrogenated with an excess of DDQ to give quantitative yields of pyrene **115,** peropyrene **116,** and teropyrene **117,** respectively.[66] As a result, this synthetic sequence has become a useful method for preparing the pyrene-type condensed aromatic hydrocarbons.

A series of pyrenophanes **(118–122)** with different stacking modes, which were designed to study the correlation between structure and fluorescence spectra of the pyrene excimer, were prepared by combining the

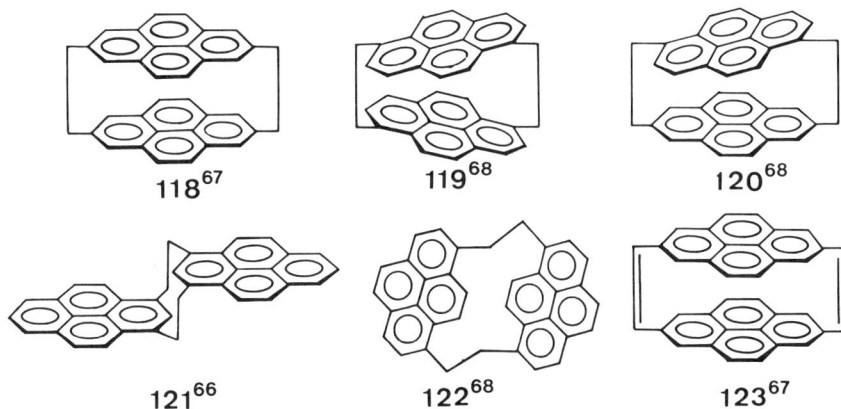

transannular reaction–dehydrogenation sequence with the direct photo-desulfurization of the disulfide.[67-69] [2.2](2,7)Pyrenophane-1,13-diene (123) was also prepared by a Stevens rearrangement of the disulfide followed by a Hofmann elimination.[68]

IV. MULTILAYERED METAPARACYCLOPHANES

[2.2]Metaparacyclophanes, in which a meta-substituted benzene ring is stacked at some angle on a para-substituted benzene ring by two ethylene bridges, have been of great interest because of their unique structure, flipping of the meta ring, and electronic interaction between both rings. This section describes the synthesis and properties of the title compounds as intermediates for multilayered para- and metacyclophanes.

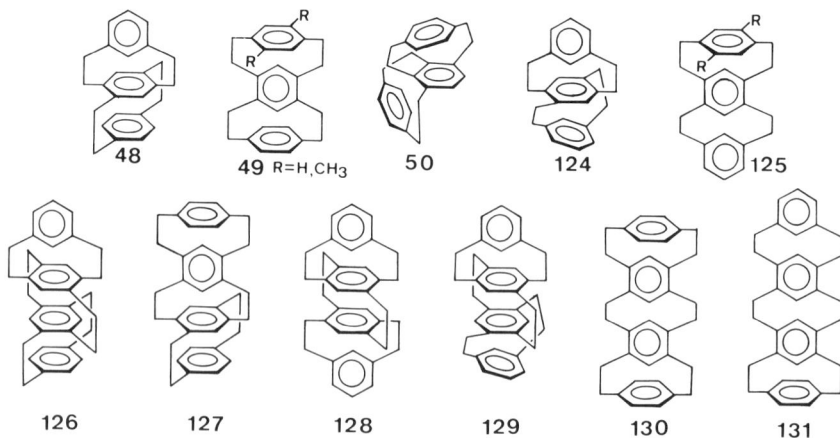

A. Synthesis and Isomerization

A series of triple- and quadruple-layered [2.2]metaparacyclophanes (48–50 and 124–131) was synthesized by a combination of general synthetic methods such as Stevens rearrangement–Raney nickel desulfurization, pyrolysis of disulfones, and photo-induced desulfurization. The syntheses of three triple-layered compounds (48, 49, and 125) are shown as examples in Scheme 14.[60]

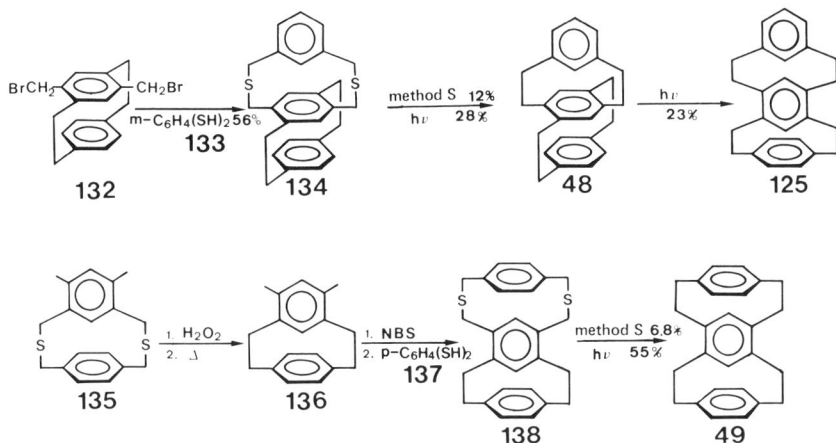

SCHEME 14. Method S: 1. $(CH_3O)_2CHBF_4$; 2. *tert*-BuOK; 3. Raney nickel.

The unexpected formation of by-product **125** in the photodesulfurization of **134** to **48** probably proceeds via a benzvalene intermediate **(139)**,
which is derived from **48,** which is initially produced during irradiation.
This assumption is supported by the fact that prolonged irradiation of **134**
resulted in an increased yield (23%) of **125** with a decrease in **48** and that

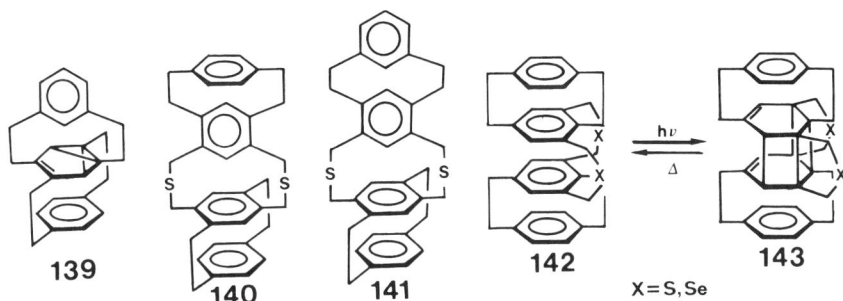

irradiation of **48** in degassed cyclohexane produced isomer **125.** Similar
skeletal rearrangement upon photodesulfurization was also observed during irradiation of disulfides **140** and **141** (to give the quadruple-layered
metaparacyclophanes **130** and **131,**[70] respectively) as well as in the photoisomerization of **96** to **99** described previously.[55]

An intriguing photoisomerization was provided by irradiation of quadruple-layered dithia- and diselena-metaparacyclophanes **(142)** (X = S
and Se) with a high-pressure mercury lamp in dry benzene to give a highly
strained cage compound **(143)** in 83 and 47% yields, respectively.[71] Com-

pounds **143** were stable at room temperature and showed a reversion to **142** in THF solution, i.e., photochromic property (X = S: quantitative yield at 90°C, half-life $\tau_{1/2}$, 30 min; X = Se: 70% at 60°C, $\tau_{1/2}$, 25 min).

B. Structures and Nuclear Magnetic Resonance Spectra

The multilayered metaparacyclophanes are expected to exhibit conformational isomerism due to the fixation of the more mobile meta-substituted benzene ring.

Two compounds (**48** and **124**) of the triple-layered metaparacyclophanes exhibit an ^{1}H-nmr pattern similar to that of [2.2]metaparacyclophane except for upfield shifts of some aromatic protons due to the anisotropy of the third benzene ring. The energy barrier for the conformational flipping of the meta ring is 18 and 19 kcal/mol for **48** (T_c 100°C) and **124** (T_c 116°C), respectively, both values being lower than that (21 kcal/mol)[72] of [2.2]metaparacyclophane. The reason for these lower energy barriers is considered to be the twisting of the central benzene ring by bridging with the two meta rings.

For compounds **49** and **125,** two conformers are possible (*anti*-**144** and *syn*-**145**). However, compound **125** shows no temperature dependence in

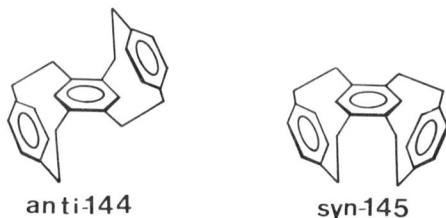

anti-**144**   syn-**145**

its nmr spectrum, suggesting that it is composed of a single conformer. The syn structure, assumed from nmr spectral analysis, was confirmed by X-ray crystallography (Fig. 11). The inner benzene ring is bent into a boat form, as seen in the structure of u,d conformer **81** of triple-layered metacyclophane in Fig. 9. In contrast, the dimethyl derivative **125** (R = CH$_3$) consists of two conformers in a 1 : 1 ratio according to nmr analysis. This result is explained by considering that the methyl substituent equalizes the stabilities of the syn and anti conformers by rendering the syn conformer less stable than the parent compound **125** (R = H).

In contrast to the preferential syn structure for **125,** the anti conformer **144** was assigned to **49** (R = H, CH$_3$) by nmr analysis. Its detailed structure by X-ray analysis shows a partial overlapping of three benzene rings,

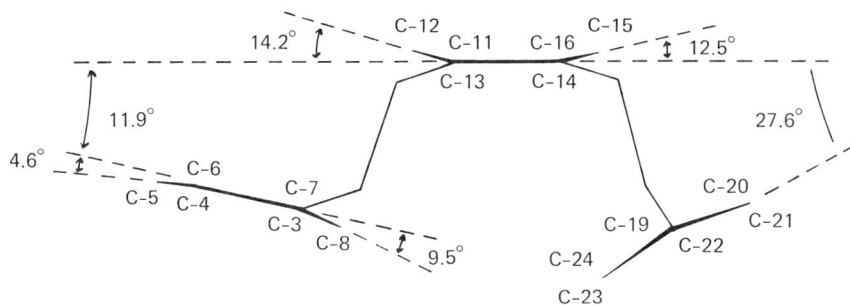

Fig. 11. Side view of triple-layered [2.2]metaparacyclophane **125** (R = H).

as seen in Fig. 12. An interesting conformational flipping of **49** was observed at elevated temperature. Thus, two types of aromatic protons of the two para-substituted rings coalesce at 85°C and appear as a singlet in the middle of their initial positions at higher temperature (140°C), indicating a conformational change between the two anti forms by means of double flipping of the two para rings.[73]

By comparison with the nmr spectra of the triple-layered compounds, both quadruple-layered compounds **130** and **131** were assigned a zigzag type of structure like that of quadruple-layered metacyclophane conformer **91,** whereas the structure of **127** contained the anti form **144** as a subunit.

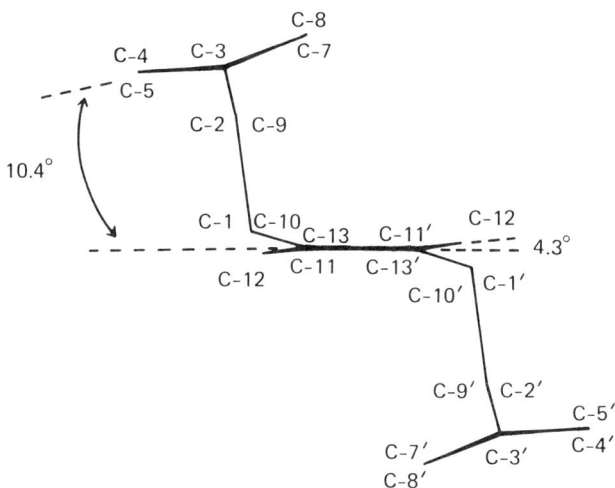

Fig. 12. Side view of triple-layered [2.2]metaparacyclophane **49** (R = H).

V. OTHER MULTILAYERED CYCLOPHANES

A number of intriguing properties of multilayered cyclophanes composed of only benzene rings are described in the preceding three sections. Some multilayered cyclophanes containing other unsaturated systems are described in this section.

A. Multilayered Heterophanes

A series of multilayered cyclophanes (146–151) containing heteroaromatic rings such as furan and thiophene were prepared to study the transannular electronic interaction and other properties. The Hofmann elimination method is also very convenient for the synthesis of

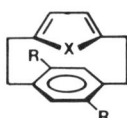

146 X=O
147 X=S

148 X=O
149 X=S

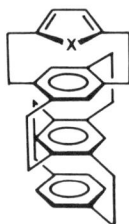

150 X=O
151 X=S

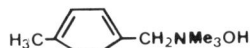

152 X=O
X=S

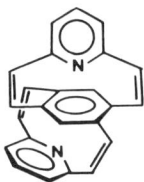

153

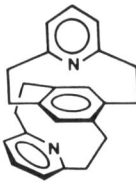

154

paracycloheterophanes. By cross-breeding pyrolysis of ammonium hydroxide **152** with quaternary bases (**1, 5, 11, 12,** and **17**) of the corresponding layered cyclophanes, double- to quadruple-layered paracycloheterophanes were prepared and were purified by column chromatography on silica gel with silver nitrate or gel permeation liquid chromatography.[74]

The most remarkable feature of their nmr spectra is that all the aromatic protons appear upfield, as in the case of the multilayered paracyclophane series, but the upfield shifts are relatively small and rather irregular. Another marked feature of their nmr spectra is that the aromatic protons of the benzene ring facing the heteroaromatic ring appear to be nonequivalent in the thiophene series and equivalent in the furan series. This result can be explained by the fixation of the thiophene ring in the former series and the rapid interconversion of the furan ring in the latter series on the basis of variable-temperature nmr spectroscopy[74] and X-ray crystallography[75] of **148** (R = H) and **149** (R = CH₃)(Fig. 13).

The two series of multilayered paracycloheterophanes show analogous tendencies in their electronic spectra, and the characteristic absorptions of the heteroaromatic rings disappear successively as the number of layers increases. The absorption curves of the thiophene series show an unexpected resemblance to those of the multilayered paracyclophanes, and the spectra of the furan series are likewise unexpectedly similar to those of the corresponding multilayered metaparacyclophanes. These spectral properties seem to be partly attributable to the similarity of their stacking modes.

In order to study a direct interaction between two pyridine nitrogens through the π-electron cavity of a benzene ring, the synthesis of triple-layered pyridinophanetetraene **(153)** was carried out by pyrolysis of the tetraphenyl sulfoxide.[76] This compound shows very little basicity in the formation of its monoprotonated derivative, an ion of interest for its potential symmetry. There is a relatively short N—N distance (5.04 Å)

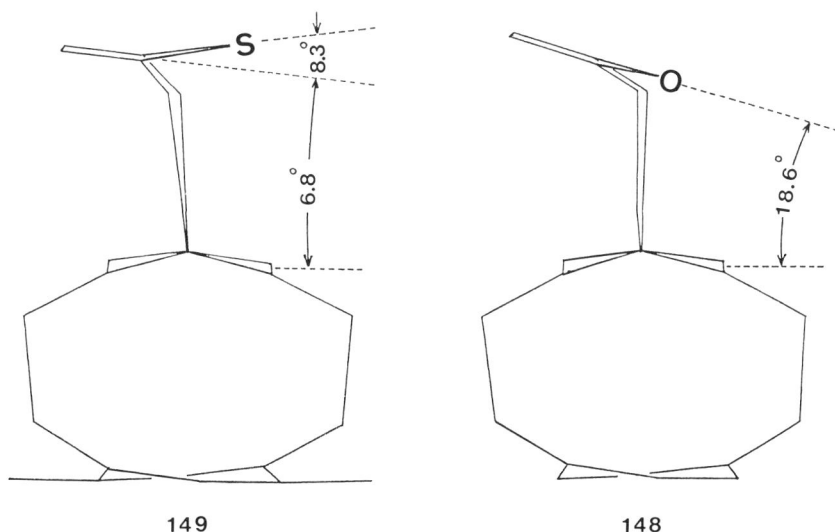

Fig. 13. Profiles of triple-layered paracycloheterophanes **148** (R = H) and **149** (R = CH₃).

according to an X-ray crystallographic analysis.[77] By nmr analysis, the behavior of the triple-layered pyridinophane **154** was shown to be remarkably similar to that of the simple [2.2](2,6)pyridinoparacyclophane.[76]

B. Multilayered Donor–Acceptor Cyclophanes

In this section multilayered donor–acceptor (DA) cyclophanes with both donor and acceptor moieties in the same molecule are described. Two series of multilayered DA cyclophanes having either a parabenzoquinone or a tetracyanoquinodimethane moiety as an acceptor have been reported.

1. MULTILAYERED PARACYCLOPHANEQUINONES

Three isomeric triple-layered dimethoxyquinones **(156–158)**[78–80] and two isomeric quadruple-layered dimethoxyquinones **(159** and **160)**[80,81]

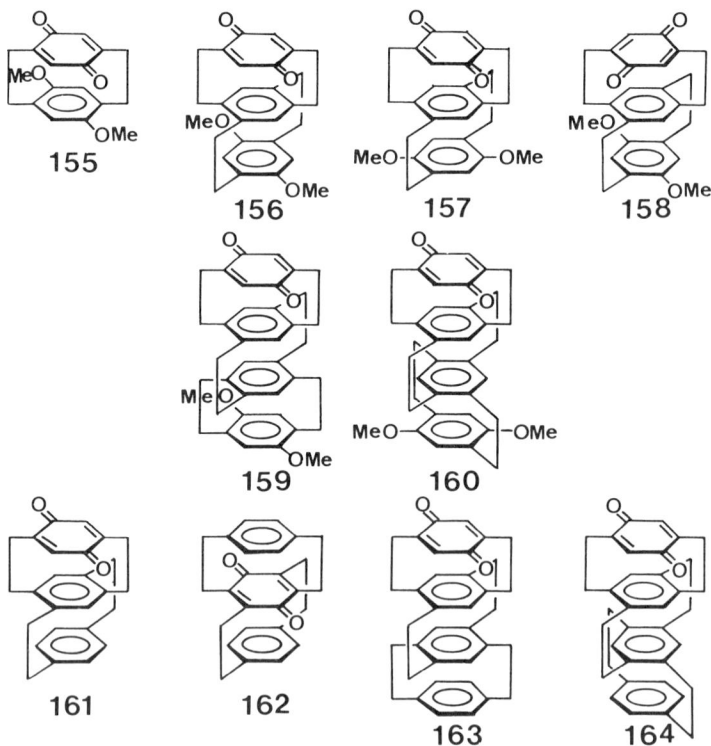

were synthesized by two groups using photodesulfurization of disulfides and cross-breeding 1,6 Hofmann elimination of quaternary ammonium bases, as shown in Scheme 15. The triple-layered dimethoxyquinones **156–158** were obtained in good yields by demethylation of **167** and **168** with boron tribromide, followed by oxidation with silver oxide. Com-

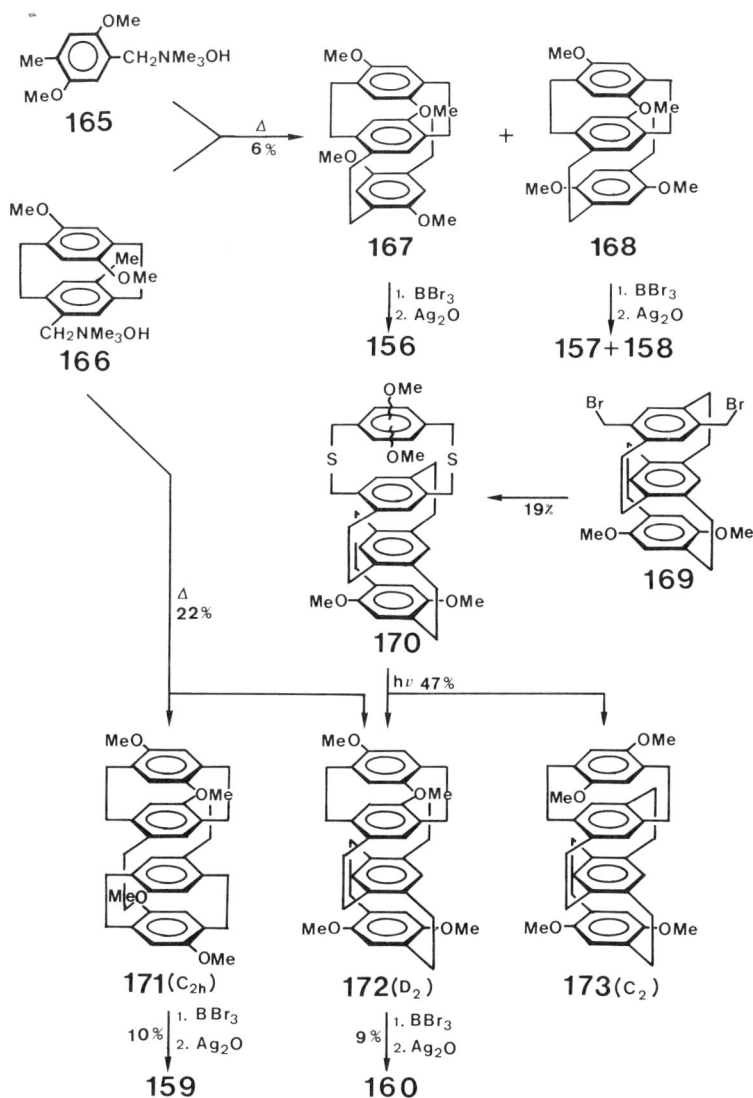

SCHEME 15

pounds **167** and **168** were obtained by cross-breeding pyrolysis of two ammonium bases (**165** and **166**) and were separated from concurrently produced tetramethoxy double- and quadruple-layered compounds.[80]

Pyrolysis of ammonium base **166** in boiling xylene gave a mixture of isomeric tetramethoxy compounds (**171** and **172**), which were tentatively assigned on the basis of their ^{1}H- and ^{13}C-nmr spectra.[80] This assignment was confirmed by an alternative synthesis of isomer **172.** Thus, photodesulfurization of disulfide **170,** derived from dibromide **169** of well-determined structure, gave a mixture of isomeric tetramethoxy compounds (**172** and **173**) in good yield.[80] The less soluble isomer was identical in all respects (as determined by various spectral data) with the isomer **172** obtained by the pyrolytic method; undoubtedly, **172** has the D_2 symmetric structure. The other two isomers (**171** and **173**) are consequently assigned to C_{2h} and C_2 symmetric structures, respectively. Also, the C_{2h} structure of **171** was confirmed by an X-ray crystallographic determination.[81] The reference quinonophanes **161–164** were prepared in the same way.[80,82]

All of the DA cyclophane compounds **155–164** show broad and structureless CT bands in the longer-wavelength region of their electronic spectra (see Fig. 14 for **156–158**). It is known that CT transitions of various double-layered DA cyclophanes are markedly affected by the

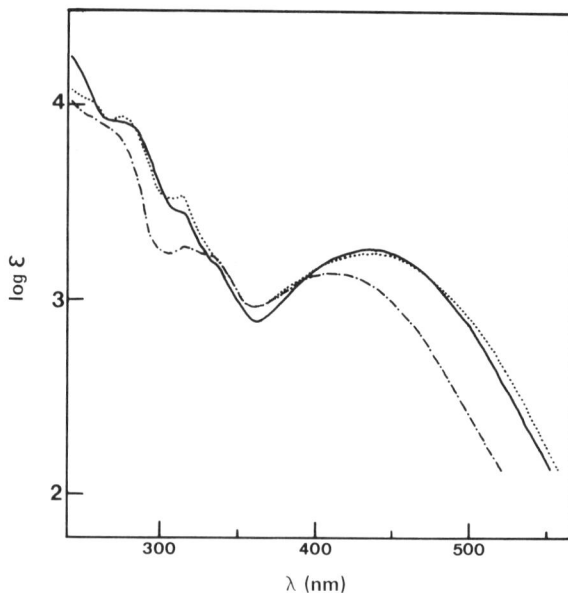

Fig. 14. Electronic spectra of three isomers of triple-layered quinonophanes in CH_2Cl_2: ·····, **156**; ---, **157**; ——, **158**.

relative orientation of the donor and the acceptor moieties. However, the isomeric pairs (**156** versus **158** and **159** versus **160**) of triple- and quadruple-layered DA cyclophanes do not follow such a pattern. Thus, these isomeric pairs show nearly superimposable absorption curves (as in Fig. 14), indicating no orientational dependence between the presumed main donor (dimethoxybenzene) and bezoquinone acceptor.

For the series of dimethoxyquinones **155, 156,** and **159,** the CT band maxima do not shift in parallel with the increase in the number of layers, that is, **155 > 159 > 156.** In compound **155,** the dimethoxybenzene moiety, having a relatively lower ionization potential, is so closely placed to the quinone moiety that the CT band of **155** shifts to longer wavelength. Compared with **155,** multilayered compounds **156** and **159** have two types of donors: the durene moiety (the inner benzene ring) and the dimethoxybenzene moiety. For **159,** the durene moiety is considered to function as the main donor from the fact that the absorption curves of **159** and **163** are roughly superimposable. In other words the two methoxy groups show no contribution to the CT transition. For this reason there is no orientational dependence of the quinone and dimethoxybenzene moieties on the CT transitions of **156–160** mentioned before.

2. MULTILAYERED PARACYCLOPHANES HAVING A TETRACYANOQUINODIMETHANE UNIT

Four triple-layered cyclophanes (**175–178**)[83] containing a tetracyanoquinodimethane moiety as an acceptor were prepared by photodesulfurization of the corresponding disulfide and a modification of Wheland–

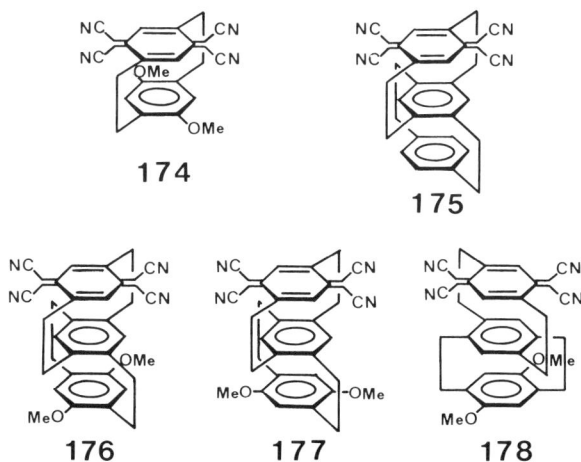

174

175

176   **177**   **178**

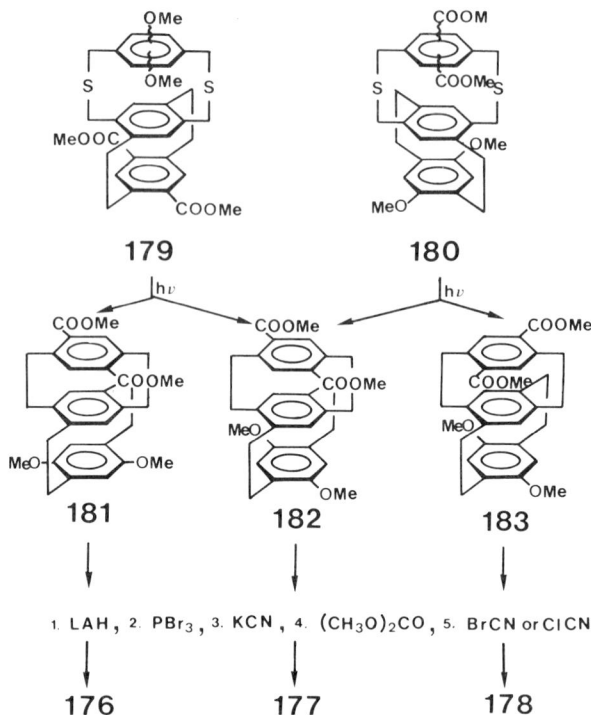

SCHEME 16

Martin's tetracyanoquinodimethane (TCNQ) synthesis,[84] as was the double-layered compound **174**.[85] Thus, three isomeric dimethoxy diesters **(181–183)** were prepared by photodesulfurization of **179** and **180** and as-signed configurationally as seen in Scheme 16. All the TCNQ-phanes **(174–178)** were obtained *in situ* without successive alkaline hydrolysis, decarboxylation, and oxidation, which were required for the synthesis of the sterically uncrowded TCNQ derivatives.

The electronic spectra of the TCNQ-phanes exhibit markedly broad CT bands (tailing to over 1000 nm). As shown in Fig. 15 the absorption maximum of double-layered TCNQ-phane **174** shifts to longer wavelength than do the maxima of the triple-layered compounds **176–178**. The fact that the CT bands of **176–178** show a marked red shift compared with that of **175** clearly indicates that the dimethoxybenzene moiety functions as a strong π donor through the electron cavity of the sandwiched inner ben-zene ring.

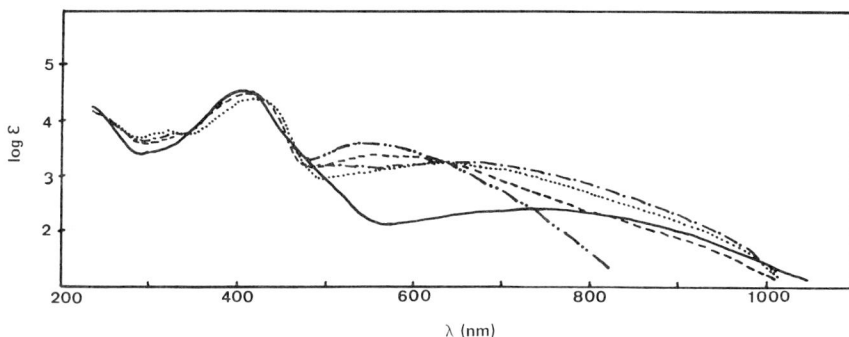

Fig. 15. Electronic spectra of TCNQ-phanes in CH_2Cl_2: ———, **174**; —··—, **175**; ----, **176**; —·—, **177**; ·····, **178**.

3. Intramolecular Exciplex: Dicyano Triple-Layered [2.2]Paracyclophane

To study excited termolecular complexes formed with the methyl-substituted benzenes and dicyano- or tetracyanobenzene in the singlet excited state, three isomers **(184–186)** of the dicyano triple-layered [2.2]paracyclophane were synthesized in the usual way.[86] By a fluorescence and nanosecond laser photolysis study, the intermolecular 2:1 complex was determined to be the structure (DD) $^+A^-$ in the excited state.[87]

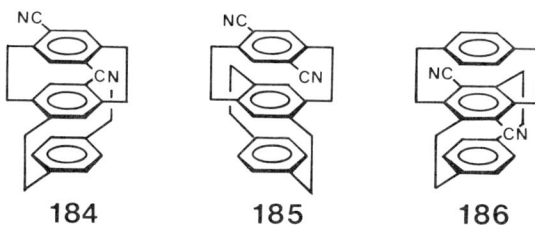

184　　　**185**　　　**186**

C. Triple-Layered Cyclophanes Containing an Anthracene Ring and a Diacetylene Group

A series of anthracenophanes[88–90] and paracyclophadiynes[91,92] was prepared to study the electronic interaction of the benzene ring with the anthracene and diacetylene chromophores, respectively. The transannu-

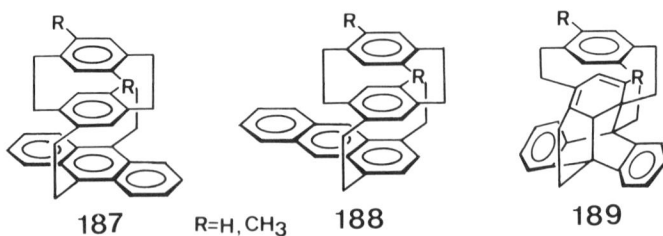

187   R=H,CH3   188   189

lar electronic interaction of triple-layered anthracenophane **187** is expected to be stronger than that of the isomeric anthracenophane **188** on
the basis of the electronic spectra of the corresponding double-layered
anthracenophane pair. For this purpose the synthesis was carried out by
the cross-breeding pyrolysis of quaternary ammonium hydroxides to give
compounds **187** and **188**.[90]

However, an unexpected product (**189**: R = H, CH₃) was obtained
instead of the desired cyclophane **187**. The structure of **189** was determined by electronic and ¹H-nmr spectra and finally by X-ray crystallographic analysis.[93] The formation of **189** was interpreted to be the result of
an intramolecular Diels–Alder reaction of the highly strained inner benzene ring with the anthracene ring in compound **187**. This unusual cycloaddition is, to our knowledge, the first example in which benzene
reacts as a dienophile under ordinary reaction conditions, and strain relief
is obviously a driving force for the reaction.

A triple-layered paracyclophadiyne (**191**) was prepared by intramolecular oxidative coupling of the diethynyl compound **190,** which was obtained from dimethyl[2.2]paracyclophane (**10**) using an eight-step sequence (Scheme 17).[94] As shown by X-ray analysis (Fig. 16),[95] the

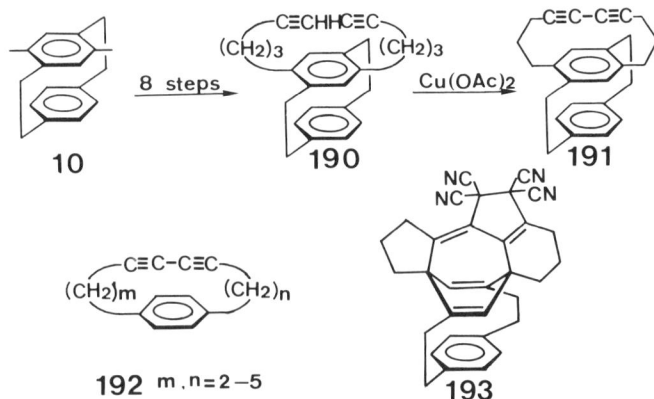

SCHEME 17

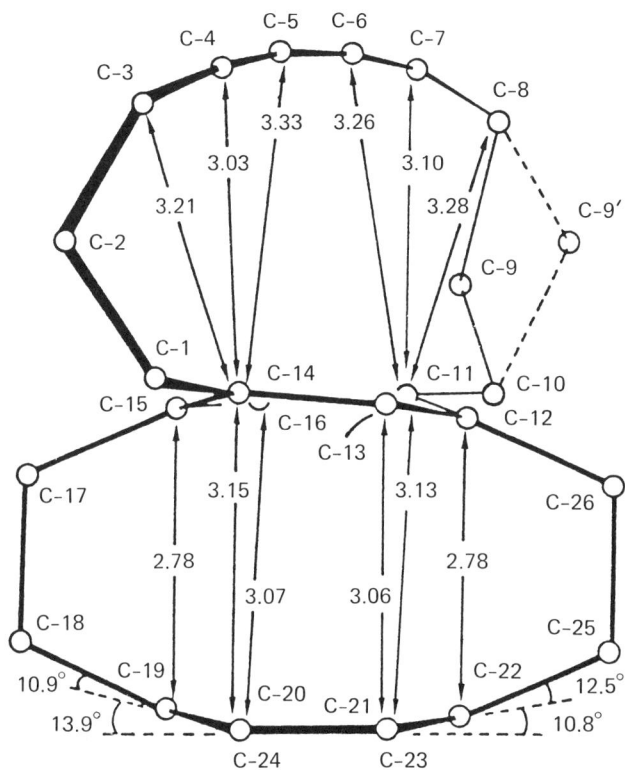

Fig. 16. Molecular structure of paracyclophadiyne **191**.

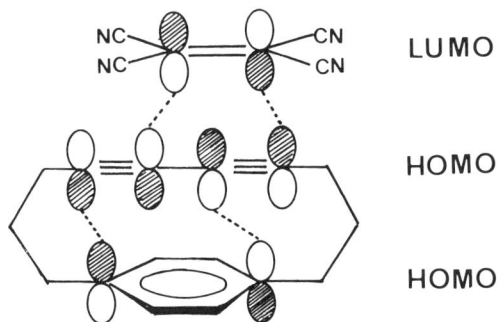

Fig. 17. Three π-system interaction orbital set for multicycloaddition of paracyclopha-diyne with TCNE.

diacetylene bond is bent into a bow shape, and the two benzene rings into a boat and a twisted shape, indicating severe electronic repulsion among the three π systems.

During a CT spectrum measurement of the **191**–TCNE complex, the initial blue color of the CT complex disappeared rapidly at room temperature, and a colorless solid was obtained. The product was determined to be a 1 : 1 cycloadduct by its mass, ir, and nmr spectra.[94] The other paracyclophadiynes **(192)** also gave similar 1 : 1 cycloadducts at higher temperature and in good yields ($m = 2$, $n = 4$, 130°C, 76%; $m = n = 3$, 140°C, 73%; $m = 3$, $n = 4$, 140°C, 34%). These reactions were interpreted by means of MO theory as being a three π-system cycloaddition, that is, the interaction of an HOMO of benzene, an HOMO of diacetylene, and an LUMO of TCNE (Fig. 17), provided that they proceed in a concerted manner.[96]

NOTES ADDED IN PROOF

(Section II,A,1) Selenium extrusion from diselena[3.3]cyclophanes and the desulfurization of dithia[3.3] system was developed for the synthesis of [2.2]cyclophanes using three methods: (1) benzyne hydrogenolysis, (2) flash pyrolysis, and (3) photolysis. The photodeselenative ring contraction in $P(NMe_2)_3$ is preferred over all of chalcogen-atom extrusion methods so far studied, e.g., 80% yield of triple-layered [2.2]paracyclophane **13**.[15a]

(Section II,A,6,b) The spin density distribution and the energy stabilization in the cation radicals generated by the electrochemical oxidation of some multilayered [2.2]paracyclophanes were reported and interpreted with a simple theoretical model.[32a]

REFERENCES

1. R. W. Griffin, Jr., *Chem. Rev.* **63**, 45 (1963); D. J. Cram, *Rec. Chem. Prog.* **20**, 71 (1959); D. J. Cram and J. M. Cram, *Acc. Chem. Res.* **4**, 204 (1971); T. Sato, *Nippon Kagaku Kaishi* **92**, 277 (1971); *Kagaku no Ryoiki* **23**, 672, 765 (1969); F. Vögtle, *Chem.-Ztg.* **95**, 668 (1971); *Angew. Chem.* **84**, 75 (1972); *Chimia* **26**, 64 (1972); *Synthesis* p. 85 (1973); T. Kai, *J. Crystallogr. Soc. Jpn.* **18**, 269 (1976); J. Kleinschroth and H. Hopf, *Angew. Chem.* **94**, 485 (1982).

1a. S. Misumi, and Y. Sakata, *J. Synth. Org. Chem., Jpn.* **29**, 114 (1971); *Hyomen* **17**, 239
    (1979); S. Misumi, *Kagaku no Ryoiki* **28**, 927 (1974); **32**, 651 (1978); S. Misumi and T.
    Otsubo, *Acc. Chem. Res.* **11**, 251 (1978); *Chem. Educ. Tokyo* **28**, 249 (1980); Y. Sakata,
    Kagaku no Ryoiki **28**, 947 (1974); *J. Synth. Org. Chem., Jpn.* **38**, 164 (1980).
2. H. O. House, D. G. Koepsell, and W. J. Campbell, *J. Org. Chem.* **37**, 1003 (1972).
3. R. H. Martin and M. Baes, *Tetrahedron* **31**, 2135 (1975); cf. R. H. Martin, *Angew.
    Chem., Int. Ed. Engl.* **13**, 649 (1974).
4. A. Salzer and H. Werner, *Angew. Chem.* **84**, 949 (1972); H. Werner, *Angew. Chem.*, **92**,
    758 (1980).
5. S. P. Kolesnikov, J. E. Dobson, and P. K. Skell, *J. Am. Chem. Soc.* **100**, 999 (1978).
6. W. Siebert, J. Edwin, and M. Bockmann, *Angew. Chem., Int. Ed. Engl.* **17**, 868 (1978);
    W. Siebert, C. Böhle, and C. Krüger, *ibid.* **19**, 746 (1980); J. Moraczewski and W. E.
    Geiger, Jr., *J. Am. Chem. Soc.* **100**, 7429 (1978).
7. H. E. Winberg, F. S. Fawcett, W. E. Mochel, and C. W. Theobald, *J. Am. Chem. Soc.*
    82, 1428 (1960); H. E. Winberg and F. S. Fawcett, *Org. Synth. Collect. Vol.* **5**, 883
    (1973).
8. D. T. Longone and H. S. Chow, *J. Am. Chem. Soc.* **86**, 3898 (1964); **92**, 994 (1970).
9. T. Otsubo, S. Mizogami, I. Otsubo, Z. Tozuka, A. Sakagami, Y. Sakata, and S. Misumi,
    Bull. Chem. Soc. Jpn. **46**, 3519 (1973); T. Otsubo, H. Horita, and S. Misumi, *Synth.
    Commun.* **6**, 591 (1976).
10. H. Horita, Ph.D. Thesis, Osaka University, Osaka, Japan (1978).
11. T. Otsubo, S. Mizogami, Y. Skata, and S. Misumi, *Chem. Commun.* p. 678 (1971).
12. T. Otsubo, S. Mizogami, Y. Sakata, and S. Misumi, *Tetrahedron Lett.* p. 4803 (1971).
13. H. Mizuno, K. Nishiguchi, T. Otsubo, S. Misumi, and N. Morimoto, *Tetrahedron Lett.*
    p. 4981 (1972); H. Mizuno, K. Nishiguchi, T. Toyoda, T. Otsubo, S. Misumi, and N.
    Morimoto, *Acta Crystallogr., Sect. B* **33**, 329 (1977).
14. T. Otsubo, S. Mizogami, Y. Sakata, and S. Misumi, *Tetrahedron Lett.* p. 2457 (1973).
15. T. Otsubo, Z. Tozuka, S. Mizogami, Y. Sakata, and S. Misumi, *Tetrahedron Lett.* p.
    2927 (1972).
15a. H. Higuchi and S. Misumi, *Tetrahedron Lett.* **23**, 5571 (1982); H. Higuchi, M. Kugi-
    miya, T. Otsubo, Y. Sakata, and S. Misumi, *Tetrahedron Lett.* **24**, 2593 (1983).
16. Y. Koizumi, T. Toyoda, H. Horita, and S. Misumi, *32nd Nat. Meet. Chem. Soc. Jpn.*
    Abstract I, p. 184 (1975).
17. C. J. Brown, *J. Chem. Soc.* p. 3265 (1953); K. Lonsdale, H. J. Milledge, and K. V. K.
    Rao, *Proc. Soc. London, Ser. A* **255**, 82 (1960); H. Hope, J. Bernstein, and K. N.
    Trueblood, *Acta Crystallogr., Sect. B* **28**, 1733 (1972).
18. K. Nishiyama, M. Sakiyama, S. Seki, H. Horita, T. Otsubo, and S. Misumi, *Tetrahe-
    dron Lett.* p. 3739 (1977); *Bull. Chem. Soc. Jpn.* **53**, 869 (1980).
19. R. H. Boyd, *Tetrahedron* **22**, 119 (1966); *J. Chem. Phys.* **49**, 2574 (1968); C. F. Sieh, D.
    McNally, and R. H. Boyd, *Tetrahedron* **25**, 3653 (1969).
20. D. L. Rodgers, E. F. Westrum, Jr., and J. T. S. Andrews, *J. Chem. Thermodynam.* **5**,
    733 (1973).
21. N. L. Allinger, L. A. Freiberg, R. B. Hermann, and M. A. Miller, *J. Am. Chem. Soc.*
    85, 1171 (1963).
22. S. Iwata, K. Fuke, M. Sasaki, S. Nagakura, T. Otsubo, and S. Misumi, *J. Mol. Spec-
    trosc.* **46**, 1 (1973), and references cited therein.
23. R. E. Merrifield and W. D. Philip, *J. Am. Chem. Soc.* **80**, 2778 (1958).
24. D. J. Cram and R. C. Helgeson, *J. Am. Chem. Soc.* **88**, 3515 (1966).
25. T. Otsubo, S. Mizogami, Y. Sakata, and S. Misumi, *Bull. Chem. Soc. Jpn.* **46**, 3831
    (1973).

26. B. V. Cheney, *J. Am. Chem. Soc.* **90,** 5386 (1968), and references cited therein.
27. H. J. Reich and D. J. Cram, *J. Am. Chem. Soc.* **91,** 3534 (1969).
28. T. Kaneda, T. Otsubo, H. Horita, and S. Misumi, *Bull. Chem. Soc. Jpn.* **53,** 1015 (1980).
29. M. Nakazaki, K. Yamamoto, and S. Tanaka, *J. Chem. Soc., Chem. Commun.* p. 433 (1972); M. Nakazaki, K. Yamamoto, S. Tanaka, and H. Kametani, *J. Org. Chem.* **42,** 287 (1977).
30. M. Nakazaki and K. Yamamoto, *J. Synth. Org. Chem., Jpn.* **33,** 9 (1975).
31. M. Iwaizumi, S. Kita, T. Isobe, M. Kohno, T. Yamamoto, H. Horita, T. Otsubo, and S. Misumi, *Bull. Chem. Soc. Jpn.* **50,** 2074 (1977).
32. F. Gerson, W. B. Martin, Jr., and G. Wydler, *Helv. Chim. Acta* **59,** 1365 (1976).
32a. H. Ohya-Nishiguchi, A. Terahara, N. Hirota, Y. Sakata, and S. Misumi, *Bull. Chem. Soc. Jpn.* **55,** 1782 (1982).
33. D. J. Cram, W. J. Wechter, and R. W. Kierstead, *J. Am. Chem. Soc.* **80,** 3126 (1958).
34. S. J. Cristol and D. C. Lewis, *J. Am. Chem. Soc.* **89,** 1476 (1967).
34a. T. Otsubo, M. Kitasawa, and S. Misumi, *Chem. Lett.* p. 977 (1977); *Bull. Chem. Soc. Jpn.* **52,** 1515 (1979).
35. T. Otsubo, H. Horita, Y. Koizumi, and S. Misumi, *Bull. Chem. Soc. Jpn.* **53,** 1677 (1980).
36. D. J. Cram, R. C. Helgeson, D. Lock, and L. A. Singer, *J. Am. Chem. Soc.* **88,** 1324 (1966); D. T. Hefelfinger and D. J. Cram, *ibid.* **93,** 4754 (1971).
37. H. Horita, Y. Koizumi, T. Otsubo, Y. Sakata, and S. Misumi, *Bull. Chem. Soc. Jpn.* **51,** 2668 (1978).
37a. H. Horita, N. Kannen, T. Otsubo, and S. Misumi, *Tetrahedron Lett.* p. 501 (1974).
38. R. Näder and A. de Meijère, *Angew. Chem.* **88,** 153 (1976).
39. H. Horita, T. Otsubo, Y. Sakata, and S. Misumi, *Tetrahedron Lett.* p. 3899 (1976); cf. J. G. O'Connor and P. M. Keehn, *J. Am. Chem. Soc.* **98,** 8846 (1976); H. Horita, T. Otsubo, and S. Misumi, *Chem. Lett.* p. 1309 (1977).
40. H. Ohno, H. Horita, T. Otsubo, Y. Sakata, and S. Misumi, *Tetrahedron Lett.* p. 265 (1977).
41. D. J. Cram and D. I. Wilkinson, *J. Am. Chem. Soc.* **82,** 5721 (1960); E. Langer and H. Lehner, *Tetrahedron* **29,** 375 (1973).
42. Y. Kai, N. Yasuoka, and N. Kasai, *Acta Crystallogr., Sec. B* **34,** 2840 (1978).
43. T. Otsubo, T. Kohda, and S. Misumi, *Tetrahedron Lett.* p. 2507 (1978); *Bull. Chem. Soc. Jpn.* **53,** 512 (1980).
44. M. W. Haenel, A. Flatow, V. Taglieber, and H. A. Staab, *Tetrahedron Lett.* p. 1733 (1977); D. T. Longone, S. H. Küsefoglu, and J. A. Gladysz, *J. Org. Chem.* **42,** 2787 (1977); L. Rossa and F. Vögtle, *J. Chem. Res., Synop.* p. 264 (1977).
45. C. J. Brown, *J. Chem. Soc.* p. 3278 (1953).
46. N. L. Allinger, M. A. Da Rooge, and R. B. Hermann, *J. Am. Chem. Soc.* **83,** 1974 (1961).
47. T. Sato, *Nippon Kagaku Zasshi* **92,** 27⟨ ⟨1971), and references cited therein.
48. V. Boekelheide and R. H. Mitchell, *in* "Aromaticity, Pseudo-Aromaticity, and Anti-Aromaticity" (E. D. Bergmann and B. Pullman, eds.), p. 150. Jerusalem Acad. Press, Israel, 1970.
49. R. H. Mitchell and V. Boekelheide, *J. Am. Chem. Soc.* **92,** 3510 (1970).
50. T. Umemoto, T. Otsubo, and S. Misumi, *Tetrahedron Lett.* p. 1573 (1974).
51. T. Umemoto, Ph.D. Thesis, Osaka University, Osaka, Japan (1976).
52. T. Umemoto, T. Otsubo, Y. Sakata, and S. Misumi, *Tetrahedron Lett.* p. 593 (1973).
53. H. Lehner, *Monatsh. Chem.* **107,** 565 (1976).

54. T. A. Hylton and V. Boekelheide, *J. Am. Chem. Soc.* **90,** 6887 (1968); V. Boekelheide, P. H. Anderson, and T. A. Hylton, *ibid.* **96,** 1558 (1974).
55. D. Kamp and V. Boekelheide, *J. Org. Chem.* **43,** 3470 (1978).
56. T. Otsubo, D. Stusche, and V. Boekelheide, *J. Org. Chem.* **43,** 3466 (1978).
57. Y. Kai, N. Yasuoka, and N. Kasai, *Acta Crystallogr., Sect. B* **33,** 754 (1977).
58. F. Hama, Y. Kai, N. Yasuoka, and N. Kasai, *Acta Crystallogr., Sect. B* **33,** 3905 (1977); Y. Kai, F. Hama, N. Yasuoka, and N. Kasai, *ibid.* **34,** 3422 (1978).
59. H. W. Geschwend, *J. Am. Chem. Soc.* **94,** 8430 (1972).
60. N. Kannen, T. Umemoto, T. Otsubo, and S. Misumi, *Tetrahedron Lett.* p. 4537 (1973); N. Kannen, T. Otsubo, Y. Sakata, and S. Misumi, *Bull. Chem. Soc. Jpn.* **49,** 3307 (1976).
61. H. Iwamura, H. Kihara, S. Misumi, Y. Sakata, and T. Umemoto, *Tetrahedron Lett.* p. 615 (1976); *Tetrahedron* **34,** 3427 (1978).
62. T. Sato, E. Yamada, Y. Okamura, T. Amada, and K. Hata, *Bull. Chem. Soc. Jpn.* **38,** 1049 (1965).
63. T. Kawashima, Ph.D. Thesis, Osaka University, Osaka, Japan (1979).
64. K. Nishiyama, K. Hata, and T. Sato, *Tetrahedron* **31,** 239 (1975).
65. T. Umemoto, T. Kawashima, Y. Sakata, and S. Misumi, *Tetrahedron Lett.* p. 463 (1975).
66. T. Umemoto, T. Kawashima, Y. Sakata, and S. Misumi, *Tetrahedron Lett.* p. 1005 (1975).
67. T. Umemoto, T. Kawashima, Y. Sakata, and S. Misumi, *Chem. Lett.* p. 837 (1975).
68. T. Umemoto, S. Satani, Y. Sakata, and S. Misumi, *Tetrahedron Lett.* p. 3159 (1975).
69. T. Kawashima, T. Otsubo, Y. Sakata, and S. Misumi, *Tetrahedron Lett.* p. 5115 (1978).
70. N. Kannen, T. Otsubo, and S. Misumi, *Bull. Chem. Soc. Jpn.* **49,** 3208 (1976).
71. H. Higuchi, K. Takatsu, T. Otsubo, Y. Sakata, and S. Misumi, *Tetrahedron Lett.* **23,** 671 (1982); H. Higuchi, M. Kugimiya, Y. Sakata, and S. Misumi, *47th Natl. Meet. Chem. Soc. Jpn.* Summary II, p. 1007 (1983).
72. D. T. Hefelfinger and D. J. Cram, *J. Am. Chem. Soc.* **92,** 1073 (1970).
73. N. Kannen, T. Otsubo, Y. Sakata, and S. Misumi, *Bull. Chem. Soc. Jpn.* **49,** 3203 (1976).
74. S. Mizogami, T. Otsubo, Y. Sakata, and S. Misumi, *Tetrahedron Lett.* p. 2791 (1971); N. Osaka, S. Mizogami, T. Otsubo, Y. Sakata, and S. Misumi, *Chem. Lett.* 515 (1974); T. Otsubo, S. Mizogami, N. Osaka, Y. Sakata, and S. Misumi, *Bull. Chem. Soc. Jpn.* **50,** 1841 (1977).
75. Y. Kai, J. Watanabe, N. Yasuoka, and N. Kasai, *Acta Crystallogr., Sect. B* **36,** 2276 (1980).
76. I. D. Reingold, W. Schmidt, and V. Boekelheide, *J. Am. Chem. Soc.* **101,** 2121 (1979).
77. A. W. Hanson, *Acta Crystallogr., Sect. B* **33,** 257 (1977).
78. H. A. Staab, U. Zapf, and A. Gurke, *Angew. Chem.* **89,** 841 (1977).
79. H. Machida, H. Tatemitsu, Y. Sakata, and S. Misumi, *Tetrahedron Lett.* p. 915 (1978).
80. H. Machida, H. Tatemitsu, T. Otsubo, Y. Sakata, and S. Misumi, *Bull. Chem. Soc. Jpn.* **53,** 2943 (1980).
81. H. A. Staab and U. Zapf, *Angew. Chem.* **90,** 807 (1978).
82. H. Tatemitsu, T. Otsubo, Y. Sakata, and S. Misumi, *Tetrahedron Lett.* p. 3059 (1975).
83. H. Tatemitsu, B. Natsume, M. Yoshida, Y. Sakata, and S. Misumi, *Tetrahedron Lett.* p. 3459 (1978); M. Yoshida, Y. Tochiaki, H. Tatemitsu, Y. Sakata, and S. Misumi, *Chem. Lett.* p. 829 (1978).
84. R. C. Wheland and E. L. Martin, *J. Org. Chem.* **40,** 3101 (1975).

85. H. A. Staab and H.-E. Henke, *Tetrahedron Lett.* p. 1955 (1978).
86. M. Yoshida, H. Tatemitsu, Y. Sakata, S. Misumi, H. Masuhara, and N. Mataga, *J. Chem. Soc., Chem. Commun.* p. 587 (1976).
87. H. Masuhara, N. Mataga, M. Yoshida, H. Tatemitsu, Y. Sakata, and S. Misumi, *J. Phys. Chem.* **81,** 879 (1977).
88. J. H. Golden, *J. Chem. Soc.* p. 3741 (1961); D. J. Cram, C. K. Dalton, and G. R. Knox, *J. Am. Chem. Soc.* **85,** 1088 (1963); H. Wynberg and R. Helder, *Tetrahedron Lett.* p. 4317 (1971).
89. T. Toyoda, I. Otsubo, T. Otsubo, Y. Sakata, and S. Misumi, *Tetrahedron Lett.* p. 1731 (1972); A. Iwama, T. Toyoda, T. Otsubo, and S. Misumi, *ibid.* p. 1725 (1973); *Chem. Lett.* p. 587 (1973); cf. T. Hayashi, N. Mataga, Y. Sakata, S. Misumi, M. Morita, and J. Tanaka, *J. Am. Chem. Soc.* **98,** 5910 (1976); T. Toyoda and S. Misumi, *Tetrahedron Lett.* p. 1479 (1978); M. Morita, T. Kishi, M. Tanaka, J. Tanaka, J. Ferguson, Y. Sakata, S. Misumi, T. Hayashi, and N. Mataga, *Bull. Chem. Soc. Jpn.* **51,** 3449 (1978); R. Nemoto, K. Ishizu, T. Toyoda, Y. Sakata, and S. Misumi, *J. Am. Chem. Soc.* **102,** 654 (1980).
90. A. Iwama, T. Toyoda, M. Yoshida, T. Otsubo, Y. Sakata, and S. Misumi, *Bull. Chem. Soc. Jpn.* **51,** 2988 (1978).
91. T. Matsuoka, Y. Sakata, and S. Misumi, *Tetrahedron Lett.* p. 2549 (1970); T. Matsuoka, T. Negi, and S. Misumi, *Synth. Commun.* **2,** 87 (1972); T. Matsuoka, T. Negi, T. Otsubo, Y. Sakata, and S. Misumi, *Bull. Chem. Soc. Jpn.* **45,** 1825 (1972).
92. T. Kaneda and S. Misumi, *Bull. Chem. Soc. Jpn.* **50,** 3310 (1977).
93. T. Toyoda, A. Iwama, Y. Sakata, and S. Misumi, *Tetrahedron Lett.* 3203 (1975); T. Toyoda, A. Iwama, T. Otsubo, and S. Misumi, *Bull. Chem. Soc. Jpn.* **49,** 3300 (1976).
94. T. Kaneda, T. Ogawa, and S. Misumi, *Tetrahedron Lett.* p. 3373 (1973).
95. T. Toyoda, Ph.D. Thesis, Osaka University, Osaka, Japan (1976).
96. S. Inagaki, H. Fujimoto, and K. Fukui, *J. Am. Chem. Soc.* **98,** 4693 (1976).

CHAPTER 11

Cyclophanes in Host–Guest Chemistry

KAZUNORI ODASHIMA AND
KENJI KOGA

Faculty of Pharmaceutical Sciences
University of Tokyo
Tokyo, Japan

I. INTRODUCTION

Cyclophanes, or "bridged aromatic compounds" as designated by the title of an earlier monograph,[1] exhibit many interesting physical and chemical properties.[1–3] Those of particular interest are molecular deformation and transannular effects, which are most marked in [2.2]paracy-

ISBN 0-12-403002-5

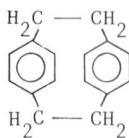

$$H_2C — CH_2$$

1

clophane **(1)**,* one of the smaller cyclophanes. These abnormal properties are related to various kinds of intramolecular interactions and are unique to smaller cyclophanes in which the aromatic rings are fixed in appropriate spatial relationships.

Unlike these smaller cyclophanes, the larger cyclophanes are generally quite normal in their benzene ring planarity, chemical reactivity, π basicity, etc., and therefore are not interesting in these respects. However, these compounds have one structural feature that their smaller analogs do not have. They possess intramolecular cavities of considerable size. This structural feature confers on the larger cyclophanes a remarkably interesting and significant property that cannot be expected in the smaller cyclophanes. A foreign molecule may be accommodated in the intramolecular cavity of a cyclophane and an *inclusion complex* formed. Thus, the unique property of the larger cyclophanes is related not to intramolecular but to intermolecular interactions, and on this basis a new field of cyclophane chemistry is developing, as will be described.

II. CYCLOPHANES AS ARTIFICIAL INCLUSION HOSTS

Before describing the complexation chemistry of cyclophanes, let us briefly consider the significance of the larger cyclophanes in the field of organic chemistry. One of the fundamental problems organic chemists face today is how to design and carry out highly selective reactions. There have been a considerable number of successful approaches to this fundamental problem using nonenzymatic systems, and very often the formation of some simple noncovalent complex before the reaction plays a significant role in high selectivity. In these cases metal complexes are

---

* For convenience the conventional nomenclature is used throughout this chapter.[4] The *Chemical Abstracts* nomenclature of [2.2]paracyclophane is tricyclo[8.2.2.2.^{4,7}]hexadeca-4,6,10,12,13,15-hexaene.

most frequently involved, but certain other kinds of complexes may also be involved, for example, hydrogen bond complexes.

These artificial reactions are invaluable because of the simplicity of the system and the possibility of wide application, especially when appropriate functional groups near the reaction site are available to form a complex that effectively generates the favorable transition state geometry. However, in many reactions the complexes formed by simple molecules or ions are not effective. For example, reactions such as those shown in Eqs. (1) and (2) are quite difficult to effect with high selectivity using simple artificial systems.[6–8,12]

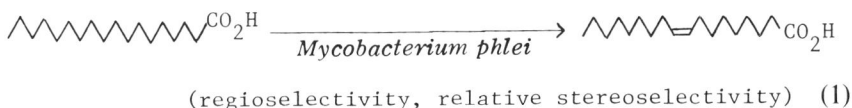

$$\text{(regioselectivity, relative stereoselectivity)} \quad (1)$$

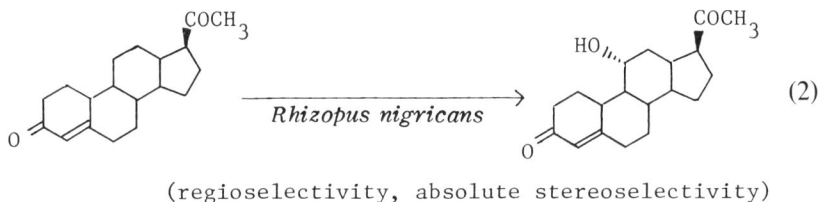

$$\text{(regioselectivity, absolute stereoselectivity)}$$

In contrast, biological systems carry out such reactions with high selectivity, often with "perfect" selectivity that is far beyond the present capacity of artificial chemistry. Here again, noncovalent complexes, that is, enzyme–substrate complexes, play a significant role. These complexes, however, are somewhat different from those formed between simple molecules or ions. The whole (or at least a considerable part) of the substrate molecule is accommodated (or included) within the enzyme cavity, so that a large area of the substrate surface is in contact with the enzyme through the multiple binding sites that are appropriately located and oriented. As a result, the substrate is bound strongly and fixed tightly in the enzyme active site.

Such a complex is possible only if the complexing molecule possesses multiple and convergent[10,10a] binding sites that are well located and oriented in a well-defined structure. This complexing molecule is called the *host,* whereas the complexed (included) molecule is called the *guest,* and the resulting complex is called a *host–guest complex.*[10,10a] Therefore, a host–guest complex is generally a stoichiometric inclusion complex of definite structure.

Enzymes are typical biological host molecules, and their well-defined macromolecular structure confers two factors that are most important for high selectivity.[11],* The first is substrate discrimination. Only substrates that fit well (sterically and electronically) within the enzyme cavity are complexed; substrates that do not fit well are not complexed and therefore are never brought to the reaction site. The second factor consists of proximity and orientation effects. The bound substrate is immobilized and its reaction site assumes a well-defined positional and orientational relationship with the reactive group of the enzyme, so that the reaction is promoted when this relationship is favorable and is inhibited when it is not favorable. Thus, extremely high speed and selectivity are conferred on enzymatic reactions through the formation of highly structured molecular complexes.

If we could imitate and generalize such biological reactions with simpler organic systems by the use of artificial host compounds, there might arise a novel and effective approach to highly selective reactions. The artificial host compounds would be much simpler than the natural ones such as enzymes, antibodies, and receptors but might be somewhat larger and more complicated than the usual reagents or catalysts. Whereas complexation by simple molecules or ions controls only the local geometry in the vicinity of the binding site(s), complexation by artificial hosts would control the entire geometry of the bound substrate, so that the reaction would be governed by the (well-defined) geometry of the whole complex.† Therefore, this novel approach should offer wider possibilities for dealing with the problem of selectivity and, in principle, would be especially effective for the sorts of reactions that are difficult to carry out either by simple artificial systems (*vide supra*) or by natural enzymes. (Note that natural enzymes are effective only for a limited number of substrates. There are many reactions in which natural enzymes do not display high selectivity.) Thus, the field of host–guest chemistry[10,10a] has been developing over the past 15 years as an important branch of biomimetic chemistry‡ to increase the ability of chemists to control organic reactions.

---

* Other important factors are conferred by the enzymatic structure, such as induced fit, strain effect, and multifunctional catalysis.[5] However, the two factors noted in the text are ubiquitous in enzymes and, more importantly, may possibly be designed and realized in relatively simple artificial systems.

† There are some examples in which the reactions are governed by the geometry of the whole complex by the use of complexing agents that are much simpler than artificial host molecules.[6,12-14]

‡ The term *biomimetic chemistry* was coined by Breslow to describe all efforts to imitate either the result or the style of biochemical reactions, in particular reactions catalyzed by enzymes.[6,7]

Because the first step in enzymatic reactions is host–guest complex formation between enzyme and substrate, the first problem in host–guest chemistry is undoubtedly the design and synthesis of artificial host compounds having inclusion cavities of definite structure. Several groups of artificial hosts have been developed so far*: (*a*) cycloamyloses (cyclodextrins),[15] (*b*) cycloalkanes,[16] (*c*) cyclophanes, (*d*) cyclic peptides,[17] (*e*) cyclic neutral polyligands[10a,18–21,21a] (crown ethers,[10a,18,20,21,21a] cryptands,[19–21,21a] etc.), (*f*) cyclic polyions[22–25] (cyclic polyanions[22] and cyclic polycations[23–25]), and (*g*) acyclic neutral polyligands.[26] All of these groups except (*g*) are composed of macrocyclic compounds. They are intrinsically suitable as artificial hosts, because they generally contain stable and well-defined inclusion cavities,† even though they are not macromolecules. Of these macrocyclic host groups, cyclodextrins, crown ethers, cryptands, and recently cyclic polyions have been most widely and systematically studied.

Crown ethers and cryptands are usually employed in organic solvents, and their major driving forces for complexation are charge–dipole interaction and/or hydrogen bonding.‡ Cyclodextrins and cyclic polyions, however, are usually employed in water, and their major driving forces for complexation are hydrophobic and charge–charge interactions, respectively. Crown ethers, cryptands, and cyclic polyions are completely synthetic hosts, whereas cyclodextrins are semisynthetic hosts in which cavity structure is already defined but in which some chemical modifications are still possible.

Cyclophanes are characterized by their aromatic ring(s), which may in principle act as rigid structural units, as hydrophobic and/or van der Waals binding sites, and as π donors or acceptors. These aromatic rings confer on the inclusion cavities well-defined structure and sufficient depth. (For example, inclusion cavities of paracyclophanes would be as

---

* Some related systems (micelles, reversed micelles, synthetic polymers, etc.) are frequently used in biomimetic chemistry to imitate enzymatic action.[8] However, these are intrinsically mobile systems with no definite binding site, so that the substrate may be bound strongly but not immobilized. Therefore, these systems may be suitable for effecting large rate enhancement,[9] but they are not suitable for achieving high selectivity.[8]

† Acyclic polyligands are not thought to be suitable hosts because they generally do not have inclusion cavities of well-defined structure. However, there are a few examples of high selectivity exhibited by this kind of host compound.[27,27a]

‡ Complexation by crown ethers and cryptands in water is usually weak because charge–dipole interaction and/or hydrogen bonding between the host and guest compete with those between water and guest. In contrast, complexation by cyclic polyions is usually strong because charge–charge interaction between the host and guest predominates over charge–dipole interaction and/or hydrogen bonding between water and guest.

deep as those of cyclodextrins.) Cyclophanes can be employed in organic solvents as inclusion hosts for organic guests (van der Waals and/or charge-transfer interactions) or metal ions (π coordination). When they are made water soluble (by the incorporation of appropriate hydrophilic groups), they can also be employed in water as inclusion hosts for hydrophobic guests.

Both cyclodextrins and water-soluble cyclophanes possess hydrophobic cavities of well-defined structure. However, whereas the former are semisynthetic, the latter are completely synthetic and therefore subject to wide structural modification. This versatility is quite important for artificial hosts, which must be designed and synthesized arbitrarily.

Cyclophanes in organic solvents are expected to constitute an artificial host group using driving forces for complexation that are different from those of crown ethers and cryptands. In water they are expected to constitute a versatile hydrophobic host group that complements cyclodextrins.* In contrast to cyclodextrins, cyclophanes can be designed and synthesized arbitrarily. As a result, it may become possible to study the details of host–guest interactions systematically and to arbitrarily develop hosts that selectively form complexes with guests of particular interest such as important biological molecules.

At present, cyclophanes are not as widely studied and applied as cyclodextrins, crown ethers, and cryptands. However, with respect to their characteristics and possibilities as described in the preceding discussion, cyclophanes undoubtedly constitute a significant and promising host group for which vigorous studies and wide applications are expected in the future.

By definition, any macrocyclic hosts that possess aromatic rings are cyclophanes, so that a wide range of artificial hosts are included in this group. In this chapter attention is focused on cyclophanes in which the aromatic rings are (or are intended to be) major binding sites. Cyclophanes in which the aromatic rings are auxiliary binding sites or rigid structural units (crown ethers and cryptands having aromatic rings, cyclophane porphyrins, etc.) also play an important role in the field of biomimetic chemistry, and these are discussed in Chapter 12.

* The major advantages of using cyclodextrins rather than cyclophanes are that they are available on a large scale and are nontoxic because of the D-glucose unit. However, they are difficult to functionalize by design because they have a considerable number of equivalent functional groups. (The smallest cyclodextrin, α-cyclodextrin, has 18 hydroxyl groups!) In addition, their cavities are already defined and are subject only to simple modification (e.g., capping).

III. PRELIMINARY STUDIES

In the 1950s considerable attention was focused on the physical and chemical properties of the smaller cyclophanes by the pioneering studies of Cram and others.[1,2] It was during this period that the larger cyclophanes (with established structure) were used for the first time as inclusion hosts for organic guests.

Stetter and Roos obtained stable and isolable crystalline complexes that were presumed to be stoichiometric inclusion complexes.[28] They synthesized a series of paracyclophanes (2a–2c) containing two benzidine skeletons bridged by two alkyl chains. Recrystallization of 2b ($n = 3$) and 2c

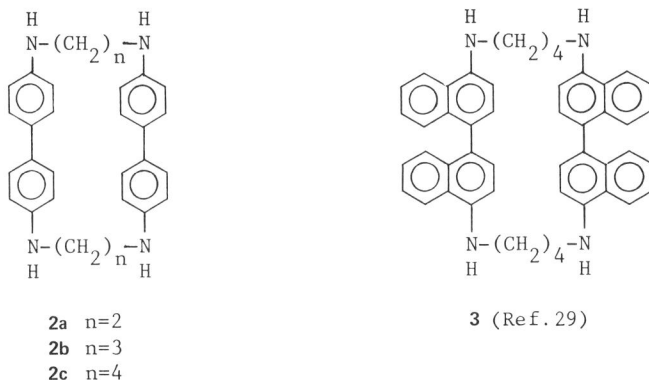

2a n=2
2b n=3
2c n=4

3 (Ref. 29)

($n = 4$) from dioxane or benzene yielded crystalline molecular complexes with these solvents, which tenaciously remained after drying *in vacuo* with heating (Table I). The most remarkable of these was the complex of 2c with dioxane, from which the solvent could not be removed after drying *in vacuo* at 150°C for 30 hr! In addition, 2c was quite stable as a dioxane complex, whereas it gradually colored on exposure to air when free of dioxane. The cavities of 2b and 2c are presumed by examination of molecular models to be large enough to include a foreign molecule. In contrast, such molecular complexes could not be obtained for 2a ($n = 2$), the cavity of which is presumed by examination of molecular models to be too small to include a foreign molecule.

On the basis of these observations themselves, one would not be able to determine whether the solvent molecule is included within the cavity or is merely stacking outside the cavity. The former possibility is supported by the unusual tenacity of the host in retaining the solvent and the observed relationship between this tenacity and the presumed cavity size; these

TABLE I

Drying Conditions and Composition of Complexes Involving hosts 2b and 2c

Host	Solvent	Drying condition (*in vacuo*)	Stoichiometry of the residual solvent[a]
2b	Dioxane	12 hr	1.25
		24 hr, 80°C	1.25
		30 hr, 150°C	0
2c	Dioxane	12 hr	0.75
		30 hr, 150°C	0.75
2b	Benzene	5 hr, 80°C	1
		5 hr, 150°C	0
2c	Benzene	5 hr, 80°C	1
		5 hr, 150°C	0

[a] The stoichiometry was determined on the basis of elemental analyses (C, H, and N).

properties are frequently observed with crystalline inclusion complexes, as will be discussed.

Both analogous and different types of paracyclophanes (**2d–2f** and **3–7**) were prepared,[29,30–33] and similar complexes were obtained for **3** and **7c**. Thus, a stoichiometric complex of **3** with dioxane was isolated by Faust and Pallas[29] and of **7c** with dioxane and benzene by Inazu and co-workers.[33] * However, no complex was isolated for other cyclophanes, indicating a quite strict structural requirement for complex formation to occur.

So far we have described complex formation by paracyclophanes that are designed to possess inclusion cavities of well-defined structure and sufficient depth. These are intramolecular cavities that are expected to capture guest molecules unimolecularly. However, there is another type of cyclophane, which also forms crystalline complexes with appropriate guest molecules but which captures the guests multimolecularly rather than unimolecularly. In this type of complex, the guest is accommodated in a cavity or channel formed in the crystal lattice by an arrangement of host molecules. That is, two or more host molecules participate in forming an inclusion cavity. This type of complex formation is observed with a wide range of host molecules but generally only in a crystalline state.[34] Among these hosts are two types of cyclophanes: type I and type II. Type I hosts are cyclic lactones of 2-hydroxybenzoic acids, and type II hosts are condensation products of formaldehyde and benzene derivatives.

The first examples of crystalline complexes formed by type I hosts, tetrasalicylide (**8**)·2 CHCl$_3$ and tetra-*o*-cresotide (**9**)·2 CHCl$_3$, were re-

* See Supplement 1 in Addendum.

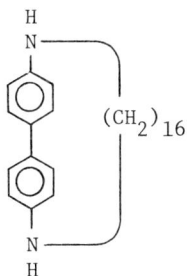

4 (Ref. 30)

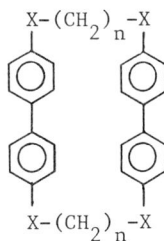

2d X=NH, n=7 (Ref. 30)
2e X=NH, n=8
2f X=NH, n=9

5a X=O, n=3 (Ref.31)
5b X=O, n=4
5c X=O, n=5

6 X=CH_2, n=4 (Ref.32)

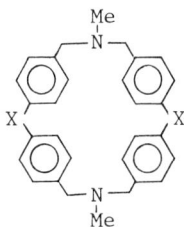

7a X = CH_2
7b X = $(CH_2)_2$
7c X = $CH_2N(CH_3)CH_2$

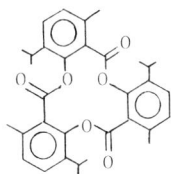

10 (tri-o-thymotide)

Type I

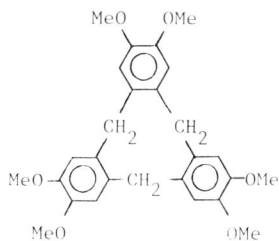

11 (cyclotriveratrylene)

Type II

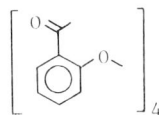

8

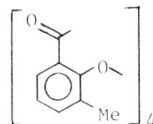

9

ported as early as 1892.[35] Later, in the 1950s, Baker and co-workers reinvestigated the previous studies and established the correct structure of many related hosts of this type.[36,36a] They also isolated a number of similar crystalline complexes with simple organic solvents such as chloroform, benzene, heptane, hexane, and ethanol. Among this type of host, special attention was directed toward tri-*o*-thymotide (**10**), first because it forms crystalline complexes with a remarkable range of guests, and second because it undergoes optical resolution on the formation of crystalline complexes (see Section IV,C).

Of type II hosts, cyclotriveratrylene (or "cycloveratryl," **11**) has been studied most widely. It was originally prepared as an acid-catalyzed condensation product of veratrole (1,2-dimethoxybenzene) and formaldehyde,[37] and its crystalline complexes with a number of organic guests were isolated.[38,44] However, it was not until the 1960s that the correct structure (cyclic trimer) was established by Lindsey[40] and others.[41,42] Until then the structure had been believed to be a cyclic dimer[37,38] or a cyclic hexamer.[39,43,44] The preferred conformations of **10** and **11** were shown to be "propeller"[45] and "crown,"[41,46] respectively, as shown in Fig. 1.

Hosts of both type I and type II are orthocyclophanes in which a planar conformation is not preferred. The nonplanar shape may explain the tendency of these hosts to form inclusion complexes because a molecule of this type is very likely to pack in the crystal in such a way as to leave open space and to include smaller molecules.[44] However, the structural requirement for such a phenomenon seems to be quite strict because many analogous compounds having three or more benzene rings failed to form inclusion complexes.[36,47,48]

Thus far, preliminary studies of the complexation chemistry of the larger cyclophanes have been described. In these studies sufficient evidence supporting the formation of inclusion complexes (and not simple stacking complexes) was not obtained. Information on the exact structure

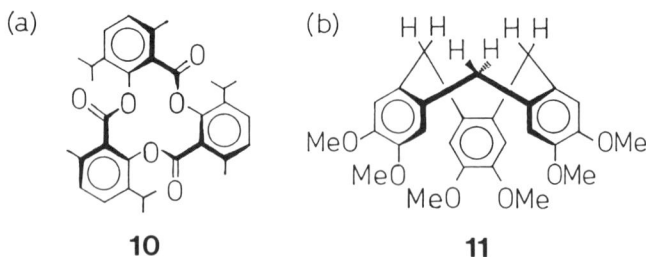

10   **11**

Fig. 1. Preferred conformations: (a) "propeller" conformation of tri-*o*-thymotide (**10**); (b) "crown" conformation of cyclotriveratrylene (**11**).

of host–guest complexes either in the crystalline state (X-ray) or in solution (nmr) was lacking, so there was no reliable basis for the broad use of these compounds as inclusion hosts. However, these preliminary studies provided impetus for the development of the complexation chemistry of the larger cyclophanes.

IV. CYCLOPHANE APPLICATIONS IN NONAQUEOUS SOLUTION

A. Preferred Conformation in Solution

As described in the previous section, the larger cyclophanes, especially paracyclophanes, may contain stable unimolecular cavities of well-defined structure and sufficient depth. However, the larger ring structure tends to make their conformations less rigid than those of their smaller analogs. Therefore, it is important to determine which of the possible conformations is preferred in solution, and for this purpose various spectroscopic studies have been carried out, mainly in nonaqueous solution.

The first problem is planarity of the aromatic rings. Ultraviolet spectroscopy gives important information in this case because electronic energy levels are very sensitive to structural deformation. Many larger cyclophanes have been shown to have planar aromatic rings on the basis of the close similarity of their uv spectra to those of the acyclic reference compounds whose aromatic rings are undoubtedly planar.[31,49–53] For example, Tabushi and co-workers showed the aromatic rings in the higher homologs of **1 (12a–12e)**[50,51] to be planar, in contrast to **1** itself, which has benzene rings that are deformed from planarity by about 13°.[2,54]

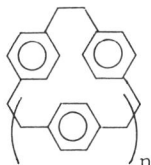

12a n=1 ($[2^3]$paracyclophane)

12b n=2 ($[2^4]$paracyclophane)

12c n=3 ($[2^5]$paracyclophane)

12d n=4 ($[2^6]$paracyclophane)

12e n=6 ($[2^8]$paracyclophane)

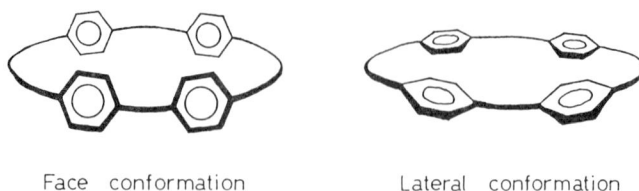

Face conformation   Lateral conformation

Fig. 2. Two extreme conformations of paracyclophanes.

The second problem is conformational freedom of the macrocyclic structure. Many larger cyclophanes exhibit ^{1}H-nmr spectra that appear as the average of the spectra of all conformers. Usually, the signals due to discrete conformers are not observed, indicating that conformational exchange is very rapid on the nmr time scale. Therefore, the larger cyclophanes are generally not restricted conformationally at room temperature.[31-33,49-53,55-64] However, broadening and separation of the signals are often observed at lower temperatures, indicating restricted conformational mobility.[51,53,57,58,64]

The third and most important problem is the determination of the preferred conformation of the aromatic rings. In paracyclophanes there are two extreme conformations: *face* and *lateral* (or *edge*) (Fig. 2).[51,61] The face conformation allows inclusion cavities with a sufficient depth, whereas the lateral conformation fills up the cavities so that the inclusion of guest molecules is not favored.

This conformational problem can be examined by observing ^{1}H-nmr chemical shift(s) of the aromatic protons. If the face conformation is preferred, the protons of each aromatic ring will be in the shielding region of the other aromatic ring(s). As a result, the signals of the aromatic

TABLE II

Chemical Shifts of the Aromatic Protons in 1, 5, 12, 13, and 14

Compound	$\Delta\delta$ (ppm)[a]	Compound[b]	$\Delta\delta$ (ppm)[c]
5a ($n = 3$)	6.80	**12b** ([2^4]PCP)	6.65
5b ($n = 4$)	6.87	**12c** ([2^5]PCP)	6.68
5c ($n = 5$)	6.98	**12d** ([2^6]PCP)	6.75
13 (acyclic)	7.23	**12e** ([2^8]PCP)	6.82
1 ([2^2]PCP)	6.30[c]	**14** (acyclic)	6.94
12a ([2^3]PCP)	6.62[c]		

[a] Measured in CDCl$_3$; TMS as an internal reference.[31]
[b] PCP, paracyclophane.
[c] Measured in CCl$_4$; TMS as an internal reference (Tabushi *et al.*,[51] and references cited therein).

protons of the cyclic compound will appear at higher field than those of the acyclic reference compound having aromatic rings that are presumed to rotate freely. That is, there is a negative cyclization shift (Eq. 3) in the aromatic protons:

$$\Delta\delta_{cyc} = \delta(\text{cyclophane}) - \delta(\text{acyclic reference}) \quad (\text{ppm}) \qquad (3)$$

This is actually the case for a number of the larger cyclophanes, and moderate $\Delta\delta_{cyc}$ values of -0.1 to -0.5 ppm have been observed.[31,32,49,51,55,59–61,61a]

In addition, this shielding effect should be stronger with decreasing size of the macroring. As shown in Table II this trend is clearly seen in the series of paracyclophanes **5a–5c** and **12a–12e**, which were prepared by Inazu *et al.*[31] and Tabushi *et al.*,[51] respectively.

12b	X=H
15a	X=NO$_2$
15b	X=CN
15c	X=COCH$_3$
15d	X=Br
15e	X=OH
15f	X=OCOCH$_3$
15g	X=NEt$_2$

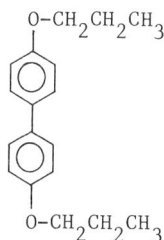

13

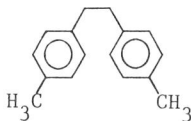

14

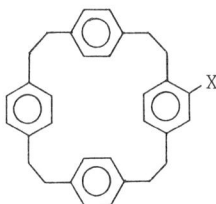

Detailed studies were carried out by Tabushi and co-workers[51,57] for [2⁴]paracyclophane **(12b)** and its derivatives **(15a–15g)**. The signal for the ethylene protons of **12b**, a singlet at room temperature, splits into two doublets at temperatures below about $-85°$C. This splitting is attributed to a constricted twisting of the ethylene chains, and the two signals are assigned to the protons frozen in axial and equatorial positions (H$_a$ and H$_e$, respectively) (Fig. 3). In contrast, the signal of the aromatic protons shows only little line broadening, indicating that the motions of the benzene rings are not greatly restricted. In such circumstances the chemical shifts of the two ethylene protons would reflect the *average* orientation of the benzene rings. Thus, the observed chemical shift difference of H$_a$ and H$_e$ ($\Delta\delta_{obs} = 0.51$ ppm) was compared with the theoretical values, which were calculated to be as follows for each of the three extreme states of the

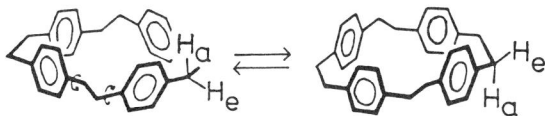

Fig. 3. Twisting of the ethylene chains.

benzene ring orientation: all-face conformation, $\Delta\delta_{calc} = 1.0$ ppm; all-lateral conformation, $\Delta\delta_{calc} = 0$ ppm; nonrestricted conformation with freely rotating benzene rings, $\Delta\delta_{calc} = 0.12$ ppm. Comparison of the observed value with the calculated values may suggest that the benzene rings still vibrate or rotate to produce a statistically averaged shielding effect but that the face conformation is greatly favored in a statistical sense.[51] When the benzene ring has a bulky or electron-withdrawing substituent (15a–15d, 15g), the lateral conformation tends to be favored, possibly by steric repulsion and by dipole–π interaction, respectively.[57]

As described so far, there are a number of examples in nonaqueous solution in which a face conformation is preferred. However, this is not always the case, as exemplified by compounds 15a–15d and 15g as well as by compounds 16 and 17. In 17 and related compounds, a planar conformation tends to be favored because of the conjugated structure.[63,64] More information about the conformations of the larger cyclophanes in both aqueous and nonaqueous solution is necessary for the design of stable and well-defined inclusion cavities.

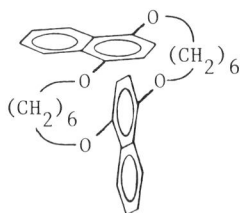

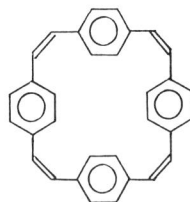

16 (Ref.61)   17 (Ref.64)

B. Complex Formation with Metal Ions

Although there are a large number of "cyclophanes" that form inclusion complexes with metal ions in organic solvents, most of these are members of the classes of host compounds in which the donor heteroatoms (oxygen, nitrogen, sulfur, etc.) are the major binding sites (crown ethers, cryptands, or their analogs having aromatic rings).[10a,18–21,21a] In these hosts the aromatic rings act as auxiliary binding sites or rigid structural units. As is well known such classes of "cyclophanes" have been widely studied as important artificial host compounds, and these are discussed in Chapter 12.

A number of studies have also been made of cyclophanes in which the aromatic rings are intended to be the major binding sites for metal ions.

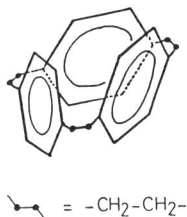

= $-CH_2-CH_2-$

Fig. 4. π-Prismand.

Pierre and co-workers[65] applied [2³]paracyclophane (**12a,** Section IV,A) as a soft donor for a heavy-metal ion. The solubility of CF_3SO_3Ag in chloroform was greatly enhanced in the presence of **12a,** and the formation of an extremely stable 1 : 1 complex in this solution was observed by ¹H-nmr spectroscopy. The stability constant K_s of this complex in CD_3OD is 1.95×10^2 M^{-1} (24°C), which is approximately 100 times greater than the known values for similar acyclic systems. In addition, this complex was isolated as a crystalline powder, and the X-ray powder diffraction studies indicated that it is a stoichiometric compound and not a mere mixture of uncomplexed cyclophane and silver salt. Furthermore, the field desorption mass spectrum showed an ion peak corresponding to [**12a**·Ag]⁺. From these observations it is presumed that **12a** acts as a "π-cryptand" that accommodates the guest ion in its cavity. Because of its characteristic shape the authors named it π-prismand (Fig. 4).

There are other examples of cyclophanes that are designed to include metal ions in their cavities, but complex formation has not been reported so far.[66–69] Additional studies of this sort of metal inclusion will be significant because, as mentioned in Section II, cyclophanes in organic solvents are expected to constitute a group of novel artificial hosts that use for complexation driving forces that are different from those of crown ethers and cryptands.

C. Complex Formation with Organic Guests

As described briefly in Section III, two types of cyclophanes (types I and II) have long been known to form inclusion complexes in which the guests are accommodated in the cavities formed in the crystal lattice. Generally, this kind of host compound forms inclusion complexes only in the crystalline state. Recent advances in X-ray crystallography have facilitated the elucidation of the detailed structure of the complexes.

Among type I hosts, special attention has been directed toward tri-*o*-thymotide **(10)**, first because it forms crystalline complexes with a remarkable range of guests, and second because it undergoes optical resolution on formation of crystalline complexes. Lawton and Powell found through an extensive X-ray study that the crystalline complexes of **10** generally belong to one of two classes: *cavity type* (or *cage type*) or *channel type*.[70] The cavity-type complexes are formed with various guests having lengths shorter than 9.5 Å, whereas the channel-type complexes are formed with longer guests. In the former the host molecules are presumed to form cavities of limited dimension, and in the latter they are presumed to form channels extending throughout the entire crystal. Both types of complex formation have been observed directly in X-ray studies.[71-76] In host–guest complex formation the dimensions of the guest molecules frequently determine whether inclusion occurs. Here, however, the dimensions determine which of the structurally different types of inclusion complex is formed; a similar tendency is also observed with host **21** (*vide infra*).

Powell and co-workers also investigated the stereochemistry of **10**.[77] Because of the substituents on the benzene rings, **10** is not coplanar and exists in enantiomeric forms that are related to each other as right- and left-handed three-bladed propellers. A careful X-ray examination suggested that, when **10** forms crystalline complexes with guests, the host molecules in any one crystal are of the same kind, that is, all in one enantiomeric form. Thus, although the crystalline complex is a racemate as a whole, optical resolution takes place in every single crystal. This must require the formation of chiral inclusion cavities, and optical resolution of racemic guests can be achieved by growing large single crystals. A systematic study was carried out by Green and co-workers in which the complexed guests showed an enantiomeric excess of up to 47%[78] (see also Gerdil and Allemand[79]). These workers also performed photoisomerization of stilbene in the crystalline complex of **10**.[75] Among the analogous hosts investigated,[80] only **18** possessed both of the properties exhibited by **10**: crystalline inclusion complex formation and spontaneous optical resolution.[81]

X-ray structural studies were also carried out for the crystalline complexes of type II hosts, that is, the complexes of **11** with benzene[82] and of **19** with 2-propanol.[83] Two cyclotriveratrylenyl units were bound together to construct a macrocage molecule, which was shown to form a 1 : 1 crystalline complex with chloroform.[84]

There are two other classes of cyclophanes, *calixarenes* **(20)** and [2⁶]metacyclophane **(21)**, that have been shown to form crystalline inclu-

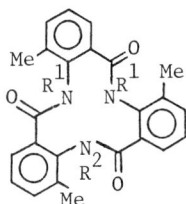

18 R^1=Me; R^2=CH$_2$Ph

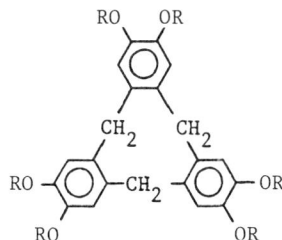

11 R=Me (cyclotriveratrylene)

19 R=H

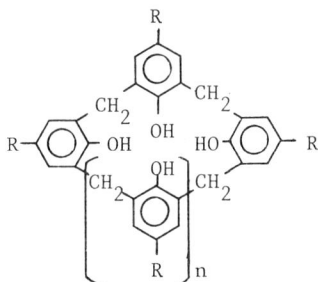

20 (calixarene)

20a n=1, R=Bu^t

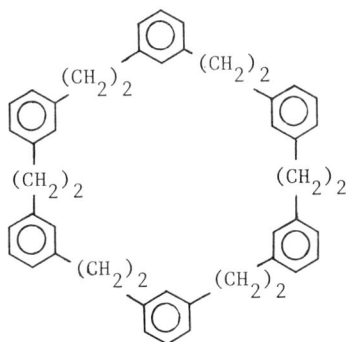

21 ([2^6] metacyclophane)

sion complex with organic guests (solvents).[85]* Both are metacyclophanes having intramolecular cavities.

Gutsche and co-workers carried out a systematic study of calixarenes, which were named after their characteristic chalicelike shape.[86,87] A number of crystalline complexes with simple organic guests were isolated, and frequently there was extreme difficulty in removing the complexed guest molecules.[87] For example, the guest molecule could not be removed from the 1 : 1 crystalline complex of *p-tert*-butylcalix[4]arene **(20a)** with toluene after heating at 207°C for 48 hr at 1 mm Hg, and 0.5 mol of the guest still remained after heating at 257°C for 144 hr at 1 mm Hg. The structure of this 1 : 1 complex was elucidated by the X-ray study of Andreetti and co-workers.[88] The chalicelike shape of the host molecule and the inclusion of the guest molecule within the chalice are apparent.

* See Supplement 1 in Addendum.

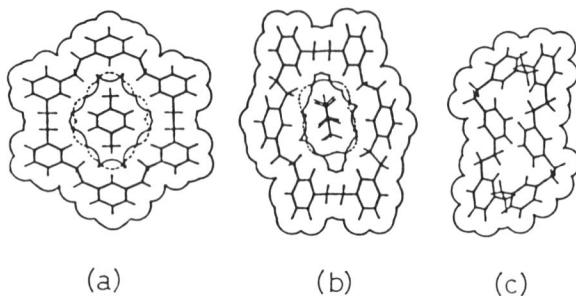

Fig. 5. Spatial shapes of the complexed and uncomplexed [2⁶]metacyclophanes **21**: (a) complexes with *p*-xylene and *o*-xylene; (b) complexes with benzene, *m*-xylene, *n*-heptane, cyclohexane, geraniol, etc.; (c) uncomplexed host.

[2⁶]Metacyclophane **(21)** was found to form crystalline complexes with a wide range of guests, and in some cases selectivity was observed.[89] For example, separation of *p*-xylene from its regioisomers and purification can be carried out by crystalline complex formation with **21** (see also MacNicol and Wilson[27a] and Ollis *et al.*[80]).* A detailed X-ray study of the structure of these complexes was performed by Itai and co-workers.[90] In these complexes, long channels are formed by the stacking of doughnut-shaped host molecules, and the guest molecules are included in these channels. Two modes of inclusion are observed. In one the guest molecules are sandwiched parallel to the macrocycle of the host (Fig. 5a), and in the other they are oriented vertically so as to penetrate the channel (Fig. 5b). The host molecules in the uncomplexed state adopt a squashed conformation, and consequently no channel is formed (Fig. 5c).

The formation of inclusion complexes with organic guests is of great significance in many respects. Considering that most organic reactions are carried out in organic solvents, additional possibilities will exist if this sort of complex formation is also effected in solution.

V. CYCLOPHANE APPLICATIONS IN AQUEOUS SOLUTION

A. Inclusion Complex Formation

The formation of inclusion complexes by hydrophobic interaction in aqueous solution is significant for the following reasons:

* See Supplement 2 in Addendum.

1. Hydrophobic interaction can be used, in principle, to form complexes with a remarkable range of organic guests, including those without any particular functional group that may facilitate binding. (Notice, for example, that the presence of a primary ammonium group is generally required for the complexation with crown ethers.)
2. With hosts having hydrophobic cavities of definite structure, the fit of steric structure between the cavity and guest affects the stability of the host–guest complexes considerably. Consequently, a strict discrimination among the guests might be possible on the basis of the steric fit.
3. Although water itself is generally not regarded as a good solvent for organic reactions, a hydrophobic cavity would create a nonpolar microenvironment that might facilitate (or retard) the reactions.
4. Host–guest complex formation in water is a fundamental process in many biological reactions such as those between substrate and enzyme, antigen and antibody, and drug and receptor. Simple artificial hosts may serve as suitable model systems for these biological reactions.

Considering these features and the possibilities for complexation in aqueous systems, it is of great significance to design and synthesize water-soluble host compounds having hydrophobic cavities of definite structure. As discussed in Section II, cyclophanes in water are expected to constitute a versatile hydrophobic host group, first because their aromatic rings can serve as rigid structural units for constructing well-defined hydrophobic cavities, and second because they are totally artificial hosts that can be designed and synthesized arbitrarily. Thus, considerable effort has been directed toward the design and synthesis of water-soluble cyclophanes, and the studies in this field have been a major part of the host–guest chemistry of cyclophanes.[59,61a,91–123]

The first problem in this field was to confirm that this class of host compounds indeed forms inclusion complexes and not simple stacking complexes. Thus, several spectral and kinetic comparisons of these cyclophanes and acyclic reference compounds were made. Tabushi and co-workers observed a marked change in the fluorescence spectrum of **22** induced by cyclophane **7c** in acidic water.[93] A large enhancement of the emission intensity and a blue shift of the emission maximum were observed, indicating that **22** was transferred into a nonpolar environment in the presence of **7c**. Because such a spectral change was negligible for the acyclic reference compound **23** under the same conditions, it was suggested that **22** is included in the hydrophobic cavity of **7c** rather than being merely stacked outside the cavity. Thus, the host **7c,** the complexing

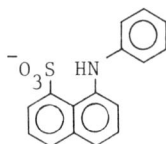

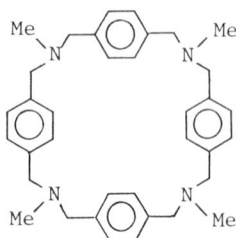

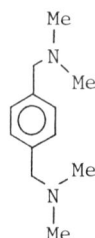

22 (ANS)

7c

23

capacity of which had been known only in the crystalline state,[33] * was successfully applied (in a protonated form) to the complexation in water. Similar spectroscopic results were also obtained for **24b**[92] and **25**,[94] which are intrinsically soluble in neutral water, in contrast with **7c**.

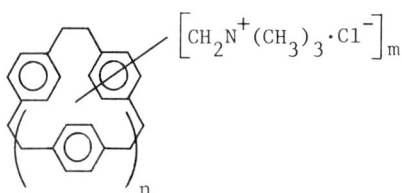

24a n=3, m_{av}=3
24b n=4, m_{av}=4

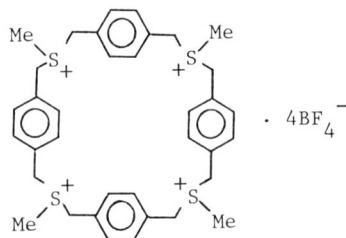

25

A study by Murakami and co-workers using kenetic data suggested the formation of an inclusion complex. Deacylation (transacylation accompanied by the release of p-nitrophenolate ion) of p-nitrophenyl dodecanoate was promoted by a factor of 7.4 in the presence of cyclophane oxime **(26)**.[97–99] The reaction was consistent with Eq. (4), involving complex formation between the host and guest before the intracomplex transacylation by nucleophilic attack of the oximate anion:

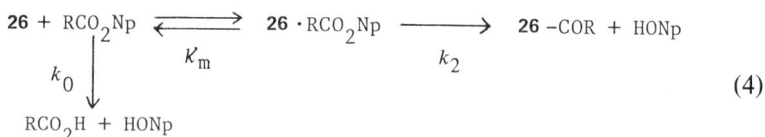

$$(4)$$

Np, p-nitrophenyl; K_m, Michaelis constant; [host] = [substrate] = 9.9×10^{-6} M; pseudo-first-order rate constant $k_\psi(\mathbf{26}) = 2.25 \times 10^{-3}$ sec^{-1}, $k_\psi(\mathbf{27}) = 0.33 \times 10^{-3}$ sec^{-1}; spontaneous hydrolysis rate constant $k_0 = 0.31 \times 10^{-3}$ sec^{-1}; k_2, intracomplex reaction rate constant, $k_2(\mathbf{26}) = 2.0 \times 10^{-3}$ sec^{-1}; 20°C, 10.9% (v/v) acetone–water, pH 12

* See Supplement 1 in Addendum.

Because the reaction rate was not affected by the reference compound (27) having no cavity, inclusion complex formation between the cyclophane and the hydrophobic guest was suggested. Inclusion complex formation was further supported by inhibition experiments. Similar kinetic results have also been obtained for 28a[103] and 29.[95,96]

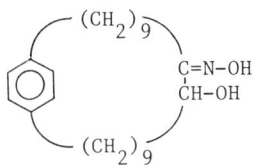

26

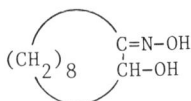

27

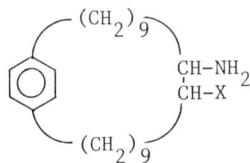

28a X=OH (Ref. 103)
28b X=H (Ref. 104)

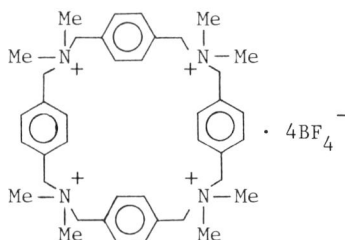

29

Although all of the above observations suggest the formation of inclusion complexes rather than simple stacking complexes, it was not until recently that direct evidence for "inclusion" was obtained with this class of host compound. 1,6,20,25-Tetraaza[6.1.6.1]paracyclophane (CP44, 30a), having two diphenylmethane skeletons bridged by two methylene

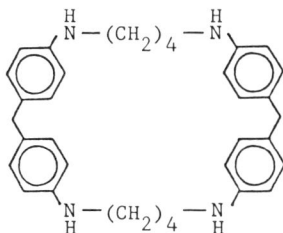

30a (CP44)

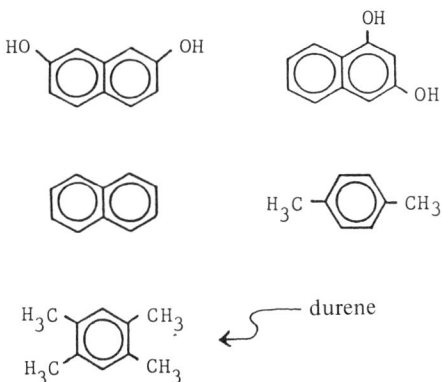

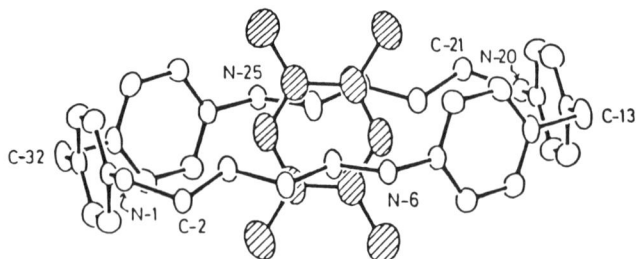

Fig. 6. Full inclusion of durene by CP44·4 HCl. The hydrogen atoms are not drawn.

chains, was designed and synthesized as a novel type of host compound.[115] Compound CP44 formed crystalline complexes from acidic water with a variety of guests having hydrophobic moieties, as exemplified here. Of these complexes, a detailed X-ray crystallographic study was carried out for the 1:1 complex with durene (CP44·4 HCl·durene·4 H_2O). The atomic positions of both host and guest could be determined definitely, showing the formation of a complex of considerable stability.

As shown in Fig. 6 a typical host–guest complex is formed, in which the guest molecule, durene, is fully included in the cavity of the host molecule, CP44·4 HCl. The guest molecule is located exactly at the middle of the cavity and is fixed tightly in close contact with the host molecule. Two characteristic features are found in the conformation of the host molecule. (a) The four benzene rings of the host are perpendicular to the mean plane of the macroring, facing one another to adopt the face conformation. (b) The two bridging chains adopt the trans-planar conformation except for the gauche conformation about the N-1—C-2 and N-20—C-21 bonds. As a result, a hydrophobic cavity that has rectangularly shaped open ends (~3.5 × 7.9 Å) and a depth of 6.5 Å is formed (Fig. 7). The inclusion geometry of the guest molecule is characteristic in that the benzene ring fits well in the cavity, being nearly parallel to the inner wall (Fig. 7). The primary basis for this good fit is the close correspondence between the thickness of the aromatic ring (3.4 Å) and the shorter width of the cavity open ends (~3.5 Å) (Fig. 8).

Thus, the formation of the inclusion complex as well as the adoption of the face conformation was confirmed directly through X-ray analysis. A similar result was obtained for the complex with naphthalene, in which the host molecule adopts nearly the same conformation as that adopted in the complex with durene.[120] It is worth noting that these crystalline complexes are formed without any participation of strong polar interactions (electrostatic interaction, hydrogen bonding) between the host and guest,

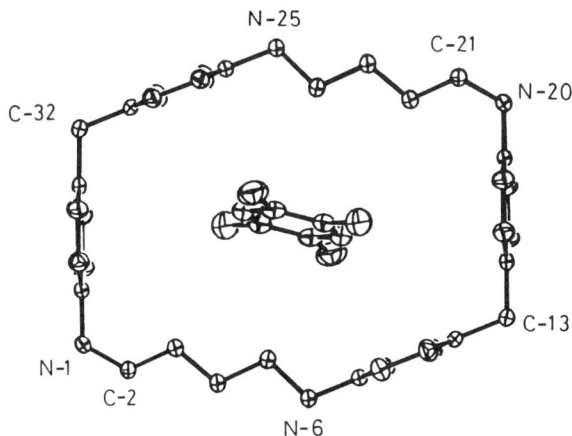

Fig. 7. Rectangularly shaped open end of the inclusion cavity of CP44·4 HCl.

because durene and naphthalene are both nonpolar guests with no functional group.

To examine whether the complexation state in crystals corresponds well to that in solution, a detailed study was carried out in acidic water[115,116] (CP44 and related compounds are all soluble in water as their amine salts below pH 2). As in **7c** and **25,** remarkable changes are induced by CP44 in the fluorescence spectra of **22** and **31** in acidic water (Fig. 9). Both of these guests are well-known hydrophobic probes, and the observed spectral changes indicate the transfer of these guests into a nonpolar environment. Similar spectral changes are observed when they are bound to the cavities of cyclodextrins and proteins or when they are dissolved in less polar solvents.[124–126] Remarkable changes are also observed in the ¹H-nmr spectrum of **32.** As shown in Figs. 10 and 11 marked upfield shifts are induced in the guest proton signals by CP44 in DCl–D₂O

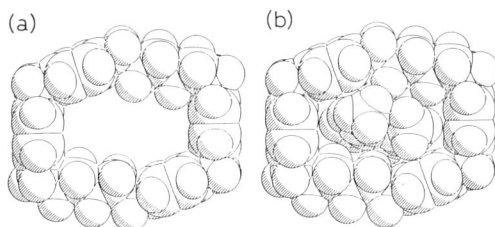

Fig. 8. Spatial shapes of (a) CP44·4 HCl and (b) CP44·4 HCl–durene complex, showing the close fit between the host and guest. The circles represent hydrogen and carbon atoms (sp^3 and aromatic) with van der Waals radii of 1.2, 1.6, and 1.7 Å, respectively.

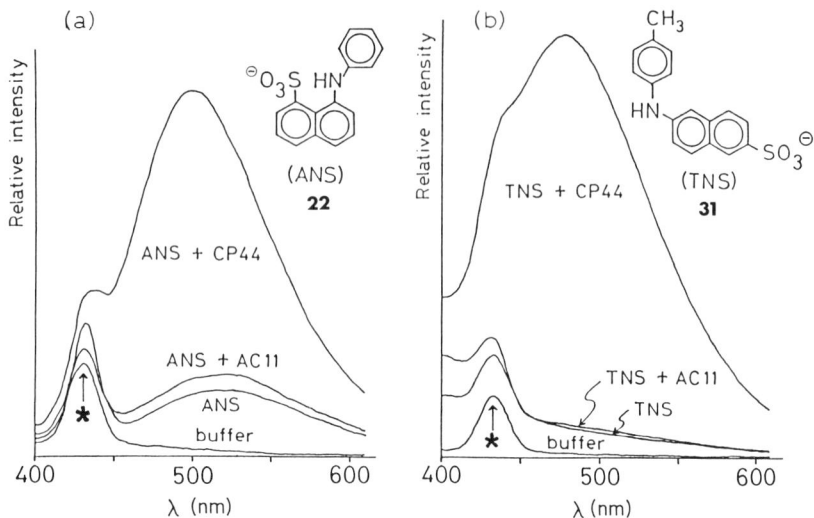

Fig. 9. Fluorescence spectra of (a) **22** and (b) **31** in the presence or absence of CP44 or AC11. Conditions: pH 2.0, 25.0 ± 0.1°C; excited at 375 nm; [CP44] = 5 × 10⁻⁵ M, [AC11] = 1 × 10⁻⁴ M, [22] = 2 × 10⁻⁶ M, [31] = 4 × 10⁻⁶ M. The asterisks indicate raman scattering of water.

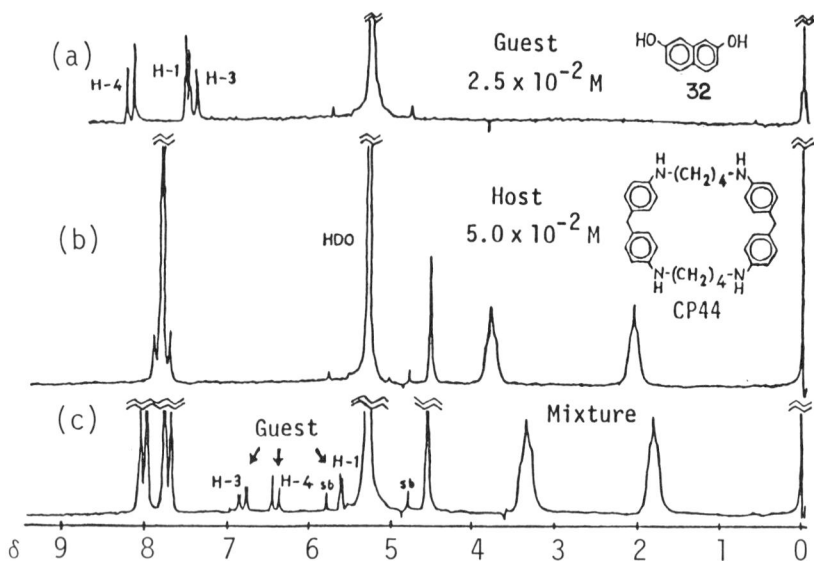

Fig. 10. ¹H-nmr spectra of (a) 2.5 × 10⁻² M **32**, (b) 5.0 × 10⁻² M CP44, and (c) their mixture in DCl–D₂O solution. Conditions: pD 1.2, room temperature; TMS as an external reference.

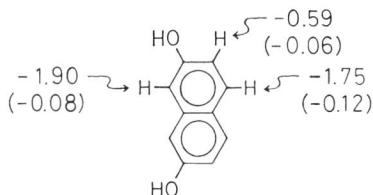

Fig. 11. Chemical shift changes of the protons of **32** induced by CP44 and AC11 (parentheses). Conditions: [AC11] = $1.0 \times 10^{-1} M$; negative values indicate upfield shifts, $\Delta\delta = \delta$(host + guest) − δ(guest) (parts per million).

solution. In ^{1}H-nmr spectra such marked upfield shifts can be attributed to a strong intermolecular ring current effect due to the aromatic rings of the host. Similar trends are observed with many of the other hydrophobic guests examined. In contrast, only small changes are induced by the acyclic reference AC11 **(33),** either in the fluorescence or the ^{1}H-nmr spectra under the same conditions. Consequently, it is suggested that inclusion complexes are formed by the protonated CP44 in water.

33 (AC11)

In addition, it is important to note that each proton signal of the host and guest shifts to a different degree, as seen in Fig. 10.[116] Again, similar trends are observed with many of the other hydrophobic guests examined. These discrete shifts indicate that complex formation occurs in a particular geometry and not in a random manner. This feature is characteristic of complex formation by hosts having well-defined structure, but it is unexpected in mobile systems such as micelles and synthetic polymers. Therefore, it is important to examine further the geometry of inclusion complexes in solution.

In the case of cyclophanes, ^{1}H-nmr spectroscopy is especially useful for this purpose because the degree of shift of the guest signals induced by the host is due primarily to the intermolecular ring current effect and is determined by the spatial relationship between the guest proton and the aromatic rings of the host. Accordingly, theoretical chemical shift

A   Pseudoaxial   inclusion

B   Axial   inclusion

C   Equatorial   contact

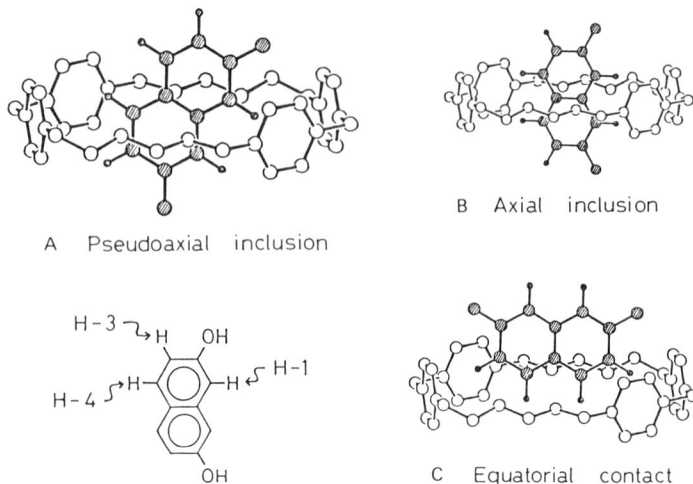

Fig. 12. Assumed geometries of the complex of CP44 with **32** in acidic water. The hydrogen atoms of the host molecule are not shown.

changes ($\Delta\delta$) for the guest signals can be calculated for each assumed geometry of the complex, and the preferred inclusion geometry can be deduced from a comparison between the observed and calculated values.

Thus, theoretical chemical shift changes of the proton signals of **32** ($\Delta\delta_{calc}$) have been calculated for several possible complexation geometries on the assumption that CP44 adopts, in acidic aqueous solution, the same conformation as it does in the crystalline complex with durene (Figs. 6 and 7).[116,121] Of the assumed geometries shown in Fig. 12, geometry A gives the best agreement between the observed and calculated values for each guest proton (Table III). The agreement of the $\Delta\delta_{calc}/\Delta\delta_{obs}$ values for all the guest protons in geometry A and the disagreement in geometries B and C also support the preference of geometry A. In this geometry the

TABLE III

Calculated and Observed Chemical Shift Changes[a] for the Possible Geometries A, B, and C for the Complex of CP44 with 32 in Acidic Water

Geometry[b]	H-1	H-3	H-4
A	−2.07 (1.09)	−0.65 (1.10)	−1.93 (1.10)
B	−2.16 (1.14)	−0.49 (0.83)	−2.12 (1.21)
C	−0.16 (0.08)	−3.06 (5.2)	−1.70 (1.0)
Observed	−1.90	−0.59	−1.75

[a] $\Delta\delta$ (parts per million). Values for $\Delta\delta_{calc}/\Delta\delta_{obs}$ are shown in parentheses.
[b] Geometries A, B, and C are depicted in Fig. 12.

TABLE IV

Stability Constants K_s of 1 : 1 Complexes between Water-Soluble Hosts and Hydrophobic Guests in Aqueous Solution

Complex with **22**			Complex with **31**		
Host[a]	K_s (M^{-1})	Reference	Host[a]	K_s (M^{-1})	Reference
CP44[b]	6.3×10^3	115	CP44[b]	9.6×10^4	118
7c[b]	5.5×10^2	93	α-CD	1.8×10	125
25	1.6×10^3	94	β-CD	1.5×10^3	125
34	1.1×10^4	108	γ-CD	1.5×10^3	125
β-CD	2.4×10	126			

[a] CD, cyclodextrin.
[b] Protonated form in acidic solution.

guest molecule is oriented with the long axis of its naphthalene ring penetrating the cavity obliquely (pseudoaxial inclusion). The preference for this geometry is consistent with ^{13}C-nmr spectra and energy calculations. In the ^{13}C-nmr spectra the largest chemical shift changes of the guest are induced at C-9 and C-10, which are presumed in geometry A to be in closest contact with the host.[116] From the energy calculations, geometry A is one of the possible geometries in which the nonbonded interaction energy between the host and guest is calculated to be lowest.[121]

Finally, two other aspects of the formation of inclusion complexes by water-soluble cyclophanes should be mentioned. One is the stability of the 1 : 1 inclusion complexes. Some examples of the stability constants of these complexes and of complexes with other water-soluble hosts are listed in Table IV. It can be seen that all of the water-soluble cyclophanes examined form quite stable complexes with the hydrophobic guests ($K_s =$ 10^2–10^4 M^{-1}),* whereas a lower stability constant is observed when the complexation mode is presumed to be a simple stacking.[59] The other aspect of inclusion complex formation by water-soluble cyclophanes is

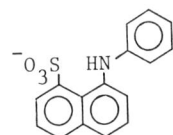

22 (ANS)

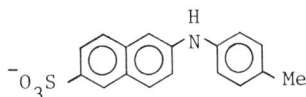

31 (TNS)

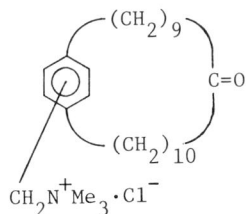

34

* See Supplement 3 in Addendum.

the degree of immobilization of the included guests. A considerable restriction of mobility of the complexed guests has been shown for a number of cyclophanes in aqueous solution on the basis of measurement of rotational correlation times by esr spectroscopy.[110,112,113] Further investigation of "dynamic coupling" between host and guest will give important information as to the fixation of the included guests.[127]

B. Acceleration of Reactions

As described in Section V,A, cyclophanes form 1 : 1 inclusion complexes with organic guests in aqueous solution. In addition, the inclusion complexes are formed in a particular geometry with some fixation of the included guests (in a statistical sense). Therefore, a large rate enhancement may result if cyclophanes are functionalized with appropriate reactive groups and employed in organic reactions in aqueous solution. The reaction will proceed via inclusion complex formation with the guests (substrates) before the intracomplex reaction. The acceleration of reaction in water based on such a mechanism is one of the important features of enzymes. In this respect, water-soluble cyclophanes may serve as preliminary models of enzymes.

Murakami and co-workers[91,97–114] carried out extensive studies on functionalized cyclophanes. As described briefly in Section V,A, deacylation of p-nitrophenyl esters is promoted by the cyclophane oxime **26** but is not affected by the acyclic reference **27**.[97,98] The reaction scheme is consistent with Eq. (4), and the rate enhancement is ascribed to the formation of an inclusion complex in which the reaction site of the substrate (C=O) and the reactive group of the host (oximate anion) are brought into close proximity. As shown in Table V the acceleration (k_ψ/k_0) is large with long-chain alkyl esters but is negligible with the corresponding acetate, which may not form an inclusion complex with **26** due to the lack of hydrophobicity.[98] Large acceleration with long-chain alkyl esters is presumably due not only to inclusion complex formation, but also to the accompanying unfolding of the alkyl moiety, which may expose the carbonyl site to the nucleophile.[105,107,113,128] A large rate enhancement by **26** also occurs with 2,4-dinitrophenyl sulfate.[102]

These reactions yield acylated oximes (**26**—COR), which are hydrolyzed very slowly under the reaction conditions. This hydrolysis rate must be large enough to regenerate the reactive group of the host and to attain the turnover behavior of a true catalyst such as the natural enzymes.[100] The stability constants K_s (approximated by $1/K_m$) of 1 : 1 com-

TABLE V

Deacylation Rates of *p*-Nitrophenyl Esters in the Presence or Absence of 26 and 27[a]

Substrate, $CH_3(CH_2)_nCO_2Np$	$k_0 \times 10^4 \, (M^{-1})$	$k_\psi \times 10^4 \, (M^{-1})$	
		26	**27**
$n = 0$	21.3	23.1	22.6
$n = 8$	4.27	16.8	4.30
$n = 10$	0.69	15.3	0.70
$n = 14$	0.10	12.6	0.11

[a] Conditions: 23.5°C, pH 9.23, 10.9% (v/v) acetone–water; [host] $= 2 \times 10^{-5} \, M$, [substrate] $= 1 \times 10^{-5} \, M$.

plexes of **26** with long-chain alkyl substrates are large (10^4–$10^5 \, M^{-1}$), indicating that the [20]paracyclophane skeleton provides an effective inclusion cavity for hydrophobic substrates.

This type of paracyclophane is a versatile model system for studying bifunctional catalysis because an additional functional group can be easily incorporated into the benzene ring that is located on the opposite side of the oxime group. Thus, functionalized [20]paracyclophanes **(35a–35c)** were prepared and their esterolytic activities toward long-chain alkyl esters were examined over the entire pH range.[105,106] Figure 13 shows the pH dependence of the deacylation rates of *p*-nitrophenyl hexadecanoate in the presence or absence of **35a–35c**.[106] This pH–rate profile is interesting in that the reactivity order of the three cyclophane oximes changes markedly in passing from alkaline to neutral and acidic conditions.

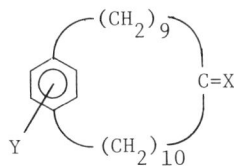

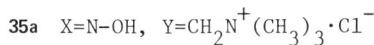

34   X=O,   Y=CH$_2$N$^+$(CH$_3$)$_3$·Cl$^-$

35a   X=N-OH,   Y=CH$_2$N$^+$(CH$_3$)$_3$·Cl$^-$

35b   X=N-OH,   Y=H

35c   X=N-OH,   Y=CO$_2$H

The reactivity order of **35a** $\gg$ **35b** > **35c** under alkaline conditions may be ascribed to the difference in stabilization of the transition state for

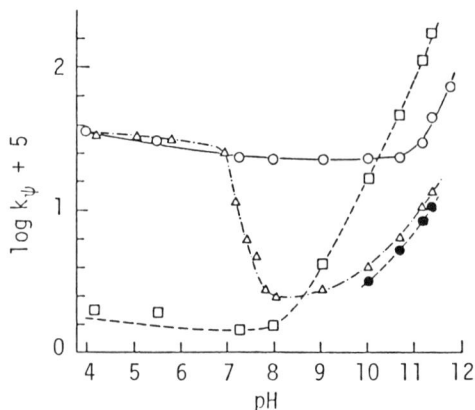

Fig. 13. The pH dependence of the deacylation rates of p-nitrophenyl hexadecanoate in the presence or absence of **35a–35c**. Conditions: $40.0 \pm 0.1°C$; [host] $= 8$–9×10^{-6} M, [substrate] $= 1 \times 10^{-5}$ M; k_ψ values are in reciprocal seconds. Key: $\square$, **35a**; $\circ$, **35b**; $\triangle$, **35c**; $\bullet$, none.

deacylation in which a negative charge develops on the carbonyl oxygen of the substrate.[105] This transition state is stabilized by the positively charged quaternary ammonium group and destabilized by the negatively charged carboxylate group. As a result, the reactivity of the oximate anion is enhanced in **35a** and reduced in **35c** compared with **35b**. Thus, a nucleophilic–electrostatic bifunctional catalysis has been attained in a simple model system (Fig. 14a). Although electrostatic interaction is generally weakened in water, it has proved to be effective in the hydrophobic microenvironment provided by the cyclophane. A related example is seen in the enhanced aminolysis of p-nitrophenyl hexadecanoate by glycine in the presence of **34**.[107] As shown in Fig. 14b the aminolysis is presumably facilitated by the anchoring of glycine on the quaternary ammonium group of the host. This may be a preliminary model for the biosynthesis carried

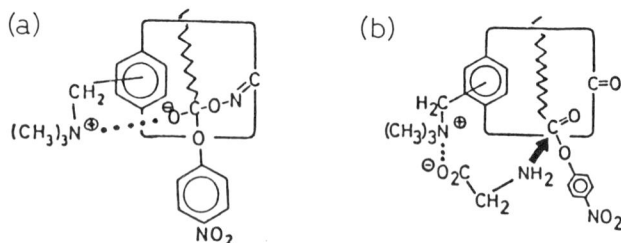

Fig. 14. (a) Nucleophilic–electrostatic bifunctional catalysis. (b) Electrostatic–hydrophobic double-field catalysis.

out by enzymes possessing dual binding sites to assemble two organic substrates, which then react with each other.

The reactivity order of **35c** ≈ **35b** >> **35a** under neutral to acidic conditions seems quite unusual. The former two compounds exhibit considerable reactivity despite the low nucleophilicity of the oxime group which must be in an un-ionized form in a neutral to acidic pH region.[106] This unusual nucleophilicity of the un-ionized oxime group may be ascribed to the structured water around the hydrophobic cavity. The basicity of these water molecules is appreciably enhanced through hydrogen bonding. Consequently, they are likely to act as favorable proton acceptors, abstracting the oxime proton in the transition state so as to promote trans-acylation (Fig. 15a). The hydrogen-bonded structure of water is disrupted by charged groups such as ammonium and carboxylate groups, so that the reactivity of **35a** as well as of **35c** (in alkaline solution) is markedly reduced, as shown in Fig. 13. Thus, the remarkable hydrophobicity of the [20]paracyclophane skeleton provides not only an effective inclusion cavity but also a hydrophobic microenvironment that may greatly affect the reactivity.

Another example of enhanced nucleophilic attack by the nonionized oxime group in neutral solution is observed in the **36**–Cu^{2+} system. In deacylation of p-nitrophenyl hexadecanoate, this system exhibits, at pH 8.12, a rate enhancement that is even larger than that of **35b** at pH 11.61.[109] As shown in Fig. 15b the coordinated Cu^{2+} may act as an effective superacid to activate the carbonyl group of the complexed substrate, so that greater reactivity is exhibited by the nonionized poor nucleophile (C=N—OH in neutral solution) than by the ionized strong nucleophile (C=N—O$^-$ in alkaline solution).

A large rate enhancement in neutral aqueous solution, as described earlier, is one of the most important problems in this field. From this

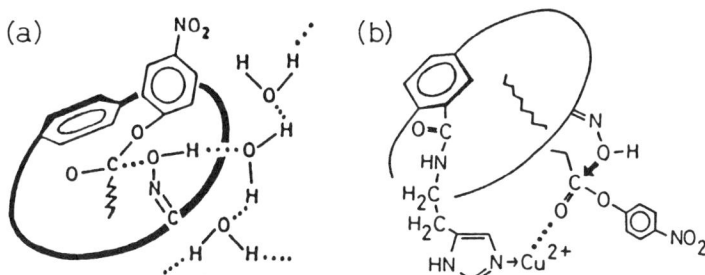

Fig. 15. (a) Unusual nucleophilic reactivity developed by structured water molecules. (b) Unusual nucleophilic reactivity developed by Cu^{2+}-assisted activation of the substrate carbonyl group.

viewpoint the introduction of a reactive group that exhibits strong nucleophilicity in a neutral pH region is significant. Imidazole is one such nucleophile, as suggested by the fact that a number of enzymes possess this reactive group in their active sites. Thus, a cyclophane bearing an imidazole group (36) was prepared, and its reactivity toward p-nitrophenyl esters was examined.[108] Cyclophane 36 was found to accelerate deacylation of various types of esters, and the more hydrophobic esters tended to be deacylated more rapidly; the largest acceleration was observed with long-chain alkyl esters. This qualitative behavior was rationalized by Hansch on the basis of a quantitative scale of hydrophobicity.[129] The involvement of the hydrophobic effect in host–guest interactions in this type of cyclophane is also supported by thermodynamic parameter analysis[98,99,105] and by the inhibitory effect of organic cosolvents such as acetone and ethanol.[97,98] Reaction with 36 yields acylated imidazoles, which are hydrolyzed so slowly that no turnover behavior is observed.

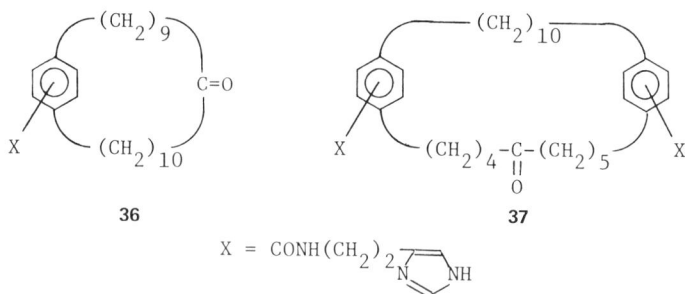

The introduction of a second functional group that facilitates hydrolysis of the acyl intermediate is necessary for this type of cyclophane to act as a true catalyst.[100] The [10.10]paracyclophane skeleton, which has also been found to be effective in forming 1 : 1 inclusion complexes,[110] may be suitable for designing such a catalyst because two functional groups can be situated on opposite sides of the cavity. Thus, a [10.10]paracyclophane bearing an imidazole group on each benzene ring (37) was synthesized, and its kinetic behavior toward a large excess of p-nitrophenyl dodecanoate in the presence of Cu^{2+} was investigated.[111] The reaction displayed burst kinetics (with respect to the release of p-nitrophenolate ion) and proceeded far beyond the stoichiometric conversion range. A detailed kinetic analysis suggested that Cu^{2+}, coordinated to one of the imidazole rings, acts as a superacid to facilitate both transacylation and hydrolysis processes (by factors of 5 to 6 and 10^2, respectively) (Fig. 16). The initial burst of the reaction corresponds to the accumulation of the acyl intermediate. After this initial stage, the transacylation and hydrolysis processes

RCO$_2$Np

Im $\longrightarrow$ Cu^{2+} $\longleftarrow$ Im

− RCO$_2$H

− HONp

Im$\rightarrow$Cu^{2+}···O=C$\longleftarrow$Im
ONp

Im$\rightarrow$Cu^{2+}···O=C $-$ Im
OH$_2$

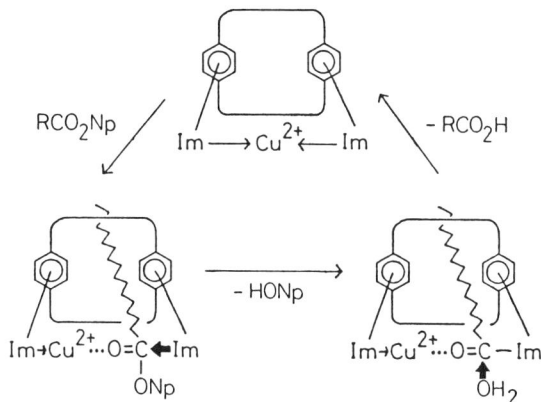

Fig. 16. Turnover behavior of **37**–Cu^{2+} involving facilitated transacylation and hydrolysis.

are balanced to give a steady-state concentration of the catalyst. This catalytic reaction is in sharp contrast to all of the reactions described above, which are stoichiometric rather than catalytic and practically never proceed beyond the stoichiometric conversion range. A true catalysis in neutral aqueous solution also occurs with **36**–Cu^{2+},[111] **28a·H**$^+$,[103] **28b·H**$^+$,[104] and **29**.[95,96]

The paracyclophanes described so far have relatively shallow cavities, with the maximum depth (~6.5 Å) corresponding to the width of a benzene ring. To include the larger guest molecules, cavities must be deepened by the introduction of appropriate units and/or substituents (see Sergheraert et al.[69]). Thus, azacyclophanes having four long-chain alkyl substituents **(38)** were prepared, and their spectral and kinetic behavior was investigated.[112,113] This type of host is effective in forming complexes with large hydrophobic guests such as organic dyes and long-chain alkyl esters. Because the complexing capacity is also markedly dependent on the charge of the guest, electrostatic interaction as well as hydrophobic interaction apparently plays a major role in host–guest complex formation. In these hosts the four hydrophobic alkyl chains may gather together like a micelle to form an octopuslike structure that deepens the hydrophobic cavity, allowing the inclusion of large guests (Fig. 17). Also, the cavity is "adjustable" because of the flexibility of the long alkyl chains. Related host molecules **(39** and **40)** have been prepared using the cyclotriveratrylene skeleton,[123,130] which adopts the "crown" conformation (Sections III and IV,C) and may facilitate the orientation of the hydrophobic alkyl chains in the same direction.

38a R = $(CH_2)_{10}CO_2H$

38b R = $(CH_2)_{10}N^+(CH_3)_3 \cdot Br^-$

38c R = $(CH_2)_{10}N^+(CH_3)_2(CH_2Im) \cdot Cl^-$

38d R = $[(CH_2)_{10}N^+H(CH_3)_2 \cdot Cl^-]_{2/4}$
     $[(CH_2)_{10}N^+(CH_3)_2(CH_2Im) \cdot Cl^-]_{2/4}$

39 R = $(CH_2CH_2O)_2CH_2CH_3$

40 R = $(CH_2)_{10}CO_2^-$

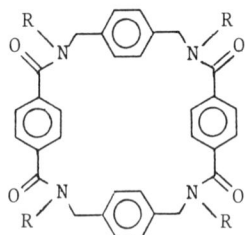

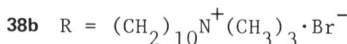

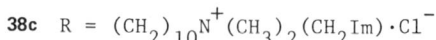

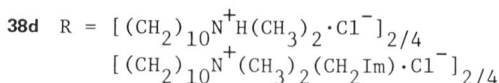

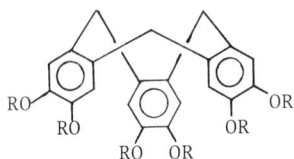

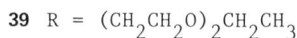

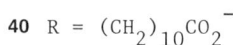

Murakami and co-workers have demonstrated a novel use of cyclophanes.[91,114] In the reduction of hexachloroacetone in nonpolar solvents (CH_2Cl_2 or CH_2Cl_2/CH_3OH), dihydronicotinamide-containing cyclophanes **41, 42,** and **43** (Zn^{2+} complex) exhibited large rate enhancement, which may be ascribed to the effective cooperation of the functional groups that are suitably fixed in the macrorings. Although the role of the macrorings here seems to be the fixation of the functional groups rather than the inclusion of the substrate, cyclophanes are employed in solutions for the first time for a reaction other than ester hydrolysis. This will surely give impetus for the application of cyclophanes to many useful

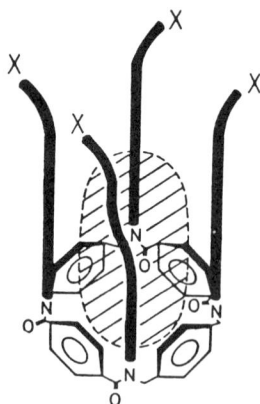

Fig. 17. "Octopus" cyclophane with a deep and adjustable hydrophobic cavity (X = $-N^+R_3$, $-CO_2^-$).

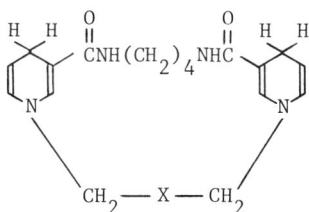

41 X = p-xylylene (Ref.114)

42 X = o-xylylene (Ref.114)

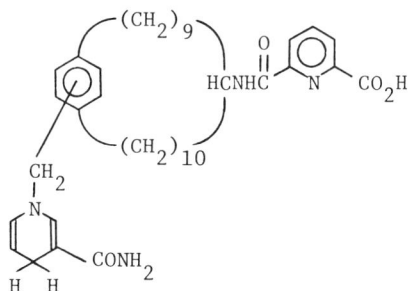

43 (Ref.114)

organic reactions, not only in organic solvents but in aqueous systems as well.

C. Substrate Selectivity

Many biological reactions proceed with extremely high selectivity, which results from complex formation by biological host molecules such as enzymes, antibodies, and receptors. These macromolecular hosts possess definite structure, which is a fundamental requirement for the formation of highly structured complexes and for strict selectivity. In order to imitate and generalize this significant characteristic of biological reactions using artificial hosts, the water-soluble cyclophanes will be among the most appropriate and promising systems, because these hosts are completely synthetic and possess cavities of well-defined structure (Section V,A). Considering that substrate-selective inclusion is the initial step in biological reactions, it is important to effect substrate-selective inclusion by water-soluble cyclophanes and to carry out a systematic study of this behavior.

For this purpose analogs of CP44 should be a fairly suitable system for the following reasons. First, the diphenylmethane skeleton can effectively form hydrophobic cavities of well-defined structure. Because there is only one intervening methylene unit, the two benzene rings will be fixed at a definite angle. In addition, Cram and co-workers demonstrated a transannular electronic coupling between the π electrons of the two benzene rings of the diphenylmethane unit.[131,132] Such a transannular coupling would tend to decrease the distance between the benzene rings and increase the population of molecules with the π orbitals of C-1 and C-1′ pointing toward each other.[131] This interaction, being especially marked in cyclic systems, may favor the face conformation for paracyclophanes, affording a deep cavity (see Section IV,A). Empirical force field and molecular orbital calculations of the molecular structure of diphenylmethane also indicate that the face (gable) conformation is the ground state for the isolated molecules.[133] Consequently, the diphenylmethane skeleton is considered to be an appropriate unit for the construction of inclusion cavities of definite structure and sufficient depth, although inclusion complex formation has not been reported for other cyclophanes containing this unit.[33,49,55,68,69] Second, the cavity structure of this type of cyclophane can be modified easily by changing the bridge or skeleton moieties. As will be seen this versatility is one of the most important features of a systematic investigation.

A number of derivatives of CP44 were synthesized and the complexation properties of these modified hosts were examined to obtain information about the relation between the nature of the cavity (size, hydrophobic area, hydrophobicity, and stability) and complexing capacity.[117] In these modified hosts the length of the bridge moieties (**30b–30g**), the unit of the bridge moieties (**44** and **45**), and the unit of the skeleton moieties (**46** and **47**) are changed.*

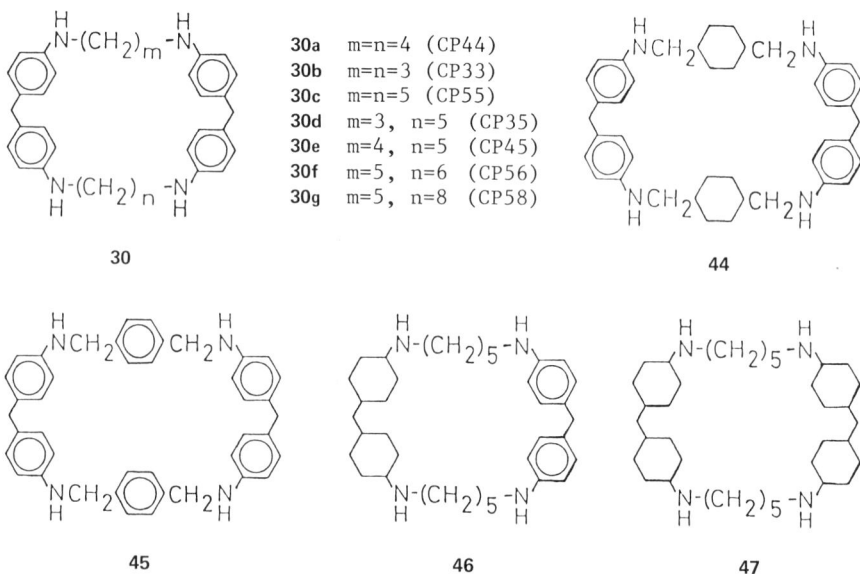

30a	m=n=4	(CP44)
30b	m=n=3	(CP33)
30c	m=n=5	(CP55)
30d	m=3, n=5	(CP35)
30e	m=4, n=5	(CP45)
30f	m=5, n=6	(CP56)
30g	m=5, n=8	(CP58)

30

44

45   46   47

The effect of modification was examined by comparing the complexing capacity of modified hosts with the hydrophobic guest **22**, which had been shown to form a stable 1 : 1 complex with CP44 in acidic water (Section

* All the 1,4-cyclohexylene units are trans.

TABLE VI

Stability Constants K_s of 1 : 1 Complexes between 22 and Modified Hosts[a]

Host	K_s (M^{-1})[b]
CP44 (**30a**)	6.3×10^3 (1.0)
CP45 (**30e**)	5.2×10^3 (0.82)
CP55 (**30c**)	9.8×10^3 (1.6)
CP56 (**30f**)	4.3×10^4 (6.8)
CP58 (**30g**)	3.9×10^4 (6.2)
44	5.0×10^5 (80)

[a] Conditions: $25.0 \pm 0.1°C$, pH 2.0. The hosts are in a protonated form.
[b] Numbers in parentheses indicate relative stability of the 1 : 1 complexes.

V,A). As shown in Table VI hosts **30c, 30e–30g,** and **44** formed stable complexes with **22,** whereas the complexes formed by **30b, 30d,** and **45– 47** were very weak, as indicated by small changes in fluorescence.

From these observations the effects of cavity modification may be described as follows:

1. Complex formation is affected by cavity size **(30a–30g).** The strongest complex formation by CP56 **(30f)** may be due to the excellent fit of its cavity with the naphthalene ring of **22,** as suggested by molecular models (see Fig. 18b). Thus, of the two hydrophobic moieties of the guest, the fit with the naphthalene ring seems to be more important than that with the benzene ring. Molecular models also suggest that the cavities of **30b** and **30d** are too small to include even a benzene ring, which is the smaller of the hydrophobic moieties of the guest.

2. The increase in the hydrophobic area of the cavity greatly enhances complex formation (**44** versus CP55 and CP56), whereas the decrease in hydrophobicity seems to be unfavorable (**45** versus CP55 and CP56). The complex between **22** and **44** ($K_s = 5.0 \times 10^5\ M^{-1}$) is by far the strongest among the corresponding complexes with other artificial water-soluble hosts (see Table IV).

3. The diphenylmethane skeleton seems to contribute to the formation of inclusion cavities, whereas the corresponding aliphatic skeleton does not (CP55 versus **46** and **47**). Despite the increased hydrophobicity, the aliphatic skeleton does not seem to be a suitable unit for constructing an inclusion cavity. One possible reason is that the face

conformation may not be favored in this nonaromatic skeleton because the transannular electronic coupling (*vide supra*) is absent. Consequently, the lateral conformation may be greatly preferred in **46** and **47** to fill up their own cavities so that the inclusion of guests is not favored (see Section IV,A).

Thus, the simple modifications that optimize cavity size and increase hydrophobic area improved complex formation 80-fold (from $K_s = 6.3 \times 10^3 \ M^{-1}$ by CP44 to $K_s = 5.0 \times 10^5 \ M^{-1}$ by **44**). This systematic cavity modification in artificial water-soluble hosts will provide important information for effecting substrate-selective inclusion in water. In addition, this may be a preliminary example of the development of hosts that selectively form complexes with guests of particular interest, such as important biological molecules.

The above-described study on modified hosts as well as the nmr study on CP44 (Section V,A) strongly supports the idea that this type of paracyclophane forms inclusion cavities of well-defined structure in acidic water. This is the fundamental requirement for effecting substrate-selective inclusion, the initial step of reactions occurring via host–guest complexation. Therefore, a systematic study of substrate-selective inclusion was carried out for three representatives of this type of host, namely, CP44, CP56, and **44,** having cavities of different size and/or hydrophobic area.[118]

Stability constants between these hosts and guests having a variety of shapes and charges are listed in Table VII, and the following tendencies have been observed for selectivity in substrate inclusion:

1. All three hosts form strong to moderate complexes with anionic guests having an aromatic ring (Table VII, **53** and **54**). In contrast, complex formation is very weak for anionic guests having a quite different structure from that of the aromatic guests (**55** and **56**).
2. Of the guests having a naphthalene ring, CP44 exhibits selectivity for the β-substituted naphthalenes (**32, 31, 50,** and **52**), whereas CP56 and **44** exhibit selectivity for the α-substituted naphthalenes (**22, 49,** and **51**).
3. With all the guests examined, host **44** forms stronger complexes than CP56 by a factor of 10 to 17.
4. For all three hosts, the complexes with the dianion guests (**51** and **52**) are stronger than those with the corresponding monoanion guests (**49** and **50**). Furthermore, only weak complex formation is observed for the aromatic guests having positive charge(s) (**57** and **58**).

As described in Section V,A, X-ray studies of the complexes of CP44 · 4 HCl with durene and naphthalene show the formation of an inclusion

TABLE VII

Stability Constants K_s of 1:1 Host–Guest Complexes[a,b]

Guest[c]		CP44	CP56	**44**
32		2.8×10^3 (11)	2.6×10^2 (1.0)	4.3×10^3 (17)
48		2.0×10^3 (2.3)	8.7×10^2 (1.0)	1.4×10^4 (16)
22 (ANS)		6.3×10^3 (0.15)	4.3×10^4 (1.0)	5.0×10^5 (12)
31 (TNS)		9.6×10^4 (2.7)	3.5×10^4 (1.0)	—
49		1.5×10^3 (0.39)	3.8×10^3 (1.0)	5.3×10^4 (14)
50		1.9×10^4 (6.4)	2.9×10^3 (1.0)	3.0×10^4 (10)
51		4.4×10^3 (0.041)	1.1×10^5 (1.0)	1.4×10^6 (13)
52		1.8×10^5 (5.5)	3.3×10^4 (1.0)	3.2×10^5 (9.6)

[a] Conditions: $25.0 \pm 0.1°C$, pH 2.0. The hosts are in a protonated form.

[b] Numbers in parentheses indicate relative stability of the complexes of CP44 or **44** compared with those of CP56, that is, K_s(CP44 or **44**)/K_s(CP56), for each guest.

[c] See Supplement 3 in Addendum for **32, 22,** and **31.**

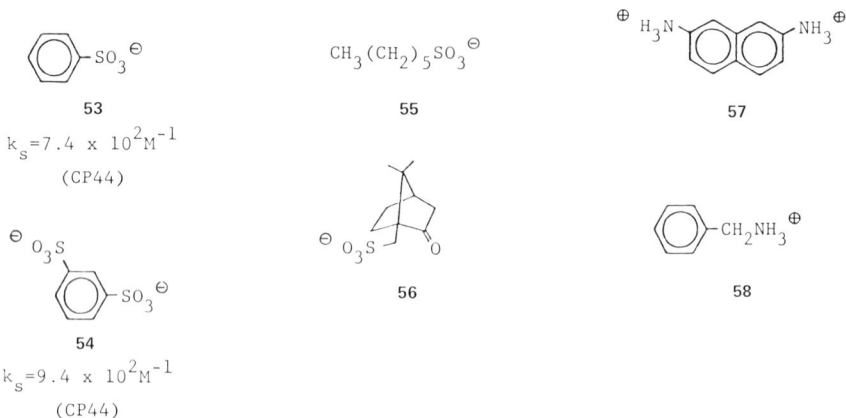

53

$k_s = 7.4 \times 10^2 M^{-1}$

(CP44)

55

57

54

$k_s = 9.4 \times 10^2 M^{-1}$

(CP44)

56

58

cavity having rectangularly shaped open ends (~3.5 × 7.9 Å), the shorter width of which is very close to the thickness of an aromatic ring. This close fit may be the primary reason for the aromatic selectivity described in point 1. Further examination has revealed that this selectivity effects remarkable discrimination between a benzene ring and a cyclohexane ring that are similar to each other in dimension except for thickness.[121a]* This kind of selectivity is not observed in cyclodextrins, which form relatively strong complexes ($K_s > 10^2 M^{-1}$) with the entire range of organic guests.[15]

In addition, the above-described studies based on the ^{1}H-nmr spectra (Section V,A) and the cavity modification (Section V,C) suggest that the small cavity of CP44 includes a naphthalene ring in the pseudoaxial geometry (Fig. 18a), whereas the large cavities of CP56 and **44** include this guest in the equatorial geometry (Fig. 18b). Consequently, steric hindrance may cause CP44 to prefer β-substituted naphthalenes, and CP56 and **44** to prefer α-substituted naphthalenes. This may be the major reason for the α and β selectivities described in point 2.

The cavity modification study also shows that complex formation by host **44** is an order of magnitude stronger than that by CP56. The same tendency is observed in complex formation with all the guests examined, as described in point 3. Considering that the open ends of the cavities of both hosts would be of similar size, the stronger complex formation by host **44** can be explained by the larger hydrophobic area of the cavity that results in a better fit with the guests. Compared with the cavity of **44**, which has cyclohexane rings, the cavities of CP44 and CP56 may be too shallow for the inclusion of an aromatic ring because of the bridging methylene chains.

---

* See Supplement 2 in Addendum.

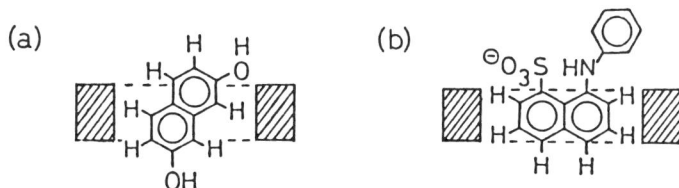

Fig. 18. Presumed inclusion geometries: (a) "Pseudoaxial" inclusion by CP44 and (b) "equatorial" inclusion by CP56 or **44**.

Thus, all the observations in points 1–3 are consistent with the results of the comprehensive study based on crystallographic and spectroscopic methods.[115–117] Therefore, it is reasonable to conclude that in this type of paracyclophane the fit of steric structure between the host and guest is one of the important factors for strong complex formation. In contrast, the observations described in point 4 clearly indicate that the *electrostatic interaction* (with hosts that are positively charged under acidic conditions) is another important factor in strong complex formation. Although electrostatic interaction is generally weakened in water, it is frequently an important binding force in the hydrophobic microenvironment provided by cyclophanes.[93,107,112,113,119] Furthermore, either factor alone is not sufficient, as evidenced by the very weak complex formation of CP44 with **55** and **56** (anionic but nonaromatic) and **57** and **58** (aromatic but cationic); for strong complex formation both these conditions must be satisfied. Thus, substrate-selective inclusion based on the host–guest recognition of *steric structure* and *charge* (multiple recognition[24,25,126,134]) is successfully effected by this type of paracyclophane in (acidic) water. Charge-transfer interaction may also be an important factor for complex formation by aromatic hosts such as these. However, evidence strongly supporting its participation has not yet been obtained[121a] (see Murakami *et al.*[110]).

The systematic investigation just described was the first example of predictable substrate-selective inclusion in water effected by totally synthetic water-soluble hosts. It is also important to note that the recognition of steric structure by these hosts is remarkably strict. For example, there are some cases of complex formation by CP44 in which the better fit of steric structure predominates over the intrinsically more favorable electrostatic interaction (**32** versus **53–56**; **50** versus **51** and **54**). Consequently, this investigation may provide a preliminary basis for the development of an effective system for strict discrimination between organic guests having slightly different steric and/or electronic structure.

Discrimination between α- and β-substituted naphthalenes was also observed by Tabushi and co-workers[95,96] in the hydrolysis of aromatic esters in the presence of **29**. The reaction is catalytic and proceeds according to Eq. (5):

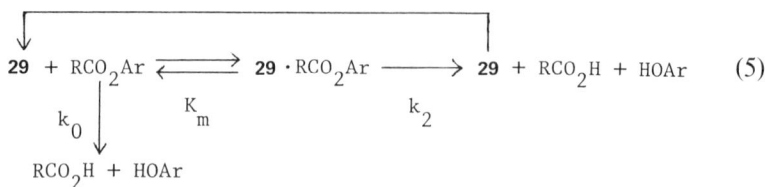

$$\textbf{29} + RCO_2Ar \rightleftarrows \textbf{29} \cdot RCO_2Ar \longrightarrow \textbf{29} + RCO_2H + HOAr \quad (5)$$

$$k_0 \downarrow \qquad K_m \qquad\qquad k_2$$

$$RCO_2H + HOAr$$

For all three substrates examined, appreciable rate enhancement is observed with the kinetic parameters, as shown in Table VIII. On the basis of a detailed study this rate enhancement was ascribed to the stabilization of the negatively charged transition state by the quaternary ammonium residue of the host (see Fig. 14a). In addition, the rate enhancement is larger for the β-naphthyl substrate **59** than for the α-naphthyl substrate **60**. It is apparent from Table VIII that this β selectivity is due to the difference in the acceleration of the intracomplex reaction (k_2/k_0) and not in the stability of the complex (approximated by $1/K_m$). Despite the stronger complexation of **60** than of **59** by the host **29**, the subsequent intracomplex hydrolysis occurs 5.4 times faster for **59** than for **60**. This difference may be due to the better spatial arrangement of the quaternary ammonium residue of the host and the carbonyl group of the guest within the inclusion complex. Thus, water-soluble cyclophanes effect substrate selectivity not only in the inclusion process but also in the intracomplex reaction process.

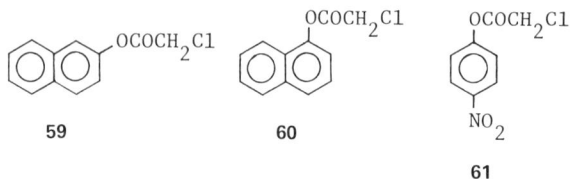

59   **60**   **61**

TABLE VIII

Kinetic Parameters for the Hydrolysis of the Aromatic Esters Catalyzed by 29[a]

Substrate	$k_0 \times 10^3$ (sec^{-1})	$k_2 \times 10^3$ (sec^{-1})	$K_m \times 10^3$ (M)	k_2/k_0
59 (β)	0.77 ± 0.01	19.2 ± 2.4	0.54 ± 0.05	25 ± 3
60 (α)	0.82 ± 0.01	4.9 ± 0.3	0.18 ± 0.02	6.0 ± 0.5
61	5.54	14.6	0.51	2.6

[a] Conditions: 20.2 ± 0.2°C, pH 8.10 (phosphate buffer).

Two major benefits, large rate enhancement and strict selectivity, can be afforded by complex formation with hosts having well-defined structure. Of these, the latter is considered to be more important for many practical purposes. From this viewpoint further efforts to effect regioselectivity[135] and stereoselectivity[136] as well will be especially important.

VI. PROSPECTS

Although host–guest complexation chemistry of cyclophanes is still in the initial stage of development, the capacity of cyclophane macrocycles to form inclusion cavities of well-defined structure and to form 1 : 1 inclusion complexes of particular geometry has been well confirmed, primarily by X-ray crystallography and nmr spectroscopy. Thus, a reliable basis has been firmly established for the application of this class of compounds as inclusion hosts. As described in Section II, cyclophanes may constitute a novel class of inclusion hosts both in organic solvents and in water. Because these are totally artificial hosts that can be designed and synthesized arbitrarily, it may become possible to study the details of host–guest interactions systematically and to arbitrarily develop hosts that selectively form complexes with guests of particular interest.

The most significant feature to be expected in the use of cyclophanes is selectivity based on the well-defined structure of these compounds. As described in Section V,C selectivity can be effected via two major processes: substrate inclusion and intracomplex reaction. Selectivity in the substrate inclusion process will be important for the use of these hosts in the physical separation of guests. One of the most fascinating applications in this area will be the utilization of this system in affinity chromatography for the separation of important biological molecules. Selectivity in the intracomplex reaction process will be important for selective organic reactions. The introduction of reactive groups at definite and appropriate positions (and orientations) is necessary for this purpose. As discussed in Section II this type of system will be most promising in reactions in which simple molecular or ionic reagents are not effective in attaining high selectivity. Because the reactions via host–guest complexation will be governed by the geometry of the whole complex, they should offer more varied possibilities for solving the problem of selectivity and therefore should effectively complement the reactions of simple artificial systems and natural enzymes. For application in synthetic reactions, it is important to develop hosts possessing dual binding sites in order to assemble selectively two organic substrates that then react with each other. Further understanding of the process and greater sophistication of the system will surely make it possible to imitate and generalize not only the style but also the selective results of biological reactions.

TABLE IX

Stability Constants K_s of the 1:1 Host–Guest Complexes in Aqueous Solutions

Host	Guest	K_s (M^{-1})	Reference
63[a]	22	5.9×10^2	61a
64	31	4.3×10^3	139
64	32	1.2×10^3	139

[a] Protonated form in an acidic condition (pH 1.5).

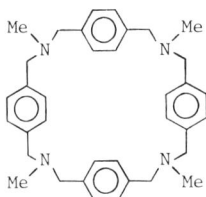

7c

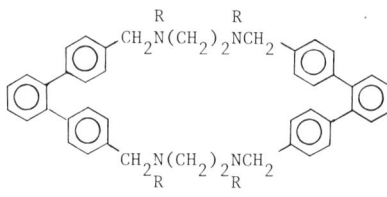

62 R = CH_2Ph

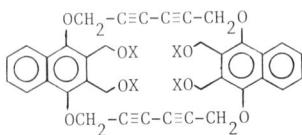

63 X = $COCH_2N(CH_3)_2$

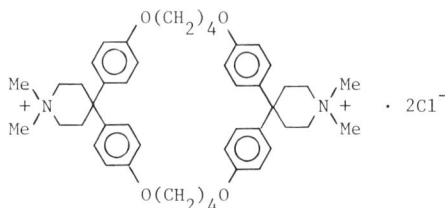

64

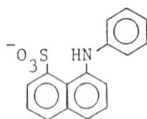

22 ANS

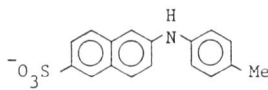

31 TNS

32

ADDENDUM

Supplement 1: The X-ray study of Barrett, Williams, and co-workers have elucidated the crystal structure of the 1 : 1 complex between **7c** and dioxane.[137] In the complex, the host molecule adopts the "dished" conformation and forms an intramolecular cavity, in which the guest molecule is included.

Supplement 2: Vögtle and Müller have shown that recrystallization of **62** from a 1 : 1 benzene/cyclohexane mixture yields only the *cyclohexane* complex.[138] This result is in striking contrast to the complexation by **30** and **44,** in which benzene is preferred to cyclohexane (Section V,C).[121a]

Supplement 3: There are some additional examples of stable 1 : 1 complexes between water-soluble paracyclophanes and hydrophobic guests. The stability constants (K_s) are shown in Table IX.

REFERENCES

1. B. H. Smith, "Bridged Aromatic Compounds." Academic Press, New York, 1964.
2. D. J. Cram and J. M. Cram, *Acc. Chem. Res.* **4,** 204–213 (1971).
3. V. Boekelheide, *Acc. Chem. Res.* **13,** 65–70 (1980).
4. D. J. Cram and J. Abell, *J. Am. Chem. Soc.* **77,** 1179–1186 (1955).
5. W. P. Jencks, "Catalysis in Chemistry and Enzymology," McGraw-Hill, New York, 1969.
6. R. Breslow, *Chem. Soc. Rev.* **1,** 553–580 (1972).
7. R. Breslow, *Adv. Chem. Ser.* **191,** 1–15 (1980).
8. H. Dugas and C. Penney, "Bioorganic Chemistry," Chapter 5. Springer-Verlag, Berlin and New York, 1981.
9. H. C. Kiefer, W. I. Congdon, I. S. Scarpa, and I. M. Klotz, *Proc. Natl. Acad. Sci. U.S.A.* **69,** 2155–2159 (1972). The term *synzyme* is introduced in this reference.
10. For these terms, see D. J. Cram and J. M. Cram, *Science* **183,** 803–809 (1974).
10a. D. J. Cram and J. M. Cram, *Acc. Chem. Res.* **11,** 8–14 (1978).
11. See, for example, G. E. Schulz and R. H. Schirmer, "Principles of Protein Structure," Chapters 10 and 11. Springer-Verlag, Berlin and New York, 1979.
12. R. Breslow, *Acc. Chem. Res.* **13,** 170–177 (1980).
13. R. Breslow, R. Rajagopalan, and J. Schwarz, *J. Am. Chem. Soc.* **103,** 2905–2907 (1981).
14. T. Endo, A. Kuwahara, H. Tasai, T. Murata, M. Hashimoto, and T. Ishigami, *Nature (London)* **268,** 74–76 (1977).
15. M. L. Bender and M. Komiyama, "Cyclodextrin Chemistry." Springer-Verlag, Berlin and New York, 1978.
16. R. Hershfield and M. L. Bender, *J. Am. Chem. Soc.* **94,** 1376–1377 (1972).
17. Y. A. Ovchinnikov and V. T. Ivanov, *in* "The Proteins, Vol. 5" (H. Neurath and R. L. Hill, eds.), Chap. 3. Academic Press, New York, 1982.

18. C. J. Pedersen, *J. Am. Chem. Soc.* **89**, 7017–7035 (1967).
19. J. M. Lehn, *Acc. Chem. Res.* **11**, 49–57 (1978).
20. R. M. Izatt and J. J. Christensen, eds., "Synthetic Multidentate Macrocyclic Compounds." Academic Press, New York, 1978.
21. S. Patai (ed.) "The Chemistry of Ethers, Crown Ethers, Hydroxyl Groups and Their Sulphur Analogues, Part 1: The Chemistry of Functional Groups, Supplement E," Chapters 1–4. Wiley, New York, 1980.
21a. F. Vögtle, (ed.) "Host–Guest Complex Chemistry (I, II): Topics in Current Chemistry **98** and **101**." Springer-Verlag, Berlin, 1981/1982.
22. J.-P. Behr, J.-M. Lehn, and P. Vierling, *Helv. Chim. Acta* **65**, 1853–1867 (1982); I. Tabushi, Y. Kobuke, K. Ando, M. Kishimoto, and E. Ohara, *J. Am. Chem. Soc.* **102**, 5947–5948 (1980).
23. F. P. Schmidtchen, *Chem. Ber.* **114**, 597–607 (1981).
24. E. Kimura, M. Kodama, and T. Yatsunami, *J. Am. Chem. Soc.* **104**, 3182–3187 (1982).
25. M. W. Hosseini and J. M. Lehn, *J. Am. Chem. Soc.* **104**, 3525–3527 (1982).
26. F. Vögtle and E. Weber, *Angew. Chem., Int. Ed. Engl.* **18**, 753–776 (1979).
27. M. Güggi, M. Oehme, E. Pretsch, and W. Simon, *Helv. Chim. Acta* **59**, 2417–2420 (1976).
27a. D. D. MacNicol and D. R. Wilson, *Chem. Ind. (London)* pp. 84–85 (1977).
28. H. Stetter and E.-E. Roos, *Chem. Ber.* **88**, 1390–1395 (1955).
29. G. Faust and M. Pallas, *J. Prakt. Chem.* **11**, 146–152 (1960).
30. G. Wittig, P. Börzel, F. Neumann, and G. Klar, *Justus Liebigs Ann. Chem.* **691**, 109–125 (1966).
30a. T. Inazu and T. Yoshino, *Bull. Chem. Soc. Jpn.* **41**, 647–652 (1968).
31. J. Nishikido, T. Inazu, and T. Yoshino, *Bull. Chem. Soc. Jpn.* **46**, 263–265 (1971).
32. R. Nagano, J. Nishikido, T. Inazu, and T. Yoshino, *Bull. Chem. Soc. Jpn.* **46**, 653–655 (1973).
33. Y. Urushigawa, T. Inazu, and T. Yoshino, *Bull. Chem. Soc. Jpn.* **44**, 2546–2547 (1971).
34. For a recent review, see D. D. MacNicol, J. J. McKendrick, and D. R. Wilson, *Chem. Soc. Rev.* **7**, 65–87 (1978).
35. R. Anschütz, *Ber. Dtsch. Chem. Ges.* **25**, 3512–3513 (1892); *Justus Liebigs Ann. Chem.* **273**, 73–93, 94–96 (1893).
36. W. Baker, W. D. Ollis, and T. S. Zealley, *J. Chem. Soc.* pp. 201–208 (1951); W. Baker, B. Gilbert, W. D. Ollis, and T. S. Zealley, *ibid.* pp. 209–213.
36a. W. Baker, B. Gilbert, and W. D. Ollis, *J. Chem. Soc.* pp. 1443–1446 (1952).
37. G. M. Robinson, *J. Chem. Soc.* **107**, 267–276 (1915).
38. V. K. Bhagwat, D. K. Moore, and F. L. Pyman, *J. Chem. Soc.* p. 443 (1931).
39. A. Oliverio and C. Casinovi, *Ann. Chim. (Rome)* **42**, 168–184 (1952).
40. A. S. Lindsey, *Chem. Ind. (London)* pp. 823–824 (1963); *J. Chem. Soc.* pp. 1685–1692 (1965).
41. H. Erdtman, F. Haglid, and R. Ryhage, *Acta Chem. Scand.* **18**, 1249–1254 (1964).
42. A. Goldup, A. B. Morrison, and G. W. Smith, *J. Chem. Soc.* pp. 3864–3865 (1965).
43. C. Casinovi and A. Oliverio, *Ann. Chim. (Rome)* **46**, 929–933 (1956).
44. V. Caglioti, A. M. Liquori, N. Gallo, E. Giglio, and M. Scrocco, *J. Inorg. Nucl. Chem.* **8**, 572–576 (1958).
45. A. P. Downing, W. D. Ollis, and I. O. Sutherland, *J. Chem. Soc. B* pp. 24–34 (1970); W. D. Ollis, J. F. Stoddart, and I. O. Sutherland, *Tetrahedron* **30**, 1903–1921 (1974).
46. B. Miller and B. D. Gesner, *Tetrahedron Lett.* pp. 3351–3354 (1965).
47. W. Baker, J. B. Harborne, A. J. Price, and A. Rutt, *J. Chem. Soc.* pp. 2042–2046 (1954).

48. J. F. Manville and G. E. Troughton, *J. Org. Chem.* **38**, 4278–4281 (1973).
49. T. Inazu and T. Yoshino, *Bull. Chem. Soc. Jpn.* **41**, 652–655 (1968).
50. I. Tabushi, H. Yamada, Z. Yoshida, and R. Oda, *Tetrahedron* **27**, 4845–4853 (1971).
51. I. Tabushi, H. Yamada, and Y. Kuroda, *J. Org. Chem.* **40**, 1946–1949 (1975).
52. E. A. Truesdale, *Tetrahedron Lett.* pp. 3777–3780 (1978).
53. H. Kondo, H. Okamoto, J. Kikuchi, and J. Sunamoto, *J. Chem. Soc., Perkin Trans. 1* pp. 3125–3128 (1981).
54. H. Hope, J. Bernstein, and K. N. Trueblood, *Acta Crystallogr., Sect. B* **B28**, 1733–1743 (1972).
55. T. Kawato, T. Inazu, and T. Yoshino, *Bull. Chem. Soc. Jpn.* **44**, 200–203 (1971).
56. I. Tabushi, H. Yamada, K. Matsushita, Z. Yoshida, H. Kuroda, and R. Oda, *Tetrahedron* **28**, 3381–3388 (1972).
57. I. Tabushi and H. Yamada, *Tetrahedron* **33**, 1101–1104 (1977).
58. L. Rosa and F. Vögtle, *Justus Liebigs Ann. Chem.* pp. 459–466 (1981).
59. E. T. Jarvi and H. W. Whitlock, Jr., *J. Am. Chem. Soc.* **102**, 657–662 (1980).
60. B. J. Whitlock, E. T. Jarvi, and H. W. Whitlock, *J. Org. Chem.* **46**, 1832–1835 (1981).
61. S. P. Adams and H. W. Whitlock, Jr., *J. Org. Chem.* **46**, 3474–3478 (1981).
61a. S. P. Adams and H. W. Whitlock, *J. Am. Chem. Soc.* **104**, 1602–1611 (1982); A. B. Brown, K. J. Haller, and H. W. Whitlock, Jr., *Tetrahedron Lett.* **23**, 3311–3314 (1982).
62. C. W. Schimelpfenig and R. R. Ford, *J. Org. Chem.* **46**, 1210–1212 (1981).
63. B. Thulin, O. Wennerström, and H.-E. Högberg, *Acta Chem. Scand., Sect. B* **B29**, 138–139 (1975); B. Thulin, O. Wennerström, I. Somfai, and B. Chmielarz, *ibid.* **31**, 135–140 (1977); H.-E. Högberg, B. Thulin, and O. Wennerström, *Tetrahedron Lett.* pp. 931–934 (1977); D. Tanner, B. Thulin, and O. Wennerström, *Acta Chem. Scand., Sect. B* **B33**, 464–465 (1979).
64. D. Tanner, B. Thulin, and O. Wennerström, *Acta Chem. Scand., Sect. B* **B33**, 443–448 (1979).
65. J.-L. Pierre, P. Baret, P. Chautemps, and M. Armand, *J. Am. Chem. Soc.* **103**, 2986–2988 (1981).
66. F. Vögtle and R. G. Lichtenthaler, *Angew. Chem., Int. Ed. Engl.* **11**, 535–536 (1972).
67. F. Bottino, S. Foti, and S. Pappalardo, *Tetrahedron* **32**, 2567–2570 (1976).
68. C. Sergheraert, P. Marcincal, and E. Cuingnet, *Tetrahedron Lett.* pp. 2879–2880 (1977).
69. C. Sergheraert, P. Marcincal, and E. Cuingnet, *Tetrahedron Lett.* pp. 4785–4786 (1978).
70. D. Lawton and H. M. Powell, *J. Chem. Soc.* pp. 2339–2357 (1958).
71. D. J. Williams and D. Lawton, *Tetrahedron Lett.* pp. 111–114 (1975).
72. S. Brunie, A. Navaza, G. Tsoucaris, J. P. Declercq, and G. Germain, *Acta Crystallogr., Sect. B* **B33**, 2645–2647 (1977).
73. R. Gerdil and J. Allemand, *Tetrahedron Lett.* pp. 3499–3502 (1979).
74. J. Allemand and R. Gerdil, *Cryst. Struct. Commun.* **10**, 33–40 (1981).
75. R. Arad-Yellin, S. Brunie, B. S. Green, M. Knossow, and G. Tsoucaris, *J. Am. Chem. Soc.* **101**, 7529–7537 (1979).
76. R. Arad-Yellin, B. S. Green, M. Knossow, and G. Tsoucaris, *Tetrahedron Lett.* **21**, 387–390 (1980).
77. H. M. Powell, *Nature (London)* **170**, 155 (1952); A. C. D. Newman and H. M. Powell, *J. Chem. Soc.* pp. 3747–3751 (1952).
78. R. Arad-Yellin and B. S. Green, *J. Am. Chem. Soc.* **102**, 1157–1158 (1980).
79. R. Gerdil and J. Allemand, *Helv. Chim. Acta* **63**, 1750–1753 (1980).
80. G. B. Guise, W. D. Ollis, J. A. Peacock, J. S. Stephanatou, and J. F. Stoddart,

J. Chem. Soc., Perkin Trans. 1, 1637–1648 (1982); A. Hoorfar, W. D. Ollis, J. A. Price, J. S. Stephanatou, and J. F. Stoddart, *J. Chem. Soc., Perkin Trans. 1,* 1649–1699 (1982); W. D. Ollis, J. S. Stephanatou, and J. F. Stoddart, *J. Chem. Soc., Perkin Trans. 1,* 1715–1720 (1982).

81. S. J. Edge, W. D. Ollis, J. S. Stephanatou, and J. F. Stoddart, *J. Chem. Soc., Perkin Trans. 1,* 1701–1714 (1982).

82. S. Cerrini, E. Giglio, F. Mazza, and N. V. Pavel, *Acta Crystallogr., Sect. B* **B35,** 2605–2609 (1979).

83. J. A. Hyatt, E. N. Duesler, D. Y. Curtin, and I. C. Paul, *J. Org. Chem.* **45,** 5074–5079 (1980).

84. J. Gabard and A. Collet, *J. Chem. Soc., Chem. Commun.* pp. 1137–1139 (1981).

85. See also F. Vögtle, H. Sieger, and W. M. Müller, *in* "Host–Guest Complex Chemistry (I): Topics in Current Chemistry **98.**" Springer-Verlag, Berlin, 1981.

86. C. D. Gutsche and R. Muthukrishnan, *J. Org. Chem.* **43,** 4905–4906 (1978); C. D. Gutsche, R. Muthukrishnan, and K. H. No, *Tetrahedron Lett.* pp. 2213–2216 (1979); R. Muthukrishnan and C. D. Gutsche, *J. Org. Chem.* **44,** 3962–3964 (1979).

87. C. D. Gutsche, B. Dhawan, K. H. No, and R. Muthukrishnan, *J. Am. Chem. Soc.* **103,** 3782–3792 (1981); C. D. Gutsche and L. J. Bauer, *Tetrahedron Lett.* **22,** 4763–4766 (1981); C. D. Gutsche and J. A. Levine, *J. Am. Chem. Soc.* **104,** 2652–2653 (1982); C. D. Gutsche and K. H. No, *J. Org. Chem.* **47,** 2708–2712 (1982); K. H. No and C. D. Gutsche, *J. Org. Chem.* **47,** 2713–2719 (1982).

88. G. D. Andreetti, R. Ungaro, and A. Pochini, *J. Chem. Soc., Chem. Commun.* pp. 1005–1007 (1979).

88a. G. D. Andreetti, R. Ungaro, and A. Pochini, *J. Chem. Soc., Chem. Commun.* pp. 533–534 (1981).

89. See *Chem. Abstr.* **87,** 134101p (1977); **88,** 121459g (1978); **89,** 6017v, 42589s, 42730f, 197137s (1978); **90,** 55144p (1979).

90. A. Itai, Y. Tanaka, and Y. Iitaka, *Am. Crystallogr. Assoc. Winter Meet., 1979* Abstract PA32 (1979).

91. Y. Murakami, Y. Aoyama, and J. Kikuchi, *J. Chem. Soc., Chem. Commun.* pp. 444–446 (1981).

92. I. Tabushi and Y. Kuroda, *Shokubai* **16,** 78–80 (1974).

93. I. Tabushi, Y. Kuroda, and Y. Kimura, *Tetrahedron Lett.* pp. 3327–3330 (1976).

94. I. Tabushi, H. Sasaki, and Y. Kuroda, *J. Am. Chem. Soc.* **98,** 5727–5728 (1976).

95. I. Tabushi, Y. Kimura, and K. Yamamura, *J. Am. Chem. Soc.* **100,** 1304–1306 (1978).

96. I. Tabushi, Y. Kimura, and K. Yamamura, *J. Am. Chem. Soc.* **103,** 6486–6492 (1981).

97. Y. Murakami, J. Sunamoto, and K. Kano, *Chem. Lett.* pp. 223–226 (1973).

98. Y. Murakami, J. Sunamoto, and K. Kano, *Bull. Chem. Soc. Jpn.* **47,** 1238–1244 (1974).

99. Y. Murakami, J. Sunamoto, H. Okamoto, and K. Kawanami, *Bull. Chem. Soc. Jpn.* **48,** 1537–1544 (1975).

100. J. Sunamoto, H. Okamoto, H. Kondo, and Y. Murakami, *Tetrahedron Lett.* pp. 2761–2764 (1975).

101. Y. Murakami, Y. Aoyama, K. Ohno, K. Dobashi, and T. Nakagawa, *J. Chem. Soc., Perkin Trans. 1* pp. 1320–1326 (1976).

102. J. Sunamoto, H. Kondo, H. Okamoto, and K. Taira, *Bioorg. Chem.* **6,** 95–102 (1977).

103. J. Sunamoto, H. Kondo, H. Okamoto, and Y. Murakami, *Tetrahedron Lett.* pp. 1329–1332 (1977).

104. Y. Murakami, J. Sunamoto, H. Kondo, and H. Okamoto, *Bull. Chem. Soc. Jpn.* **50,** 2420–2427 (1977).

105. Y. Murakami, Y. Aoyama, and K. Dobashi, *J. Chem. Soc., Perkin Trans.* 2, 24–32 (1977).
106. Y. Murakami, Y. Aoyama, and K. Dobashi, *J. Chem. Soc., Perkin Trans.* 2, 32–38 (1977).
107. Y. Murakami, Y. Aoyama, K. Dobashi, and M. Kida, *Bull. Chem. Soc. Jpn.* **49,** 3633–3636 (1976).
108. Y. Murakami, Y. Aoyama, M. Kida, and A. Nakano, *Bull. Chem. Soc. Jpn.* **50,** 3365–3371 (1977).
109. Y. Murakami, Y. Aoyama, M. Kida, and J. Kikuchi, *J. Chem. Soc., Chem. Commun.* pp. 494–496 (1978).
110. Y. Murakami, Y. Aoyama, M. Kida, A. Nakano, K. Dobashi, C. D. Tran, and Y. Matsuda, *J. Chem. Soc., Perkin Trans.* 1 pp. 1560–1567 (1979).
111. Y. Murakami, Y. Aoyama, and M. Kida, *J. Chem. Soc., Perkin Trans.* 2 pp. 1665–1671 (1980).
112. Y. Murakami, A. Nakano, R. Miyata, and Y. Matsuda, *J. Chem. Soc., Perkin Trans.* 1 pp. 1669–1676 (1979).
113. Y. Murakami, A. Nakano, K. Akiyoshi, and K. Fukuya, *J. Chem. Soc., Perkin Trans.* 1 pp. 2800–2808 (1981).
114. Y. Murakami, Y. Aoyama, and J. Kikuchi, *J. Chem. Soc., Perkin Trans.* 1 pp. 2809–2815 (1981); Y. Murakami, Y. Aoyama, J. Kikuchi, and K. Nishida, *J. Am. Chem. Soc.* **104,** 5189–5197 (1982); Y. Murakami, Y. Aoyama, and J. Kikuchi, *Bull. Chem. Soc. Jpn.* **55,** 2898–2901 (1982).
115. K. Odashima, A. Itai, Y. Iitaka, and K. Koga, *J. Am. Chem. Soc.* **102,** 2504–2505 (1980).
116. K. Odashima, A. Itai, Y. Iitaka, Y. Arata, and K. Koga, *Tetrahedron Lett.* **21,** 4347–4350 (1980).
117. T. Soga, K. Odashima, and K. Koga, *Tetrahedron Lett.* **21,** 4351–4354 (1980).
118. K. Odashima, T. Soga, and K. Koga, *Tetrahedron Lett.* **22,** 5311–5314 (1981).
119. K. Odashima and K. Koga, *Heterocycles* **15,** 1151–1154 (1981).
120. K. Odashima, A. Itai, Y. Iitaka, A. Watanabe, and K. Koga, unpublished results.
121. A. Itai, Y. Ikeda, A. Watanabe, K. Odashima, K. Koga, and Y. Iitaka, *Int. Congr. Crystallogr., 12th, 1981* Abstract C-91 (1981).
121a. K. Odashima, T. Soga, K. Matsuo, K. Mori, Y. Sasaoka, and K. Koga, unpublished results.
122. D. S. Kemp, M. E. Garst, R. W. Harper, D. D. Cox, D. Carlson, and S. Denmark, *J. Org. Chem.* **44,** 4469–4473 (1979).
123. F. M. Menger, M. Takeshita, and J. F. Chow, *J. Am. Chem. Soc.* **103,** 5938–5939 (1981).
124. L. Brand and J. R. Gohlke, *Annu. Rev. Biochem.* **41,** 843–868 (1972); F. Cramer, W. Saenger, and H.-C. Spatz, *J. Am. Chem. Soc.* **89,** 14–20 (1967).
125. H. Kondo, H. Nakatani, and K. Hiromi, *J. Biochem.* (*Tokyo*) **79,** 393–405 (1976).
126. I. Tabushi, N. Shimizu, T. Sugimoto, M. Shiozuka, and K. Yamamura, *J. Am. Chem. Soc.* **99,** 7100–7102 (1977).
127. J. P. Behr and J. M. Lehn, *J. Am. Chem. Soc.* **98,** 1743–1747 (1976); R. J. Bergeron and M. A. Channing, *ibid.* **101,** 2511–2516 (1979); J.-P. Kintzinger, F. Kotzyba-Hibert, J.-M. Lehn, A. Pagelot, and K. Saigo, *J. Chem. Soc., Chem. Commun.* pp. 833–836 (1981).
128. Y. Murakami, Y. Aoyama, and M. Kida, *J. Chem. Soc., Perkin Trans.* 2 pp. 1947–1952 (1977).

129. C. Hansch, *J. Org. Chem.* **43**, 4889–4890 (1978).
130. J. A. Hyatt, *J. Org. Chem.* **43**, 1808–1811 (1978).
131. D. J. Cram and M. F. Antar, *J. Am. Chem. Soc.* **80**, 3103–3109 (1958).
132. D. J. Cram and L. A. Singer, *J. Am. Chem. Soc.* **85**, 1084–1088 (1963).
133. J. C. Barnes, J. D. Paton, J. R. Damewood, Jr., and K. Mislow, *J. Org. Chem.* **46**, 4975–4979 (1981).
134. Y. Matsui and A. Okimoto, *Bull. Chem. Soc. Jpn.* **51**, 3030–3034 (1978); J. Boger and J. R. Knowles, *J. Am. Chem. Soc.* **101**, 7631–7633 (1979).
135. R. Breslow and P. Campbell, *Bioorg. Chem.* **1**, 140–156 (1971); I. Tabushi, K. Yama-mura, K. Fujita, and H. Kawakubo, *J. Am. Chem. Soc.* **101**, 1019–1026 (1979); R. Breslow, P. Bovy, and C. L. Hersh, *ibid.* **102**, 2115–2117 (1980); see also Breslow[6,12] and Breslow *et al.*[13]
136. Y. Chao, G. R. Weisman, G. D. Y. Sogah, and D. J. Cram, *J. Am. Chem. Soc.* **101**, 4948–4958 (1979); J.-M. Lehn and C. Sirlin, *J. Chem. Soc., Chem. Commun.* pp. 949–951 (1978); D. J. Cram and G. D. Y. Sogah, *ibid.* pp. 625–628 (1981); R. Breslow, M. Hammond, and M. Lauer, *J. Am. Chem. Soc.* **102**, 421–422 (1980); G. L. Trainor and R. Breslow, *ibid.* **103**, 154–158 (1981).
137. S. J. Abbott, A. G. M. Barrett, C. R. A. Godfrey, S. B. Kalindjian, G. W. Simpson, and D. J. Williams, *J. Chem. Soc., Chem. Commun.*, 796–797 (1982).
138. F. Vögtle and W. M. Müller, *Angew. Chem. Int. Ed. Engl.* **21**, 147–148 (1982).
139. F. Diederich and K. Dick, *Tetrahedron Lett.* **23**, 3167–3170 (1982).

CHAPTER **12**

Cyclophanes as Synthetic Analogs of Enzymes and Receptors

IAN SUTHERLAND

Department of Organic Chemistry
The University of Liverpool
Liverpool, England

I. INTRODUCTION

Although the relationship between the concept of cyclophanes, a product of the imagination of the synthetic organic chemist, and biological processes may seem rather remote, I endeavor to show in this chapter how cyclophanes are well worth consideration by chemists who wish to design and prepare synthetic analogs of enzymes and receptors.

The major role of enzymes in biological processes has long been appreciated, and the concept of biological receptors has been of increasing importance in the design of new drugs and the appreciation of structure–activity relationships. The initial step in each process is the formation of a

ISBN 0-12-403002-5

complex between a large protein molecule and the substrate (S), which in the case of an enzyme (E) leads to ultimate chemical modification of the substrate molecule (S') and, in the case of a receptor (R), to a biological response (Scheme 1). The formation of a complex between a large *host* molecule and a smaller *guest* molecule is also a feature of other biological systems, such as those of transport and immune response, and the biological activity of antibiotics, toxins, hormones, and neurotransmitters.

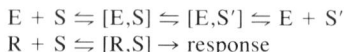

$$E + S \rightleftharpoons [E,S] \rightleftharpoons [E,S'] \rightleftharpoons E + S'$$
$$R + S \rightleftharpoons [R,S] \rightarrow response$$

SCHEME 1

The major binding forces responsible for complex formation in solution are (1) hydrogen bonding, (2) coulombic attraction, and (3) hydrophobic interactions. The first two attractive interactions are favored by solvents of low dielectric constant and are particularly important for complexes involving an anion or a cation as at least one of the components. The role of hydrophobic interactions in biological systems is well appreciated, and such interactions are of major importance in complexation by cyclodextrins[7,132] and some synthetic host molecules, particularly those of the cyclophane type discussed in Chapter 11. This chapter covers complex formation that is based primarily on the first two binding forces, which are not necessarily distinguished because recent successful models of hydrogen bonding have been based on electrostatic attractions and repulsions between an array of point changes.[91]

A synthetic host molecule having the capacity to bind guest molecules should meet a number of requirements. (1) It should be prepared by a simple route from readily available (inexpensive) starting materials; (2) it should show selectivity in complexation of guest molecules (recognition) on the basis of their constitution and configuration; (3) it should be chemically stable; (4) it should contain sites for the introduction of catalytic groups or additional recognition sites; and (5) eventual exploitation may require the possibility of attachment to polymers. The crown ethers, discovered through serendipity by Pedersen,[111,112] have excellent binding properties for cations and meet the first and third requirements. The development of crown ethers in pioneering work by Cram[24,25] and Lehn[81–84] has led to the synthesis of host molecules that satisfy, to some extent, all of the requirements.

Pedersen's work showed that simple crown ethers, such as 18-crown-6 **(1),** form crystalline 1 : 1 complexes with primary alkylammonium salts and salts of alkali metals. It has since been shown that complex formation with ammonium salts involves the hydrogen bonding shown in **2,** which is

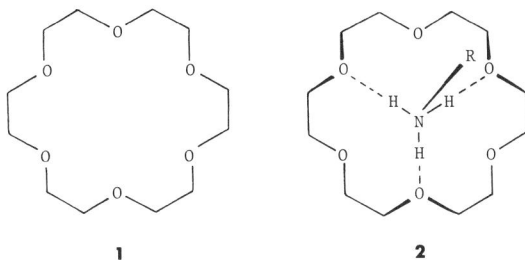

1   2

based on the crystal structures of a benzylammonium thiocyanate complex[10,134] and an ammonium bromide complex.[8,101] The conformation adopted by the macrocyclic host in these complexes is similar to that found in earlier studies of complexes with alkali metal cations.[39,120] The "face-to-face" relationship between guest and host in 2 is consistent with the ^{1}H-nmr spectrum of the complex in CD_2Cl_2 solution,[62] and the limited amount of intermolecular contact results in poor recognition of the structure of a guest alkylammonium cation by the unsubstituted 18-crown-6 host 1. Thus, the association constants K_a for a series of guest salts $R\overset{+}{N}H_3NCS^-$ in methanol at 25°C show relatively little sensitivity to the nature of the substituent R[62] (log K_a for R = H, 4.27; R = Me, 4.25; R = Et, 3.99; R = i-Pr, 3.56; R = t-Bu, 2.90). Increased recognition requires increased guest–host contact, and this has been achieved in a number of ways, including the introduction of substituents and other structural modification of the crown macrocycle. Modifications involving cyclophane systems are discussed in Section II of this chapter. Other aspects of crown ether chemistry have been the subject of a number of reviews.[13,24,31,61,82–84,116,124]

The use of metals as primary binding sites for simple guest molecules is amply exemplified biochemically, and both hemoglobin and myoglobin are well-known examples of such systems. Attempts to provide synthetic models for these hemoprotein systems have led to the synthesis of a variety of bridged porphyrins, and these are discussed in Section III with other examples of bridged metal chelates that form complexes reversibly with guest molecules.

Most biological host molecules are proteins rather than cyclophanes, but it is of interest that some of the simplest biological host molecules are based on a combination of cyclophane and peptide. Thus, antibiotics of the vancomycin group owe their biological activity to their capacity to form complexes with N-acyl derivatives of DAla-DAla, leading to the prevention of the biosynthesis of bacterial cell walls. The structure of vancomycin (3)[121,143,144] shows that the polycyclic cyclophane system included in the structure holds the binding sites in the polypeptide backbone

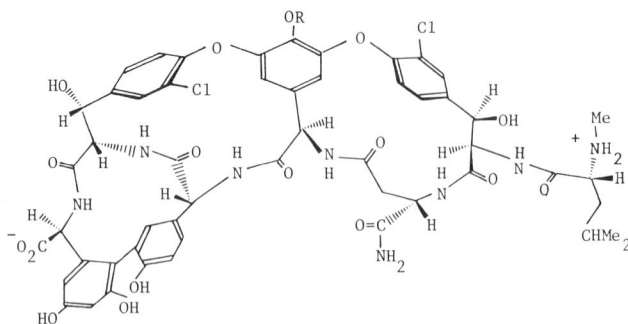

3 Vancomycin   R = Sugars

in positions that are appropriate for substrate binding. The related antibiotic avoparcin **(4)** uses a rather similar binding site,[43,93] but ristocetin **(5)** includes another cyclophane macrocycle in its structure, which presumably ensures a correct geometry for the terminal amino acid of the polypeptide backbone.[69,126,127] The details of the sugar residues are omitted from structures **3–5** because they do not appear to be involved as binding sites.

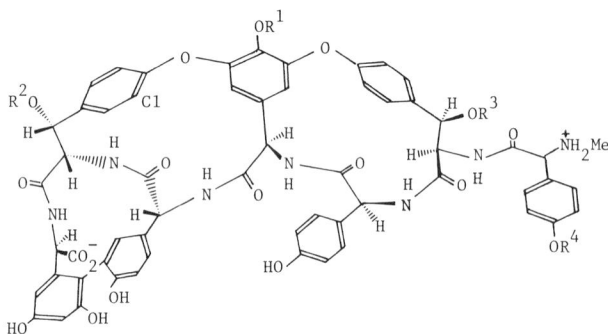

4 Avoparcin R^1, R^2, R^3, R^4 = Sugars

The antibiotics **3, 4,** and **5** may be regarded as having peptide binding sites that have their conformations controlled by cyclophane systems constructed from the side chains. These molecules of relatively low molecular weight bind dipeptide substrates selectively and strongly, even in polar solvents. Synthetic compounds of this type are not yet available, but their structures underline the advantages that can be obtained when rather flexible structures, such as small peptides, are converted to rigid structures of correct geometry and hence provide an approach to the

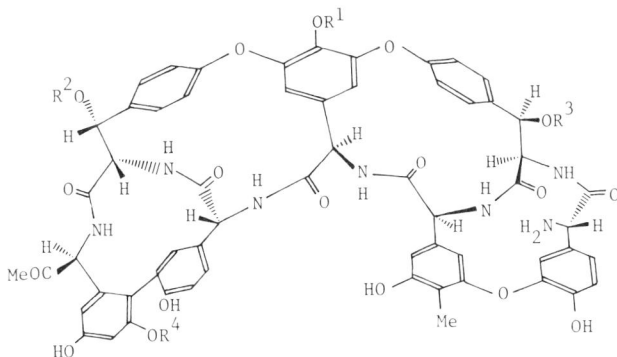

5 Ristocetin A R^1, R^2, R^3, R^4 = Sugars

incorporation of similar stereochemical control into flexible crown ether systems.

II. CROWN ETHER HOSTS BASED ON CYCLOPHANES

Cyclophanes have been used in a number of ways to control the stereochemistry of crown ether systems. Three general approaches involve (1) monocyclic systems in which the crown ether macrocycle is interrupted by a 1,3- or 1,4-disubstituted aromatic ring, (2) polycyclic systems based on two or more crown ether macrocycles held in fixed geometric relationships by a cyclophane framework, and (3) bridged systems in which the geometry of a crown ether system is controlled by a cyclophane bridge. The first device has been employed rather frequently, the second has received relatively little attention until rather recently, and the third has been used only rarely.

A. Monocyclic Systems

1. 1,3-Bridged Aromatic and Heteroaromatic Rings

The simple crown ether system (e.g., **1**) has been modified in a large number of cases by the replacement of one or more —CH_2OCH_2— units

by a 1,3-linked aromatic or heteroaromatic ring to give crown ether analogs of the types indicated by **6.** In some cases the aromatic ring provides
a binding site for a guest cation and in other cases it does not, but in all
cases, on the basis of crystal structures and the study of molecular
models, the macrocycle can adopt a conformation analogous to that commonly found in complexes of the corresponding crown ether.

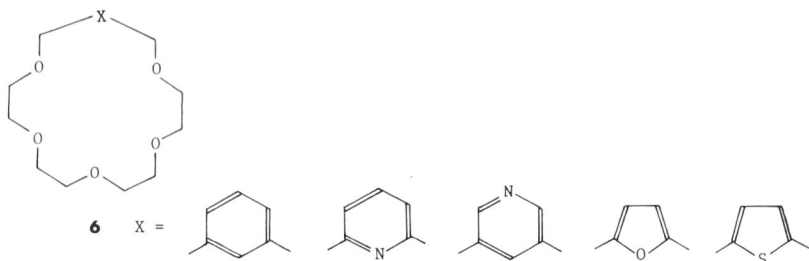

6   X =

a. 1,3-Bridged Benzene Rings

The metacyclophane **7a** was investigated as a host molecule at an early
stage in crown ether chemistry, and it was shown to be a less effective
host for primary alkylammonium cations than 18-crown-6 by a factor of
~500 in K_a as a consequence of the loss of one oxygen binding site.[130]
Subsequently, substituted systems with both *remote* **(7b)**[100] and *convergent* **(7c)**[16,102] functional groups were studied. The remote substituents
were shown to affect complexation with both alkali metal cations and
alkylammonium cations in a similar way, and a moderate correlation was
found between K_a for the substituted system **(7b)** as compared with K_a for
the unsubstituted system **(7a)** and the $\sigma_1{}^q$ values of Grob. The upfield shift

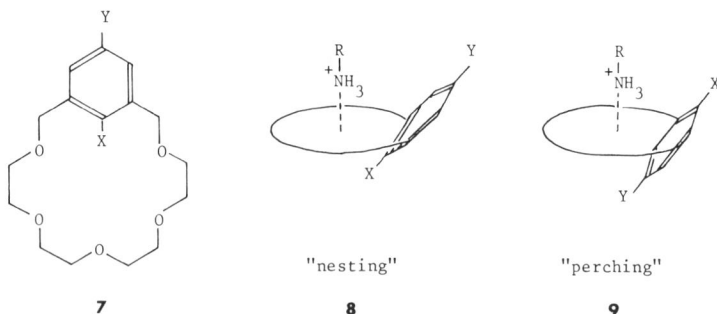

"nesting"   "perching"

7   **8**   **9**

a   X=Y=H

b   X=H, Y≠H

c   Y=H, X≠H

observed for the nmr signals of complexed *tert*-butylammonium thiocyanate as compared with the free guest suggested that these complexes select the *nesting* conformation **(8)** of the cyclophane host rather than the *perching* conformation **(9)**.

The effects of convergent functional groups were also examined, and the ester **7c** (X = CO$_2$Me) was found to be the best host for *tert*-butylammonium thiocyanate in CDCl$_3$ solution. Although the perching conformation **(9)** was found in a crystal structure[50] of the salt formed by the acid **7c** (X = CO$_2$H) with *tert*-butylamine, the high-field nmr signals of a guest *tert*-butylammonium cation in complexes of a number of different macrocycles **(7c)** suggest that the nesting conformation is important in solution. The nesting conformation has also been found[34,117] in the crystal structure of the *tert*-butylammonium thiocyanate complex of the host **7b** (Y = Br). This structure shows clearly how the 1,3-disubstituted benzene ring can be accommodated in the 18-membered ring without loss of the conformational characteristics of the macrocycle that are required for efficient complexation. The nmr spectra of 1 : 1 complexes of the macrocycle **10** and alkylammonium cations also show marked shifts to high field for the guest signals, which may be associated with a *double nesting* conformation of the host molecule **11**.[9] In this case a crystalline complex had a 2 : 1 guest/host ratio and was clearly distinct from the complex observable in solution by nmr spectroscopy.

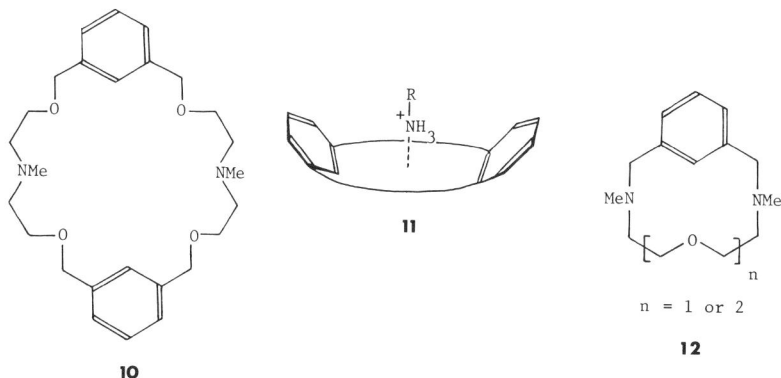

10   **11**   **12**

n = 1 or 2

This type of host molecule has been particularly suitable for studies of the kinetics of guest–host exchange, because the nmr chemical shifts of the macrocycle are dispersed by the presence of the aromatic ring. For example, it was shown in an early study[89,90] that the azacrown ethers **12** formed complexes with alkylammonium cations in CD$_2$Cl$_2$: the ^{1}H-nmr spectra of these complexes showed that face-to-face exchange of the guest cation became slow on the nmr time scale at low temperatures.

Thus, the indicated CH_2 groups in the complex, shown diagrammatically in **13a** and **13b,** were observable as an AB system at $-80°C$, which coalesced to a singlet above $-40°C$.

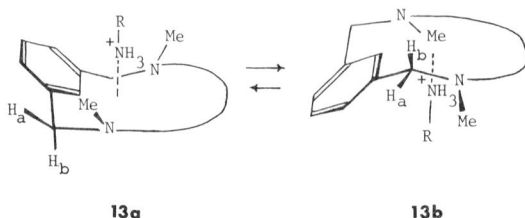

13a   **13b**

Similar temperature dependence of the nmr spectra of a number of complexes of this type was interpreted in terms of the process E + I (face-to-face exchange of guest cation and conformational inversion of host macrocycle) shown diagrammatically as **13a** ⇌ **13b.** The free energies of activation for this process were shown to vary with guest and host structure[59,60,90] and to be related to the free energy of association (ΔG_a) between the two components. This method and related[14,32–34,77–80,113] nmr techniques have been used to study a wide range of complexes and have provided useful information about the structures of complexes in solution and the binding energy ΔG_a.

The crown ether analog **7c** (X = OMe) with a convergent methoxyl substituent has led to the development of a very interesting area of cyclophane chemistry. The crown ether derivatives **14** and **15** were described by Cram's group in 1976[72] and compounds with a macrocyclic ring of suitable size (**14:** $n = 3$, and **15:** $n = 2$) were shown to be good hosts for alkali metal cations and ammonium cations. The structural variation was

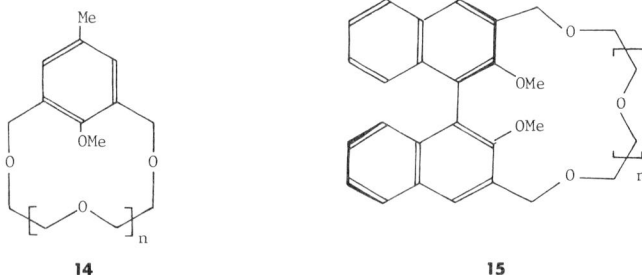

14   **15**

extended considerably in subsequent work,[73] and systems **16–19** were shown to be highly effective hosts that showed considerable discrimination between various types of alkali metal cations and also between different alkylammonium cations.

16

17

18

19

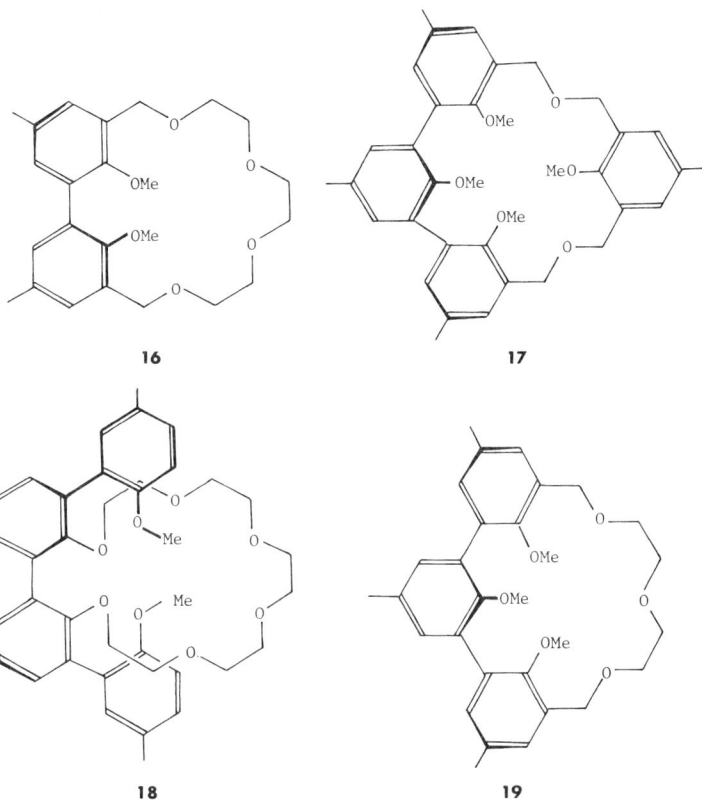

It was concluded that "anisyl units are capable of providing superb binding sites" if they are located in the correct relationship in the macrocycle (as in **16–19**), but in other cases they may be very poor. An examination of CPK models showed that anisyl units are effective when the unshared electron pairs of the convergent methoxyl groups are forced to line the cavity of the host before complexation. The spherands (e.g., **20**) are the ultimate development of this strategy.[27–29,135] The macrocycle **20** can be synthesized conveniently, as its lithium complex, by oxidative coupling of the lithiated *p*-cresol derivative **21** or the terphenyl derivative **22**. Other syntheses of spherands have also been described (Scheme 2).[28,135]

The conformation of the spherand **20** is well defined, and it represents a system that is organized for complexation and in which electron pair repulsion is relieved by interaction with a guest cation.[135] Unlike simple crown ether systems the conformation of **20** is preorganized during synthesis, and it does not require a conformational change before it can form

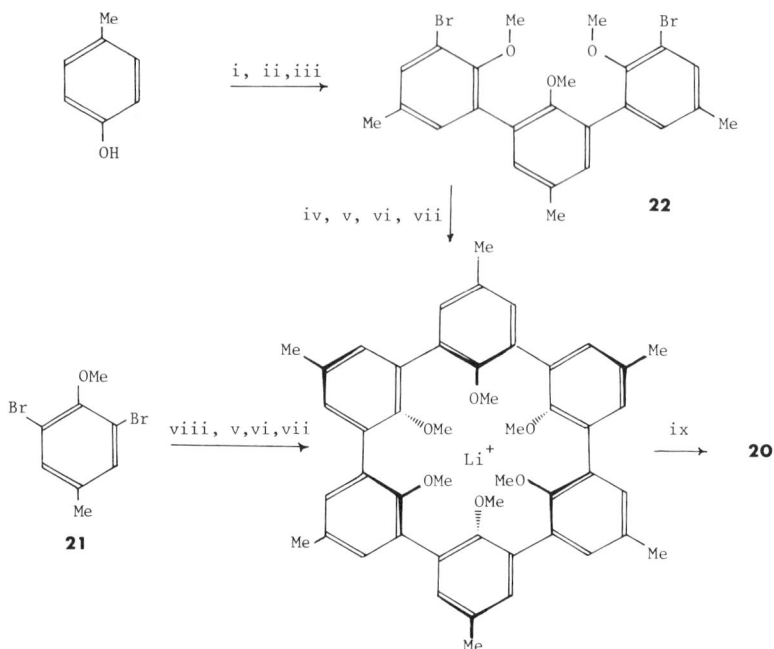

SCHEME 2. Reagents: *i*, FeCl$_3$; *ii*, Br$_2$; *iii*, Me$_2$SO$_4$; *iv*, BuLi/THF, $-80°$C; *v*, Fe(AcAc)$_3$/ C$_6$H$_6$; *vi*, H FeCl$_4$; *vii* EDTA; LiCl- *viii*, sec-Bu Li (3 mol), $-80°$C; *ix*, MeOH/H$_2$O, 150°, 5 days.

complexes with metal cations. As a consequence the spherand system **20** is a remarkably powerful and selective ligand for Li$^+$ and Na$^+$, and the crystal structures of the free spherand, the LiCl complex, and the Na-MeSO$_4$ complex show clearly the similarity of the structures of the ligand in free and complexed states.

The bridged spherand systems **23** have also been described.[27,29] They are less efficient ligands for Li$^+$ complexation than the spherand **20,** and the crystal structures of the complexes show that the ligand adopts the conformation shown in **23** with both bridges lying over the same face of the macrocyclic hexaaryl system.[27] The success of the spherand system demonstrates the advantages of such preorganized ligands based on cyclophane structures with limited conformational freedom. The extension of these ideas to the synthesis of preorganized host molecules for organic guest molecules presents an interesting and formidable challenge.

The synthesis of phenolic systems analogous to the metacyclophane **14** has also been investigated. It was shown[94] that the methyl ethers **24** could be demethylated readily by lithium iodide in pyridine to give the phenols

$X = (CH_2)_3$ or $CH_2CH_2OCH_2CH_2$

20

23

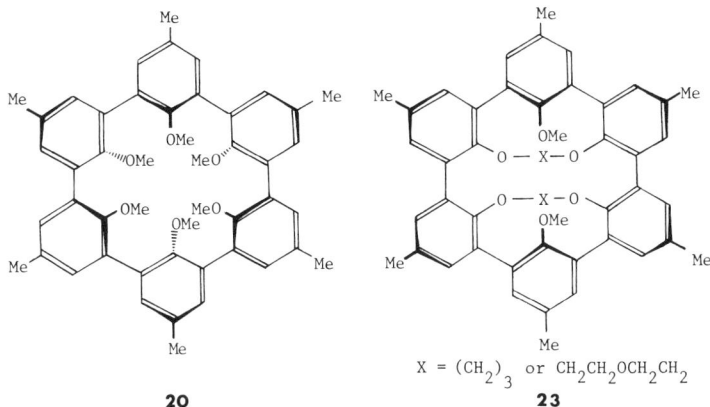

25. The 18-crown-6 analog **25b** reacted with ammonia to give the unstable ammonium salt **26** (R = H), but the corresponding *p*-nitrophenol derivative gave stable crystalline adducts with sodium salts, ammonia (**26:** R = NO$_2$), *tert*-butylamine, and 1-phenylethylamine. The allyl ethers **27** have also been used in an alternative synthetic approach to phenols **25**.[139] Oxidative demethylation of the corresponding quinol dimethyl ether gave the quinonoid cyclophane **28**.[125]

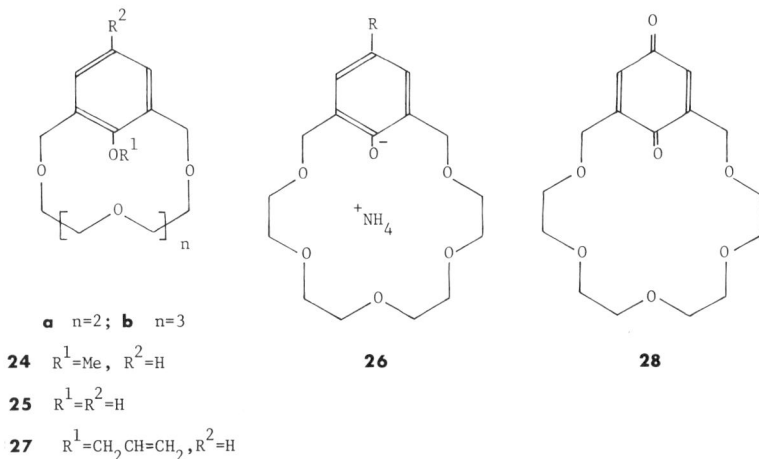

a n=2; **b** n=3

24 R^1=Me, R^2=H

25 R^1=R^2=H

27 R^1=CH$_2$CH=CH$_2$, R^2=H

26

28

b. 1,3-Bridged Heteroaromatic Rings

The incorporation of a heterocycle, such as pyridine or furan, into a crown ether system may be comparable in effect to a replaced —CH$_2$OCH$_2$— system. The two most studied examples of such cyclophane systems are those based on furan (e.g., **29**)[129,131] and pyridine (e.g., **30**).[103]

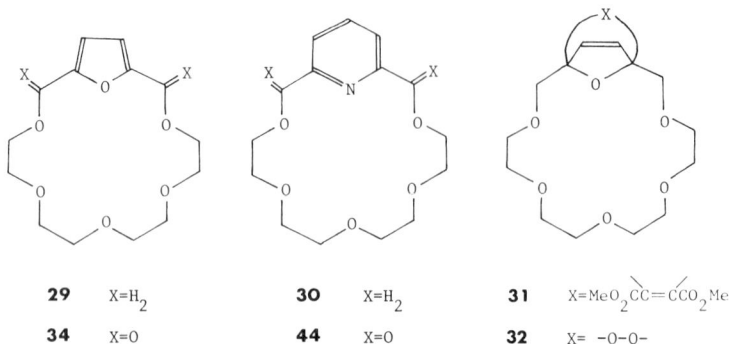

29   X=H$_2$ 30   X=H$_2$ 31   X=MeO$_2$CC=CCO$_2$Me

34   X=O   44   X=O   32   X= -O-O-

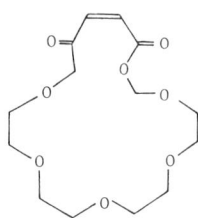   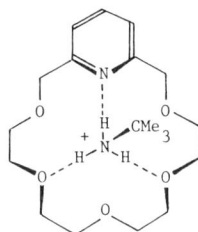

33   35

Analogs of the furanophane **29** having different macrocycle sizes have been examined as host molecules for metal cations and ammonium salts, and in addition a number of reactions of this furanophane have been described. Catalytic reduction of **29,** using a Raney nickel catalyst, gives the *cis*-tetrahydro derivative. The furanophane **29** forms a crystalline 1 : 1 complex with dimethyl acetylenedicarboxylate, which on heating gives the adduct **31.** Furanophane **29** also gives a cycloaddition product **(32)** with singlet oxygen at −60°C, which undergoes rearrangement at 9°C to give the unsaturated cyclic keto ester **33.**[46] The furanophane diesters (e.g., **34**) also form complexes with metal cations and ammonium salts.[12,14]

Pyridinophane analogs of crown ethers (e.g., **30**) form strong complexes with metal ions and ammonium salts,[103] and analysis of K_a values for a number of complexes of pyridinophanes of this type with the *tert*-butylammonium cation shows that the pyridine nitrogen atom is a slightly less effective binding site than an ether oxygen atom. The crystal structure **(35)** of the complex of **30** with *tert*-butylammonium perchlorate[97] resembles that of an 18-crown-6–alkylammonium cation complex. Hydrogen bonding (see **35**), at least in the crystal structure, involves two of the macrocycle oxygen atoms and the pyridine nitrogen atom.

A wide range of crown ether analogs based on the pyridinophane system has been prepared, but only a few examples have been selected for comment. The macrocycles **36** have been reported to catalyze the aminolysis of phenol acetates by butylamine.[49] The reason for the catalysis is not clear, but it has been suggested that the transition state for aminolysis is stabilized by complexation.

The pyridinophane crown ester **37** can be synthesized by a Hantsch reaction. Methylation (methyl fluorosulfonate) and reduction (sodium dithionite) give the bridged 1,4-dihydropyridine **38,** which forms a crystalline solvated complex with sodium perchlorate.[137] The dihydropyridine derivative **38** is also an efficient hydride donor for the reduction of a number of sulfonium salts **(39)** to a mixture of ketone R^1COCH_3 and sulfide $MeSR^2$ with simultaneous formation of the pyridinium salt **40.**[70,138] The rate of this reduction for the sulfonium salt **39** ($R^1 = R^2 = Ph$; $X = BF_4$) is approximately 500 times faster at 25°C than the reduction of the same sulfonium salt by the dihydropyridine **41.** The authors propose[70,138] that this is a consequence of a favored pathway for reduction of the sulfonium salt by the crown ester **38,** which involves the formation of a complex before hydride transfer. The system **38/40** is therefore an analog of the NAD(P)H/NAD(P) system.

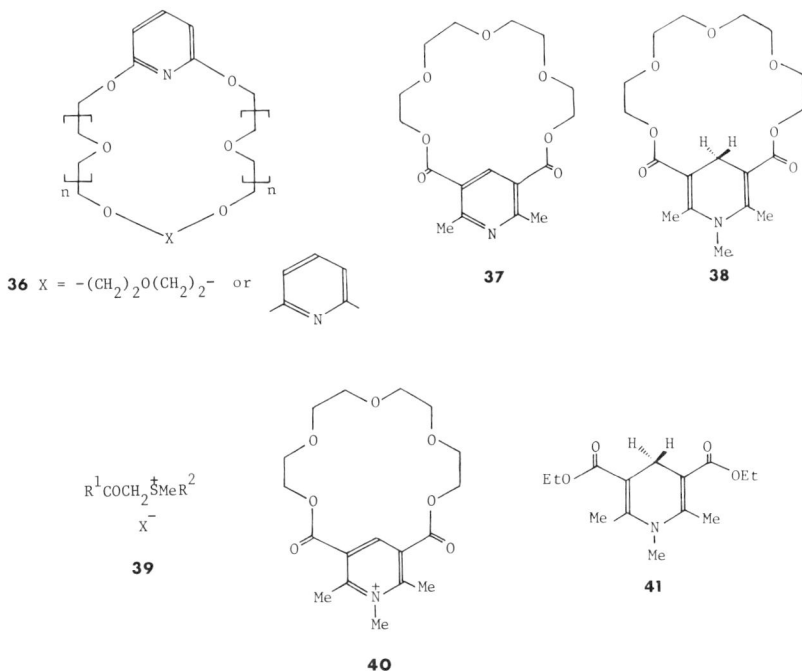

In more recent work by the same group[35,67] chiral dihydropyridine derivatives have been synthesized using amino acids as the source of chirality. The dihydropyridine **42** reduces ketones R^1COR^2 to alcohols R^1CHOHR^2 in the presence of 1 equivalent of $Mg(ClO_4)_2 \cdot 2 H_2O$ in acetonitrile at 18 to 20°C. The product alcohols were reported to be optically active, with an enantiomeric excess of the S configuration of 90% for reduction of the ketone (R^1 = Ph; R^2 = CO_2Et) by the dihydropyridine **42** (R = $CHMe_2$; X = CH_2). This corresponds to preferential addition of hydride to the re face of the carbonyl group of the ketone when the dihydropyridine **42** is derived from an L-amino acid. The authors suggest a transition state model **(43)**, based on a complex between **42**, Mg^{2+}, and

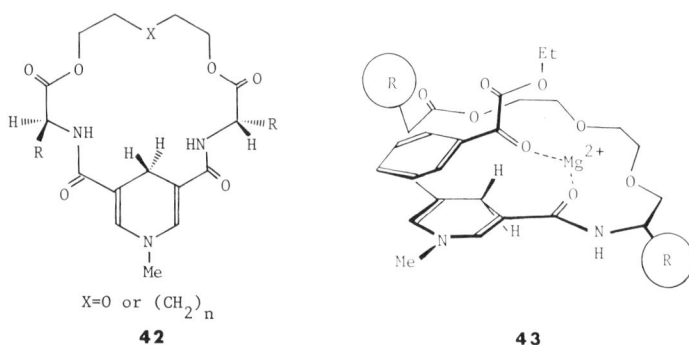

X=O or $(CH_2)_n$

42   **43**

the substrate ketone, to account for the observed enantioselectivity. The minimum requirement for high enantioselectivity is that the macrocycle in **42** be closed; the length and shape of the bridge X can be varied quite extensively, but for very long bridges [X $\geqslant$ $(CH_2)_5$, for example] enantioselectivity drops sharply.

A number of groups[11,12,15,47,104,108] have reported the synthesis of cyclic polyether diesters containing bridged pyridine (and thiophene and furan) rings (e.g., **44**; see **30**) and have described a wide range of complexes of metal cations and alkylammonium cations. The pyridinophane system has also been extended in a number of ways by the formation of cyclic diimines between 2,6-diformylpyridine and bis(primary alkylamines); in many cases the cyclization reaction is template-controlled. Examples include the formation of **45** in the presence of M^{2+} cations[40] and **46** in the presence of alkaline earth cations.[45] Binuclear systems of this type have also been described, and the formation of the Pb(II) complex **47** in the presence of $Pb(SCN)_2$ is a good example.[42] These binuclear systems are potential receptors for bridging anions, and the related Cu(II) complex **48** with an azide ion forming a bridge between two appropriately located copper atoms[41] is an interesting metalloanalog of the azide anion cryptate

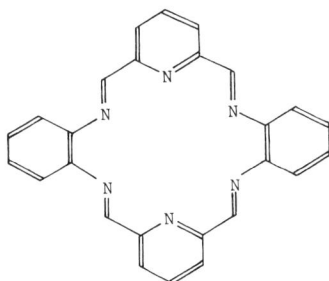

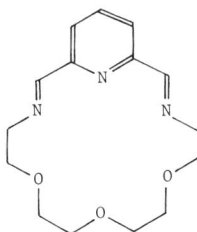

45   **46**

49, in which a purely organic receptor is used,[85,88] and related bicyclic binuclear cryptates.[96]

The binuclear copper complex bridged with an imidizolate anion **(50)** is another example that is of interest as a model for the coordination environment of bismetallated biological molecules.[23] Anion cryptates analogous to **49** have received considerable attention, but no examples have yet been reported in which the macrocyclic or macropolycyclic system is

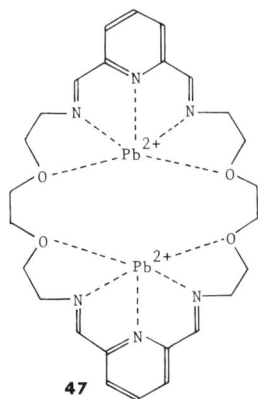

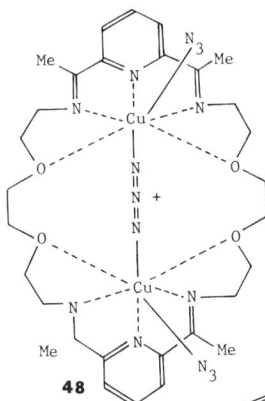

47   **48**

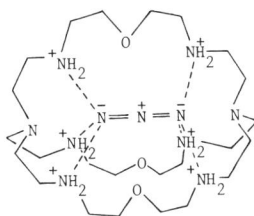

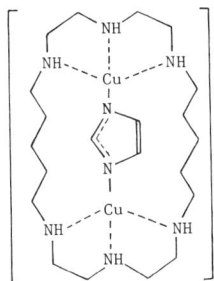

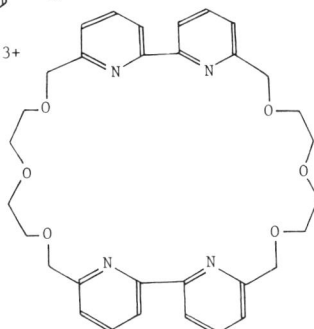

49   **50**   **51**

preorganized by the incorporation of cyclophane systems into the crypt-
and structure.[36-38,51,99,119]

The α,α-dipyridyl ligand has also been incorporated into crown ether
macrocycles, and in this case also binuclear complexes may be formed,
for example, between the ligand **51** and $CuCl_2$.[105]

2. CHIRAL SYSTEMS

The simple achiral crown ether system has been elegantly and successfully modified by Cram's group by the incorporation of a chiral 2,2'-
dioxygenated 1,1'-binaphthyl unit into the macrocycle, as, for example, in
the host **52**.[24,25] Other modifications of the crown macrocycle that have as
their primary objective the introduction of chirality include the use of
carbohydrates, tartaric acid, glycerol derivatives, and amino acids and
amino alcohols. The possible use of the plane of chirality found in suitably
substituted paracyclophanes in which the rotation of the aromatic ring is
restricted appeared to be an attractive alternative to the use of the axial
chirality of the binaphthyl systems. Unfortunately, the paracyclophanes
(53) investigated as potential chiral host molecules proved to be unsatis-
factory in that the complexes with primary alkylammonium cations were
weakly bound, and no evidence could be obtained for enantiomer discrimination using the chiral 1-phenylethylammonium cation.[110] In contrast, the
[14]paracyclophane derivative **54** has been shown to form strong complexes with primary alkylammonium salts,[5] but a system of this type in
which rotation of the aromatic ring is prevented by substitutents, as in **53,**
does not appear to have been examined.

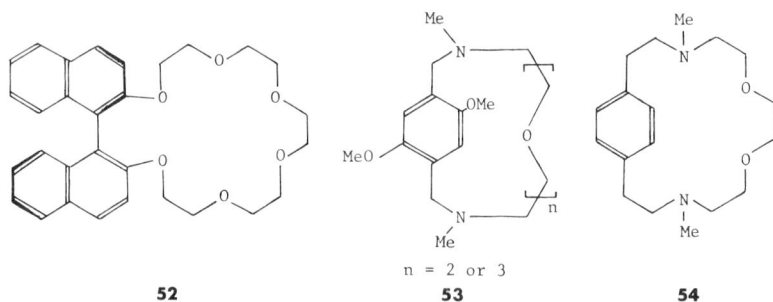

n = 2 or 3

52   **53**   **54**

Bridged 9,9'-spirobifluorene derivatives (e.g., **55**) have been investigated in some detail as chiral host molecules.[114,115] This system forms
satisfactory complexes with a wide range of metal and alkylammonium
cations and is moderately enantiomer selective for chiral alkylammonium

cations. These selectivities have been established using both distribution and electrochemical methods,[128] and they have been discussed using the crystal structure of the ammonium rhodanide complex of **55** as a guide.

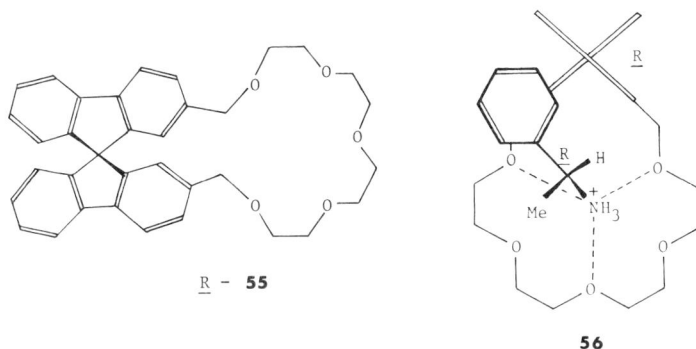

$\underline{R}$ – **55**

56

The R enantiomer of **55** forms a more stable complex with the R enantiomer of the 1-phenylethylammonium cation than with the S enantiomer. This preference was rationalized on the basis of a model for the complex in which the guest cation is situated on the upper face of the host in the orientation depicted in **56**.

B. Polycyclic Systems

1. MACROBICYCLIC SYSTEMS CONTAINING ONE CROWN ETHER MACROCYCLE

The high selectivity shown by the [k.m.n]cryptands for the complexation of different alkali metal cations was first demonstrated in the innovative work of J. M. Lehn and co-workers.[81] These studies underlined the advantages of using azacrown ethers as the starting point for the construction of macropolycyclic cryptands and have stimulated considerable interest in bridged crown ethers. A number of bicyclic cryptand systems incorporating cyclophanes in their structures have been described, and their properties as receptors for alkali metal cations have been investigated; however, there has been virtually no attempt to construct bicyclic systems that would serve as hosts for alkylammonium cations.

Most bicyclic systems have been prepared using a method based on the original procedures of Lehn's group, but the synthesis of the macrobicyclic pyridinophanes **57** is based on the rather unusual use of methyl as a protecting group for nitrogen and quaternization reactions for construct-

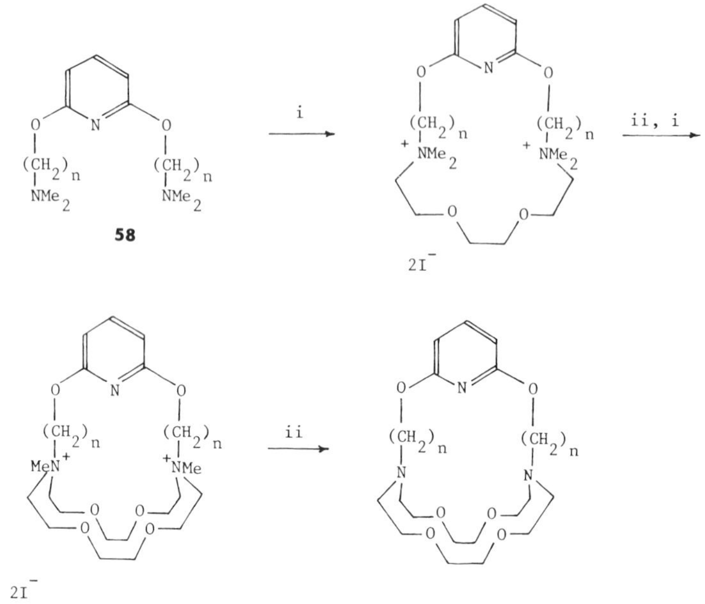

58

57

SCHEME 3. Reagents: *i*, I–(CH2)2–O–(CH2)2–O–(CH2)2–I/MeCN; *ii*, L-selectride–THF.

ing macrocycles.[106] The synthesis of **57** from the bis(tertiary amine) **58** is summarized in Scheme 3; the reductive demethylation of the bis(quaternary salt) intermediates is noteworthy. The crystal structure of a bicyclic cryptand **(57)** has been reported, as have the structures of the related cryptands **59** and **60** prepared by conventional methods from triethanolamine and 2,6-dichloropyridine or 2,6-bischloromethylpyridine, respectively.[107]

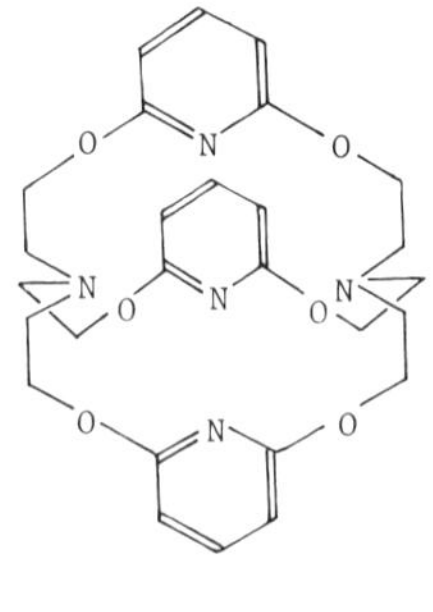

59

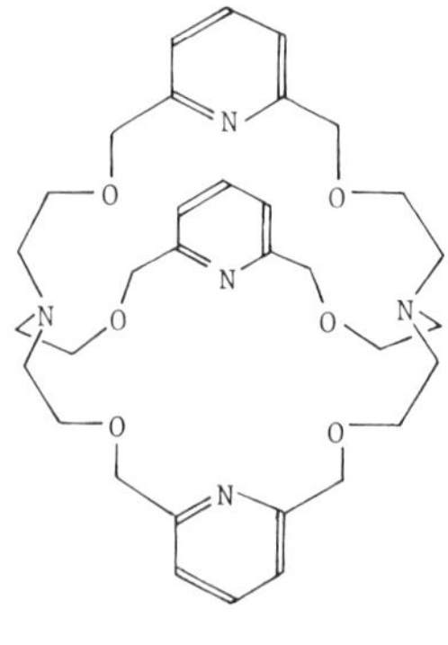

60

All these bicyclic cryptands possess potentially spherical cavities that are suitable, if of appropriate size, for metal ion complexation. Although the cavity of the uncomplexed ligand **60** is filled by a pyridine ring in its crystal structure, **60** can nevertheless form 2 : 1 complexes with $CoCl_2$ and $CuCl_2$. The crystal structure of **59** shows that the bridgehead nitrogen atoms have a planar configuration with crystallographically equivalent 120° bond angles, not ideal for complexation and in contrast with the sp^3 hybridization of the bridgehead nitrogen found for other cryptands. Similar pyridinophane cryptands (e.g., **61**) have been prepared by Vögtle's group, who have also reported the preparation of bicyclic cryptands (e.g., **62**) having a paracyclophane bridge.[18,136] Further elaborations of bicyclic cryptands by the same group have included the systems **63**,[141] **64**,[118] and the interesting ferrocenophane **65**.[109] The Na$^+$ and K$^+$ complexes of many of these cryptands have been reported.

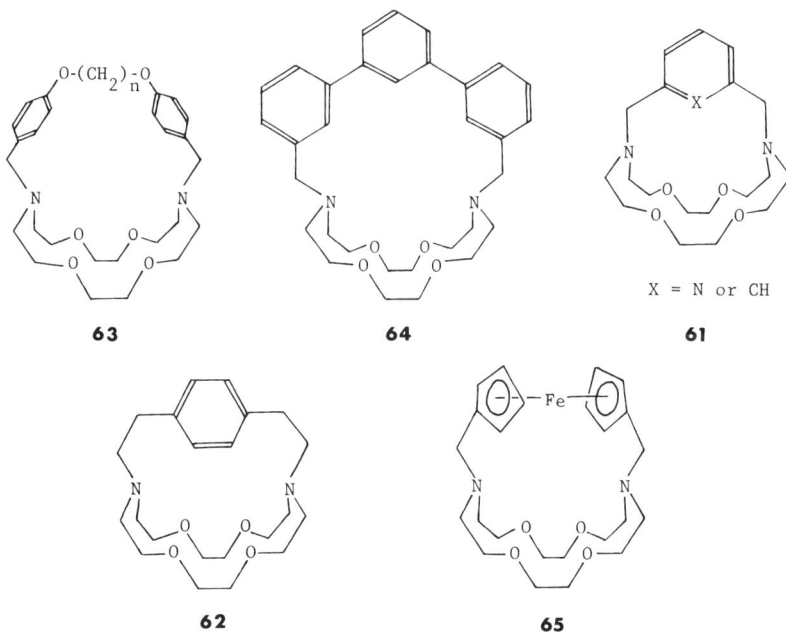

63

64

61

X = N or CH

62

65

A related bridged crown ether (**66**) has been used as a photoresponsive system.[122,123] The extraction of alkali metal salts of methyl orange from water by a solution in benzene of the crown ether **66a** having a *trans*-azobenzene bridge differs from that of a solution of the cis isomer **66b.** The trans isomer extracts Na$^+$ and K$^+$ almost equally, whereas the cis isomer extracts K$^+$ more efficiently than Na$^+$. Hence, photoirradiation of a solution of the trans isomer, which produces an equilibrium mixture of the two isomers **66a** and **66b** in 1 min, changes the extraction capacity of the organic phase in a water–benzene system. The trans isomer **66a** is also more effective for the extraction of alkylammonium cations.

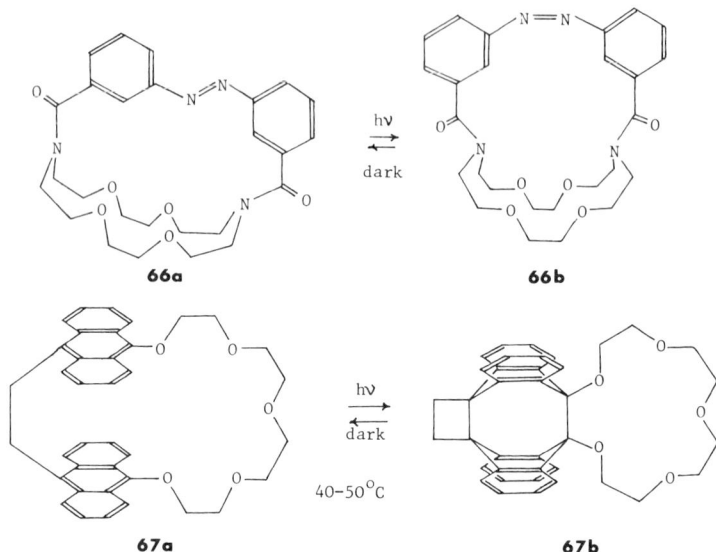

66a   **66b**

67a   **67b**

A second, and rather different, photoresponsive system is also based on a cyclophane as one component.[145] The bridged anthracene **67a** undergoes intramolecular photochemical $[_\pi 4_s + _\pi 4_s]$ cycloaddition on irradiation to give a crown ether fused to an anthracene photodimer **(67b)**. Thermal reversion **(67b → 67a)** occurs in the dark at 40 to 50°C. The half-life for this process ($t_{1/2}$ = 9.6 min) is prolonged in the presence of alkali metal cations in the order of dependence Na$^+$ > K$^+$ > Li$^+$ [$t_{1/2}$(Na$^+$) = 26 min], although the photochemical isomerization **67a → 67b** is not affected by the presence of the alkali metal cations.

2. MACROPOLYCYCLIC SYSTEMS CONTAINING MORE THAN ONE CROWN ETHER MACROCYCLE

Systems of this type, in contrast to the systems discussed in the preceding section, contain potentially two or more receptor sites for metal cations or alkylammonium cations that are held in a predetermined relationship by a cyclophane system. Three general approaches have involved (*a*) crown esters constructed on a [2.2]paracyclophane skeleton, (*b*) binaphthyl systems with multiple crown ether attachment, and (*c*) tricyclic, tetracyclic, and other polycyclic systems based on bridged azacrown ethers.

a. Systems Based on [2.2]Paracyclophane

These systems have been studied as potential two-site receptors by Cram's group.[55,56,130] The compounds studied were derived by crown ether syntheses based on the diol **68a** and the tetraacetates **68b–68d.** The derived crown ethers contained either a single macrocycle **(69)** or two macrocycles in a fixed relationship **(70–72).**

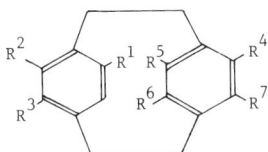

68

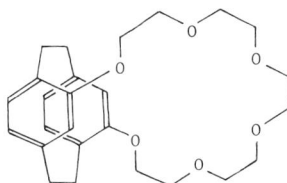

69

a $R^1 = R^7 = OH$; $R^2 = R^3 = R^4 = R^5 = R^6 = H$
b $R^2 = R^3 = R^4 = R^7 = OAc$; $R^1 = R^5 = R^6 = H$
c $R^1 = R^3 = R^5 = R^7 = OAc$; $R^2 = R^4 = R^6 = H$
d $R^1 = R^3 = R^4 = R^6 = OAc$; $R^2 = R^5 = R^7 = H$

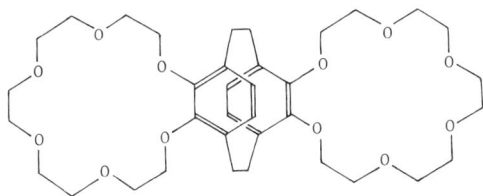

70

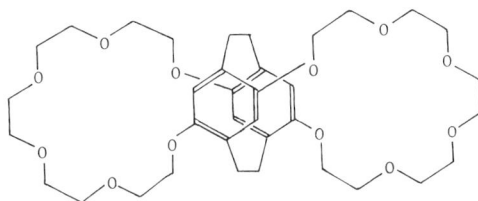

71

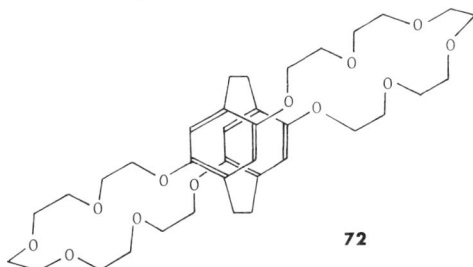

72

The aromatic rings of the [2.2]paracyclophane system serve as built-in magnetic probes for the solution structures of complexes with alkylammonium cations. Hosts having two macrocyclic receptors form 2 : 1 complexes with alkylammonium cations and with alkali metal cations. The 2 : 1 complexes formed by the host (70) are predominantly of the syn,syn type, shown diagrammatically in 73a, but the 1 : 1 complexes formed with bisalkylammonium cations appear to be polymeric (73b), even in cases such as $\overset{+}{N}H_3(CH_2)_{10}\overset{+}{N}H_3$ when the polymethylene chain might be long enough to stretch over the cyclophane system in a monomeric complex such as 73c.

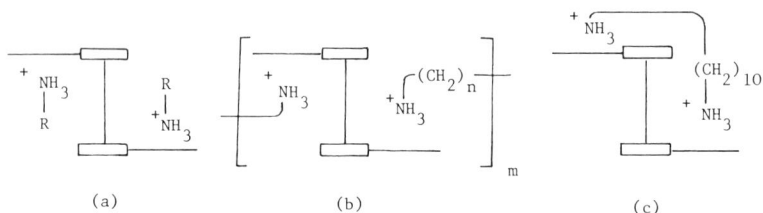

(a)   (b)   (c)

73

The complexing capacity of hosts such as 69, 71, and 72 in which the crown ether system is interrupted is considerably reduced relative to simple crown ethers, and in these systems hydrogen bonding is believed to be based on only two hydrogen bonds to each $\overset{+}{N}H_3$ group (K_a for guest tert-BuNH$_3$+ picrate of 195 and 205 for hosts 71 and 72 as compared with 434,000 for 2,3-naphtho-18-crown-6). Furthermore, the binding of NH$_4$+ picrate is ~10^3 times greater than for tert-BuNH$_3$+ picrate, but, in spite of this evidence for nonbonded interactions involving the alkyl group of alkylammonium guest cations, it was not possible to demonstrate enantiomer selectivity in complexation by the chiral hosts 71. Thus, the diastereoisomeric complexes formed between 71 and α-phenylethylammonium thiocyanate gave different chemical shifts for the aryl hydrogen atoms (δ 5.68 and 5.81), but under conditions of fast equilibration between the two diastereoisomers [(R,S) host + (R,S) guest[58,70]] a single aryl hydrogen signal was observed, with a chemical shift (δ 5.75) consistent with an approximate 1 : 1 mixture of the two diastereoisomers.

b. Binaphthyl Systems

These compounds have generally provided three-dimensional cavities rather than two related and fixed crown ether receptor sites.[54] Examples include the macrobicyclic systems 74, prepared by reaction of a polyethylene glycol dianion with the dihalide 75, and the corresponding (S,S)-bisbinaphthyl system 76, prepared by an analogous procedure. In addition, the bridged tetralin derivative **77** having two distinct cavities, each lined by eight oxygen atoms, was also reported. All of these hosts formed strong complexes with alkali metal and ammonium picrates, but although the crystal structure of the host **(77)** was reported there is no analogous structural information for the complexes.[54]

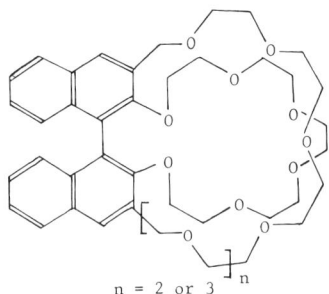

74

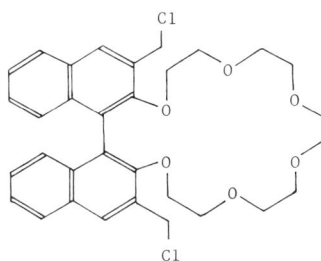

75

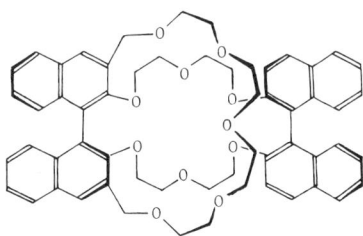

76

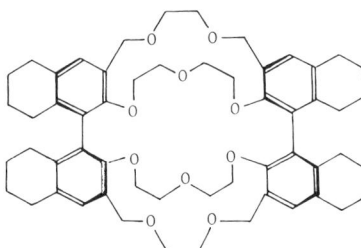

77

c. Bridged Crown Ethers

Polycyclic crown ether systems based on doubly **(78)** or triply **(79)** bridged crown macrocycles are potential hosts for two metal cations, two alkylammonium cations, or a single molecule of a bisammonium cation. Although these polycyclic compounds, shown diagrammatically in the

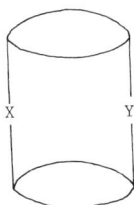

78

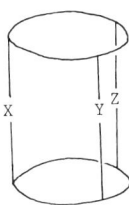

79

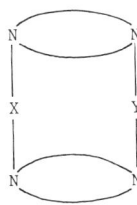

80

formulas, could in principle be based on all-oxygen crown ethers, in practice this would involve a considerable synthetic problem. All the systems of this type that have been described have therefore been based on bridged azacrown ethers (e.g., **80**), which are readily synthesized.

Tricyclic systems of type **80** have been reported in a number of publications, but their complexes have been systematically examined only relatively recently. The tricyclic hosts **81** were shown to form 1:1 and 2:1 complexes with a range of metal cations, with the two macrocyclic receptors behaving as virtually independent crown ether units.[86]

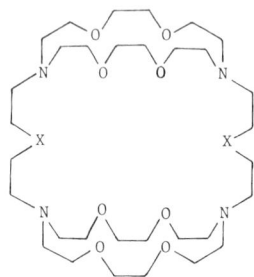 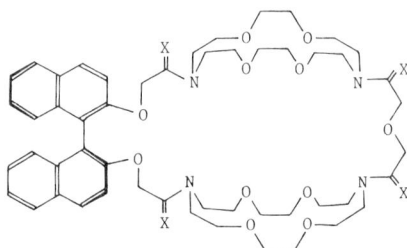

X = O or CH₂

81

a X=O;   b X=H₂

82

The chiral macrotricyclic host **82a** is of particular interest because it showed some enantiomer recognition for (±)-α-naphthylethylammonium chloride both in extraction experiments and in selective transport from H₂O through a CHCl₃ layer containing the (S) host [complexation of (−) guest preferred]. The more strongly basic host **82b** was protonated by RŃH₃ salts under transport or extraction conditions, but complexation was observed by nmr spectroscopy for solutions in CD₃OD. The alkali metal cation complexes of **82b** do show enantiomer discrimination in both anion transport and extraction experiments. For example, for mandelic acid salts, when K⁺ mandelate is used the (−) enantiomer is extracted preferentially, whereas when Cs⁺ mandelate is used the (+) enantiomer is extracted preferentially. This may reflect the differing geometries of the K⁺ and Cs⁺ complexes of the host **82b.**

Tricyclic systems of the general type **80,** in which the bridging groups contain aromatic rings, have a well-defined separation between the two macrocyclic azacrown ether receptors and might therefore be expected to have highly selective receptor properties for bisammonium cations ŃH₃(CH₂)ₙŃH₃. This has been shown to be the case for the tricyclic systems **83**[64,65,71,75,87,95] and for the related tetracyclic systems **84**.[74] These results are discussed here in some detail because they are closely related to the strategy for host design based on cyclophane systems.

In addition to ensuring structural rigidity, the aromatic ring or rings in the bridges of these polycyclic systems provide a magnetic probe for the identification of guest molecules that are enclosed in the molecular cavity lying between the two aromatic rings. Thus, guest molecules in an inclusion complex can be distinguished from guest molecules that are attached only to the outer face of the receptor macrocycles. The first compounds of this type to be examined for complexation of guest ammonium cations[64]

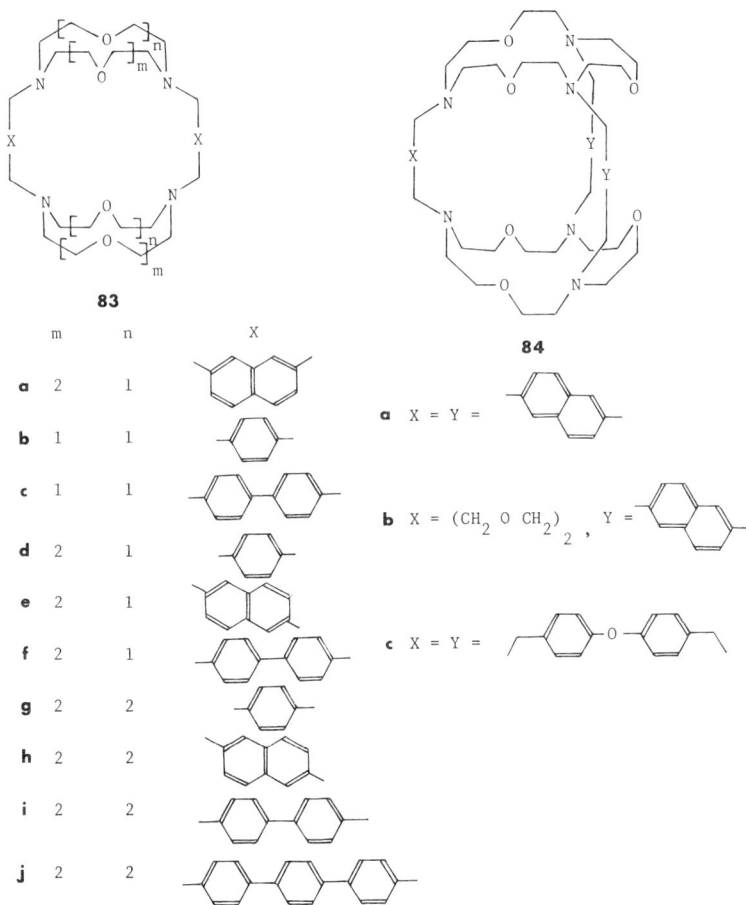

83

	m	n	X
a	2	1	
b	1	1	
c	1	1	
d	2	1	
e	2	1	
f	2	1	
g	2	2	
h	2	2	
i	2	2	
j	2	2	

84

a X = Y =

b X = $(CH_2 O CH_2)_2$, Y =

c X = Y =

were the tricyclic hosts **83a** and **83d** based on diaza-15-crown-5 macrocycles. These azacrown systems were selected because the monocyclic diaza-15-crown-5 system (e.g., **85:** $n = 2$) was known to form complexes having only the cis,cis stereochemistry shown in **86** ($n = 2$). The tricyclic hosts **83a** and **83d** were therefore expected to form inclusion complexes only, and in accord with this expectation the nmr spectrum of a mixture of **83a** and methylammonium thiocyanate (1 : 2 ratio) in CD_2Cl_2 showed two NMe signals at low temperature (δ 1.73 and 2.00 at $-20°C$) assignable to the two diastereoisomeric 1 : 2 inclusion complexes **87a** and **87b.***

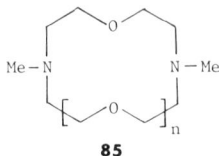

85

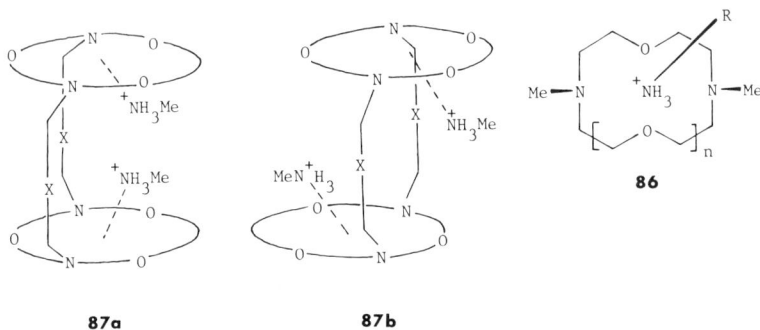

87a   **87b**   **86**

The nmr spectrum of the complex of the same host (**83a**) with the bisammonium salt $\overset{+}{N}H_3(CH_2)_6\overset{+}{N}H_3$ 2 NCS$^-$ showed high-field shifts ($\sim$1 ppm) for the α-, β-, and γ-CH_2 groups of the complexed guest cation as compared with the free guest cation, indicating the formation of an inclusion complex of a single type with a relatively high energy barrier ($\Delta G^{\ddagger} = 14.0$ kcal/mol) for the exchange of free and complexed guest biscations.† Other guest salts, $\overset{+}{N}H_3(CH_2)_n\overset{+}{N}H_3$ 2 NCS$^-$ ($n = 5$ and 7), also formed well-defined 1 : 1 inclusion complexes with host **83a,** but primary alkylammonium cations $R\overset{+}{N}H_3$ with a large group R (e.g., R = CH_2Ph) did not form well-defined complexes or show the shifts to high field characteristic of an inclusion complex. This is presumably because the limited size of the cavity did not permit inclusion complex formation, and the inward direction of complexation by the 15-membered macrocycles does not permit complexation of the guest cation to the outer face of the host.

---

* These and similar diagrams show the bridging —CH_2CH_2— groups of the macrocycles as lines for simplicity. The precise nature of the attachment of $\overset{+}{N}H_3$ to the receptor macrocycle is not specified because no precise information is available.

† Based, in the usual way, on the coalescence of the separate signals for the α-, β-, and γ-CH_2 groups of the free and complexed guest salt.

The 12-crown-4 system **85** ($n = 1$) also forms cis,cis complexes (**86:** $n = 1$) with alkylammonium cations in which the host adopts a chiral [3.3.3.3] conformation.[76,98] The incorporation of this smaller and more rigid macrocycle into the tricyclic host system **83** was therefore expected to give host molecules having a similar tendency to form inclusion complexes. The host **83b** has a very small cavity that will accommodate only the biscation $\overset{+}{N}H_3(CH_2)_2\overset{+}{N}H_3$ with substantial shifts to high field of the nmr signal for CH_2 groups of the guest cation (δ 0.89 for the CH_2 groups of the complexed cation at 25°C as compared with δ 3.08 for the free cation). At very low temperatures the nmr spectrum of the complexed cation was observed as two broad multiplet signals (δ −0.86 and 2.44), probably corresponding to a $—CH_aH_bCH_{a'}H_{b'}—$ system because the guest ^{13}C signal remained as a singlet over the entire temperature range. This result is consistent with a structure for the complex in which the host has a chiral [3.3.3.3] conformation for each 12-membered ring with slow interconversion of the two enantiomeric [3.3.3.3] conformations. Because only a single species of complex can be detected, it must be concluded that the conformations of the two macrocycles are interrelated. The free energy barrier for conformational interconversion ($\Delta G^{\ddagger} = \sim 8$ kcal/mol) is comparable to that found for the analogous process in the dibenzylammonium complex of N,N'-dimethyl-1,7-diaza-12-crown-4 (**85:** $n= 1$). The small cavity of the host **83b** will not form strong complexes with longer guest biscations $\overset{+}{N}H_3(CH_2)_n\overset{+}{N}H_3$ ($n > 2$).

The longer cavity of the biphenyl-bridged macrotricycle **83c** forms complexes with a range of bisammonium cations $\overset{+}{N}H_3(CH_2)_n\overset{+}{N}H_3$ ($n = 4, 5, 6,$ and 7), but there is no evidence for strong complexation of the shorter biscation $\overset{+}{N}H_3(CH_2)_3\overset{+}{N}H_3$ or the longer biscation $\overset{+}{N}H_3(CH_2)_8\overset{+}{N}H_3$, which are too short and too long, respectively, to fit into the host cavity. The selectivity of this host can be determined by competition experiments, because (1) the guest CH_2 signals are shifted 1–2 ppm to high field of the corresponding signals for the uncomplexed guest, and (2) exchange between free and complexed guest is slow on the nmr time scale at low temperatures so that separate signals can be seen for free and complexed guest cations. On this basis it can be shown that $\overset{+}{N}H_3(CH_2)_5\overset{+}{N}H_3$ is complexed in preference to $\overset{+}{N}H_3(CH_2)_4\overset{+}{N}H_3$, that the guests $\overset{+}{N}H_3(CH_2)_5\overset{+}{N}H_3$ and $\overset{+}{N}H_3(CH_2)_6\overset{+}{N}H_3$ are complexed almost equally readily, and that $\overset{+}{N}H_3(CH_2)_6\overset{+}{N}H_3$ is complexed in preference to $\overset{+}{N}H_3(CH_2)_7\overset{+}{N}H_3$. This selectivity has been examined in a similar manner for the series of tricyclic hosts **83b–83f** to give the results[65] that are summarized in Table I. If the length of the cavity, defined by the distance l in **88,** is compared with the length d of the extended guest $NH_3(CH_2)_n NH_3$, then optimum complex-

TABLE I

Selectivity for Guests $\overset{+}{N}H_3(CH_2)_n\overset{+}{N}H_3$ in Complexation by Hosts 83

Host	Selectivity[a]	Length (Å) of cavity, l	Length (Å) of preferred guest, d
83b	$n = 2 >> 3$	5.8	3.9
83c	$n = 3 << 4 < 5 = 6 > 7 >> 8$	10.1	7.9[b]
83d	$n = 2 > 3 > 4$	5.8	3.6
83e	$n = 3 << 4 > 5 > 6$	7.9	6.1
83f	$n = 4 < 5 < 6 > 7$	10.1	8.5

[a] The value of n refers to the formula $\overset{+}{N}H_3(CH_2)_n\overset{+}{N}H_3$. The inequality signs refer to stronger and weaker complexation as determined by competition experiments.

[b] Average length for $n = 5$ and $n = 6$.

ation is obtained for $d = \sim l{-}2$ Å, consistent with the hydrogen bonding shown diagrammatically in **88**.

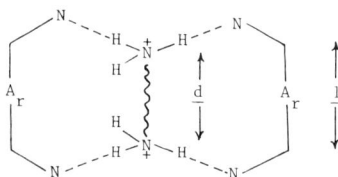

88

The nmr spectra of these complexes also provide information about the geometry of complexation, particularly on the basis of temperature dependence of the spectra, as has been discussed for the complex of **83b** with the biscation $\overset{+}{N}H_3(CH_2)_2\overset{+}{N}H_3$. Thus, the host **83d** appears to form only a single type of complex with each of the biscations $\overset{+}{N}H_3(CH_2)_n\overset{+}{N}H_3$ ($n = 2, 3,$ and 4), and the complex with $\overset{+}{N}H_3(CH_2)_2\overset{+}{N}H_3$ shows a poorly defined AA'BB' system for the nmr spectrum of the guest —CH_2—CH_2— system at very low temperatures ($\sim -100°C$), consistent with interrelated chiral conformations for both of the 15-membered rings, just as in the complex of the same guest with the host **83b**. The nmr spectra of the complex of host **83e** with the guest biscations $\overset{+}{N}H_3(CH_2)_4\overset{+}{N}H_3$ and $\overset{+}{N}H_3(CH_2)_5\overset{+}{N}H_3$ are particularly informative. There are four possible conformations for this host **(89a–89d),** which are related by rotation of the bridging naphthalene systems or of the crown macrocycles (accompanied by nitrogen inversion). The conformation **89a** is selected almost exclusively for complexation by the two guest biscations. Thus, the complexed guest $\overset{+}{N}H_3(CH_2)_4\overset{+}{N}H_3$ gives four different CH_2 signals in both the ${}^1H$- and

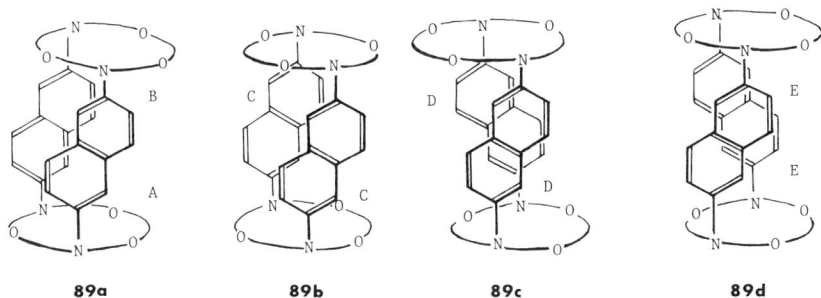

89a   **89b**   **89c**   **89d**

^{13}C-nmr spectra at low temperatures, and this would not be the case for complexation by the alternative host conformations **89b–89d**.* Furthermore, the complexed guest $\overset{+}{N}H_3(CH_2)_4\overset{+}{N}H_3$ shows eight different CH signals at very low temperatures ($< -80°C$), consistent with chiral conformations of the host macrocycles selected by the steric requirements for complexation with the guest biscation. The nmr spectra of the complex of the biscation $\overset{+}{N}H_3(CH_2)_5\overset{+}{N}H_3$ with the same host show similar, but less well defined characteristics.

This remarkably high stereoselectivity for complex formation with guest biscations $\overset{+}{N}H_3(CH_2)_n\overset{+}{N}H_3$ is not shown for complexation with a simple alkylammonium salt. Thus, the nmr spectrum of the complex of host **83e** with methylammonium thiocyanate (1 : 2 ratio) shows four $\overset{+}{N}Me$ signals at low temperatures corresponding to 1 : 2 complexes formed by at least three of the four conformations **89a–89d**. These results for the tricyclic hosts **83a–83f** are complementary to those reported for the analogous tricyclic hosts **83g–83j** based on bridged diaza-18-crown-6 systems.[71,75] These hosts form complexes with simple alkylammonium cations ($Me\overset{+}{N}H_3$ and $Et\overset{+}{N}H_3$) in a 2 : 1 ratio with a small upfield shift of the guest ^{1}H-nmr signals, but because the diaza-18-crown-6 system **90** forms cis,cis (**91a**), cis,trans (**91b**), and trans,trans (**91c**) complexes with alkylammonium cations[58] it is possible that some of the guest salt is attached to the outside of the 18-membered macrocyclic receptors of **83g–83j** in addition to being attached inside the cavity. These hosts also form 1 : 1 complexes with larger guest cations, such as $PHCh_2CH_2\overset{+}{N}H_3$. The ^{1}H-nmr spectra of the 1 : 1 complexes formed between hosts **83h** and **83i** and bisammonium cations $\overset{+}{N}H_3(CH_2)_n\overset{+}{N}H_3$ show the same marked upfield shifts for the guest CH_2 signals as for the complexes of hosts **83a–83f**, indicating very clearly that the guest biscation is confined within the central cavity of the macrotricycle.

---

* The letters A–E in **89** denote five different environments for the macrocyclic receptor sites in the four conformations.

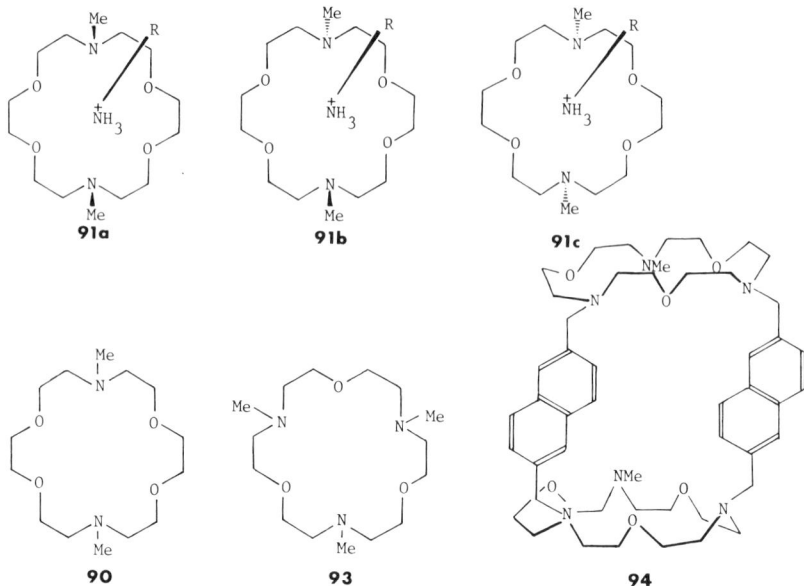

91a   91b   91c

90   93   94

The complexes between hosts **83g–83j** and biscations $\overset{+}{N}H_3(CH_2)_n\overset{+}{N}H_3$ have also been examined using ^{13}C correlation times (τ_c) as a guide to molecular motion[71] within the complex, as defined by the dynamic coupling coefficients χ. For a $^{13}CH_n$ group the ^{13}C relaxation time T_1 is given by

$$T_1^{-1} = 3.60 \times 10^{10}nr_{CH}^{-6}\tau_c$$

where r_{CH} is the C—H separation in angstroms (1.10 ± 0.2 Å). The dynamic coupling coefficient is the ratio of substrate correlation time to ligand correlation time, the latter being defined in this case by the correlation time for the ArC$\underline{H_2}$N ^{13}C nucleus because this is the host carbon atom with the longest correlation time. The value of χ is a good indication of the extent to which the motions of host and guest are coupled and hence the closeness of the fit between them.[6,17] For hosts **83g–83j** complexation in general slows molecular motion by a factor of 2 or 3. Three types of behavior have been distinguished for the values of χ, which are summarized in Table II. (*a*) Strong dynamic coupling between the guest–host pairs **83g** + $\overset{+}{N}H_3(CH_2)_3\overset{+}{N}H_3$, **83h** + $\overset{+}{N}H_3(CH_2)_5\overset{+}{N}H_3$, **83i** + $\overset{+}{N}H_3(CH_2)_7\overset{+}{N}H_3$, and **83j** + $\overset{+}{N}H_3(CH_2)_{10}\overset{+}{N}H_3$ is evident from values of χ approaching 1 for all carbon atoms, but the guests tend to gain increasing flexibility (decreasing χ) as the length of the $(CH_2)_n$ chain increases. (*b*) When the substrate is too long or too short, χ decreases, indicating in-

TABLE II

Dynamic Coupling Coefficients χ for Complexes of Hosts 83g–83j with Guests
$\overset{+}{N}H_3(CH_2)_n\overset{+}{N}H_3$

Host	Guest n^a	\multicolumn{6}{c}{χ for substrate carbon atoms[b]}					
		C-1	C-2	C-3	C-4	C-5	C-6
83g	3	0.95	0.95	—	—	—	—
83h	5	1	1	0.94	—	—	—
	6	0.77	0.71	0.69	—	—	—
	7	0.60	0.53	0.60	0.67	—	—
83i	7	0.91	0.79	0.88	0.82	—	—
	8	0.77	0.65	0.81	0.71	—	—
	9	0.39	0.48	0.41	0.30	0.35	—
83i	8	0.47	0.34	0.38	0.32	—	—
	9	0.77	0.45	0.42	0.52	0.52	—
	10	0.77	0.67	0.67	0.80	0.75	—
	11	0.80	0.50	0.45	0.45	0.45	0.55
	12	0.77	0.42	0.42	0.42	0.57	0.45

[a] The value of n refers to the formula $\overset{+}{N}H_3(CH_2)_n\overset{+}{N}H_3$. The underlined value refers to the guest biscation showing maximum average chemical shifts to high field for the guest $C\underline{H}_2$ signals.

[b] The carbons are numbered as $\overset{+}{N}H_3C(1)H_2C(2)H_2$, etc.

creased mobility of the guest and less efficient fit of the guest within the cavity. (c) Finally, particularly low values of χ are accompanied by small induced chemical shifts in the guest ^{1}H-nmr spectrum, and high values of χ are accompanied by maximal chemical shifts to high field in the guest ^{1}H-nmr spectrum.

The use of isodynamic character as a criterion for good guest–host fit in this type of complex provides an alternative to methods based on relative association constants, and the high values for χ in these complexes contrasts with the low values for χ found for the loosely fitting cyclodextrin complexes examined in earlier work by the same group.[6,17]

A number of macrotetracyclic compounds (84) analogous to the macrotricycles 83 have also been examined by Lehn and co-workers.[74] These tetracyclic systems were synthesized by the route outlined in Scheme 4 from monoprotected triaza-18-crown-6 (92).

The macrotetracycles 84 have a well-defined central cavity that is tighter and more insulated from the environment than the cavity of the analogous macrotricycles 83. It is not surprising that this leads to even greater discrimination in guest selection. Hosts 84a and 84b form 1:1 complexes with the picrates of 1,5-diaminopentane and 1,6-diaminohexane. The ^{1}H-nmr spectra of the guests show remarkably large shifts to

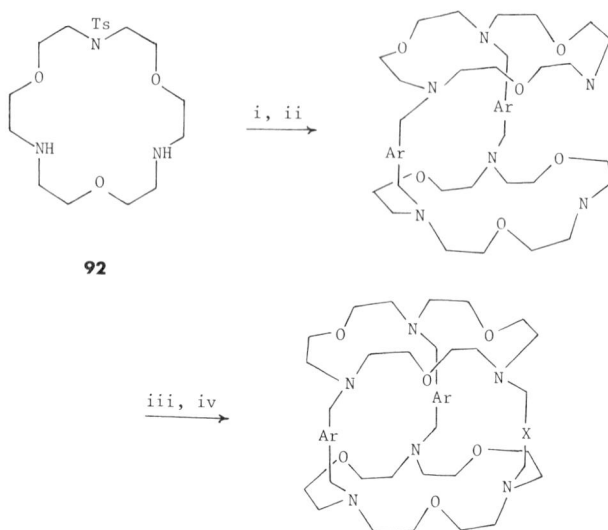

SCHEME 4. Reagents: *i*, ClCOArCOCl; *ii*, B_2H_6 then HBr–AcOH–PhOH or LiAlH$_4$–THF; *iii*, ClCOXCOCl; *iv*, B_2H_6.

high field as compared with the corresponding complexes of the triaza-18-crown-6 system **93**. Typical results are summarized in Table III. Receptors **84a** and **84b** also form complexes with MeN̄H$_3$ picrate, which show upfield shifts for the C$\underline{H}_3$ signals, indicating the occurrence of some internal binding but not ruling out the possibility of competing external binding. The hexamines **84** are strongly basic, but protonation can readily be distinguished from complexation on the basis of ^{1}H-nmr chemical shifts[74];

TABLE III

^{1}H-nmr Spectra for $\overset{+}{N}H_3(CH_2)_n\overset{+}{N}H_3$ (n = 5 and 6) in Complexes with Macrotetracyclic Hosts 84 and Related Azacrown Ether Derivatives[74]

	^{1}H-nmr spectra of guests (δ)					
	$\overset{+}{N}H_3(CH_2)_5\overset{+}{N}H_3$			$\overset{+}{N}H_3(\dot{C}H_2)_6\overset{+}{N}H_3$		
Host	C-1—H$_2$	C-2—H$_2$	C-3—H$_2$	C-1—H$_2$	C-2—H$_2$	C-3—H$_2$
84a	1.47	−1.35	−1.13	1.51	−0.76	−1.85
84b	1.62	−0.23	−0.90	—	—	—
83h	1.95	−0.22	−1.10	~2.5	0.31	−0.28
94	2.06	0.67	−0.52	—	—	—
93	~2.5	1.58	1.43	~2.5	1.53	1.41

protonation was found to occur for guest cations $R\overset{+}{N}H_3$ having a group R that is too large to fit within the host cavity.

The macrotetracycle **84a** showed strong chain length discrimination for the dications $\overset{+}{N}H_3(CH_2)_n\overset{+}{N}H_3$; complexes were formed for $n = 5$ and $n = 6$, but complexes were not formed for $n = 4$ and $n = 8$, and there appeared to be a total discrimination against internal complexation of a more bulky guest than $Me\overset{+}{N}H_3$ or $\overset{+}{N}H_3(CH_2)_n\overset{+}{N}H_3$. Guest exchange rates were slow, and it was possible to carry out competition experiments; in particular, the tetracycle **84a** and the analogous tricycle **83h** appeared to complex the dication $\overset{+}{N}H_3(CH_2)_5\overset{+}{N}H_3$ almost equally readily.[74]

These studies[64,65,71,74,75,87,95] of the rigid macropolycyclic receptors **83** and **84** show great promise as an approach to highly selective complexation, and other possible functions as coreceptors and metalloreceptors, cocatalysts, and cocarriers have been suggested.[75] The rigidity results from the cyclophane bridges. It was therefore of interest to examine less rigid systems in which one or more of the cyclophane bridges had been replaced by a more flexible $(CH_2)_n$ or $CH_2(CH_2OCH_2)_nCH_2$ linkage. Only one such study has been reported.[65] The asymmetric macrotricyclic hosts **95** were prepared by the route outlined in Scheme 5.

The macrotricycles **95** form complexes with dications $\overset{+}{N}H_3(CH_2)_n\overset{+}{N}H_3$, and the ^{1}H-nmr spectra of the complexes show the typical shifts to high

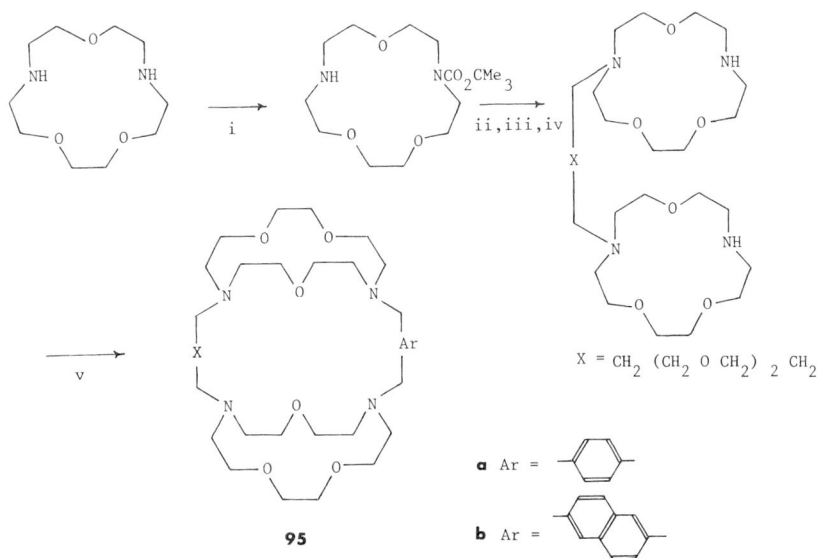

95

a Ar =

b Ar =

X = CH_2 $(CH_2\ O\ CH_2)_2$ CH_2

SCHEME 5. Reagents: *i*, $(Me_3CO—CO)_2O$; *ii*; $ClCO(CH_2OCH_2)_2COCl$; *iii*; $CF_3CO_2H–H_2O$; *iv*, $LiAlH_4$; *v*, $BrCH_2ArCH_2Br$.

field for the signals of the included guest cations. The protons of the —CH$_2$— groups of the guest are also geminally nonequivalent at temperatures at which guest exchange is slow but inversion of the conformational chirality of the macrocycles is fast. This is clearly a consequence of the two different bridges. Host **95a** shows an interesting difference in guest selectivity as compared with the more rigid host **83b.** The dications $\overset{+}{N}H_3(CH_2)_2\overset{+}{N}H_3$ and $\overset{+}{N}H_3(CH_2)_3\overset{+}{N}H_3$ are equally readily complexed, and although, as expected, the complexation of longer dications is weaker the dications $\overset{+}{N}H_3(CH_2)_4\overset{+}{N}H_3$ and $\overset{+}{N}H_3(CH_2)_5\overset{+}{N}H_3$ are also equally readily complexed. This result suggests that cations of this type may have the capacity to provide cavities of differing lengths as a result of conformational changes in the flexible $CH_2(CH_2OCH_2)_2CH_2$ bridge.

Finally, as a possible guide to further developments, the macropolycyclic crown ether analogs **96**[48] and **97**[142] have been reported. Both of these systems form complexes with alkali metal cations and suggest the possibility of constructing analogs of the bridged azacrown ethers **83,** in which three crown ether macrocycles provide receptors for up to three guest cations.

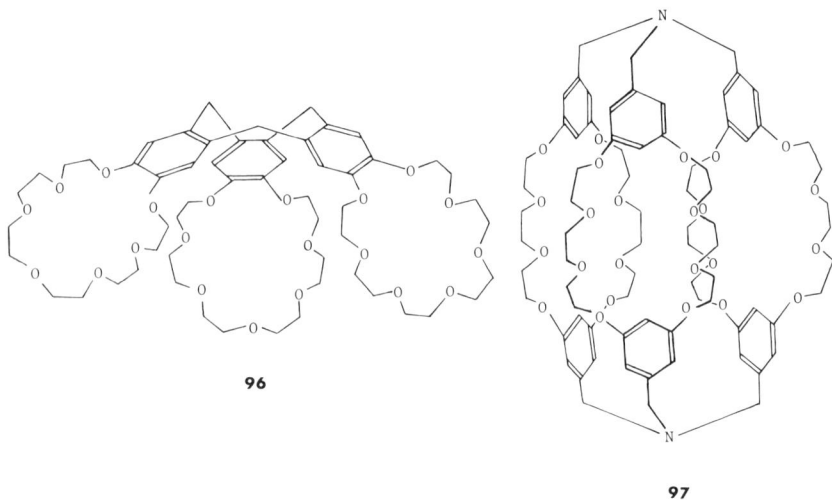

96

97

III. BRIDGED AND CAPPED PORPHYRINS AND RELATED SYSTEMS

The search for synthetic analogs of hemoglobin and myoglobin has led to the synthesis of a wide variety of substituted, bridged, and capped porphyrins. These studies have been reviewed elsewhere.[19,66] The short review given here emphasizes the bridged, face-to-face, and capped systems, because these are cyclophanes, as well as their complexation of oxygen and carbon monoxide. The "picket fence" porphyrins are not included, because with these compounds successful strategy for controlled complexation[19,20,63] does not involve the use of cyclophanes.

The features of the oxygen-binding site in hemoglobin or myoglobin that the models systems are designed to simulate include (1) an Fe(II) porphyrin situated in a hydrophobic cleft in a protein; (2) five-coordinate Fe(II), the fifth ligand being the imidazole of a histidine residue in the natural systems; (3) reversible binding of oxygen at the sixth coordination site, which is situated in a cavity that is only large enough to accept small molecules; (4) stability of the oxygen complex toward the irreversible autoxidation (Scheme 6) that is found for the denatured natural systems, which leads to the formation of a ψ-oxo dimer; and (5) selective binding of oxygen rather than carbon monoxide. Consideration of these requirements leads to the selection of models based on an Fe(II) porphyrin in which one face is sterically hindered with limited ligand accessibility and in which the other face is unhindered. This required encumbrance at one face can be provided by three types of model: (1) the picket fence model shown diagrammatically in **98,** in which the large meso substituents provide the required steric hindrance; (2) the "strapped" or "bridged" model **99,** in which one face of the porphyrin system is protected by a bridge; and (3) the capped model **100,** in which the protection is given by a cap over one face of the porphyrin system. The face-to-face porphyrins **101** may be regarded as bridged or capped and are discussed separately.

A. Bridged Porphyrins

Most bridged or strapped porphyrin models for hemoglobin have been based on porphyrin dicarboxylic or tetracarboxylic acid with the bridges introduced by ester or amide linkages. The bridge is required to direct the ligand L (Scheme 6 and **99**) to the unbridged face of the porphyrin, leaving a cavity, defined by the bridge, of a suitable size to accommodate oxygen and prevent the formation of the μ-oxo dimer.

The singly bridged systems **102** are examples of such porphyrins; the anthracene bridge excludes the bulky dicyclohexylimidazole ligand, but permits both oxygen and carbon monoxide binding.[133] The kinetics of oxygen and carbon monoxide complexation to the bridged face of the porphyrin were studied by flash photolysis. The longer bridge in the porphyrin **102b** showed relatively little effect on the rates of complexation,

98   **99**   **100**   **101**

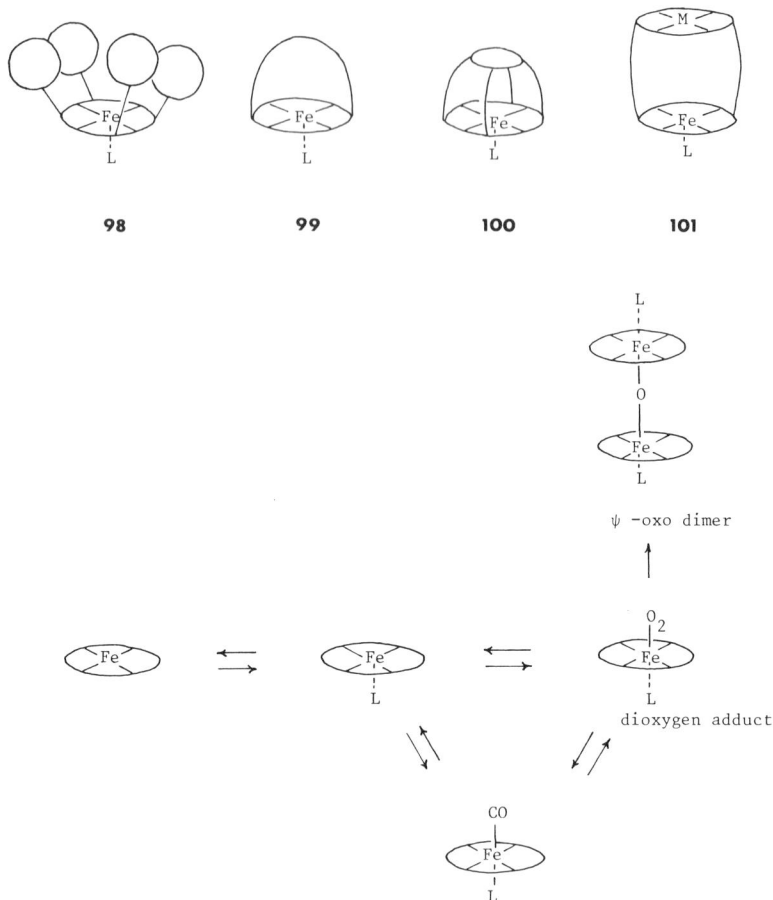

SCHEME 6. Reactions of Fe(II) porphyrins with oxygen and carbon monoxide. The circle represents a porphyrin system, and L a ligand such as N-methylimidazole.

but the shorter bridge in **102a** slowed down the rates of binding significantly for both oxygen and carbon monoxide. However, the steric effects for both guests were similar, in contrast with the natural hemoprotein systems. The analogous singly bridged systems **103** gave rather similar results.[140] The bridge was adequate in preventing attachment of the imidazole ligand to the bridged face but did not provide discrimination between oxygen and carbon monoxide to match that of the natural receptors.

The fifth ligand attached to the Fe(II) can also be built into the porphyrin system, either by using a second bridge containing a pyridine ligand, as in **104,** or by using a ligand attached to a single side chain on the porphyrin system, as in **105.**

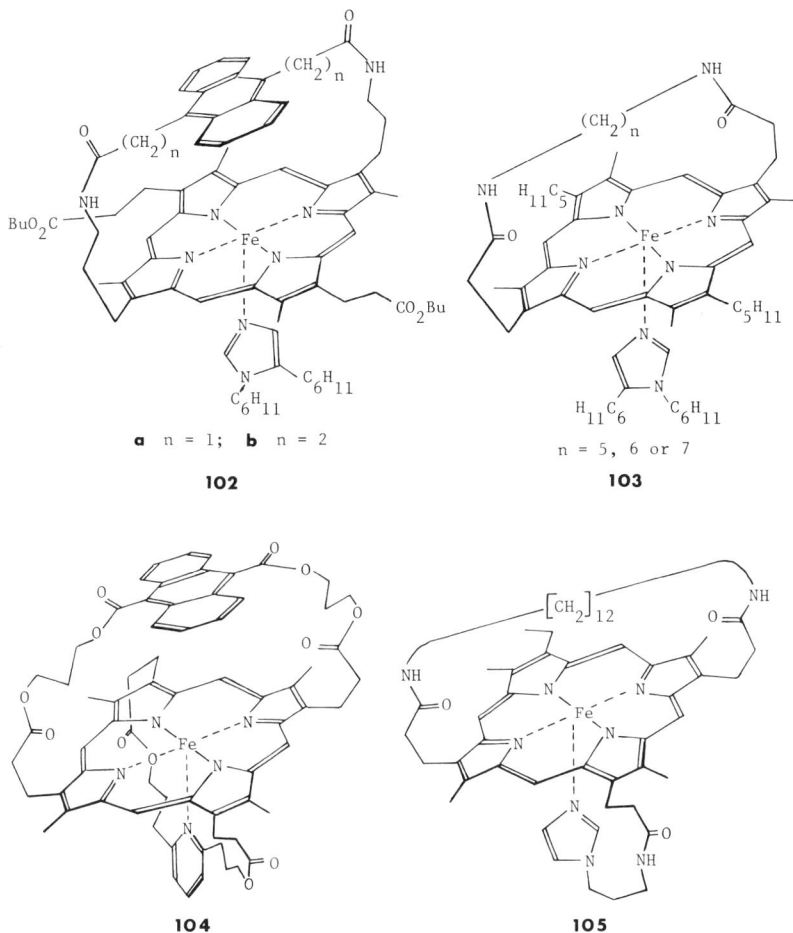

a n = 1; b n = 2

102

n = 5, 6 or 7

103

104

105

The doubly bridged system 104[3,4,30] has been shown to be a reasonably good model for hemoglobin, although the anthracene bridge does not provide adequate steric protection for the prevention of irreversible oxidation ($t_{1/2}$ = 15 min at 20°C for the dioxygen complex of 104). However, model 104 does show reversible binding for both carbon monoxide and oxygen.

Reversible oxygen uptake is not restricted to encumbered porphyrin systems, and it has also been noted for metal complexes of bicyclic (lacunar) ligands that contain a "persistent void."[57] At least one such system (106) has cyclophane character, although this is obviously not a prerequisite for activity in a system of this type. The bridging in 106 is evidently not adequate to ensure stability of the dioxygen complex, which decomposes at 20°C.

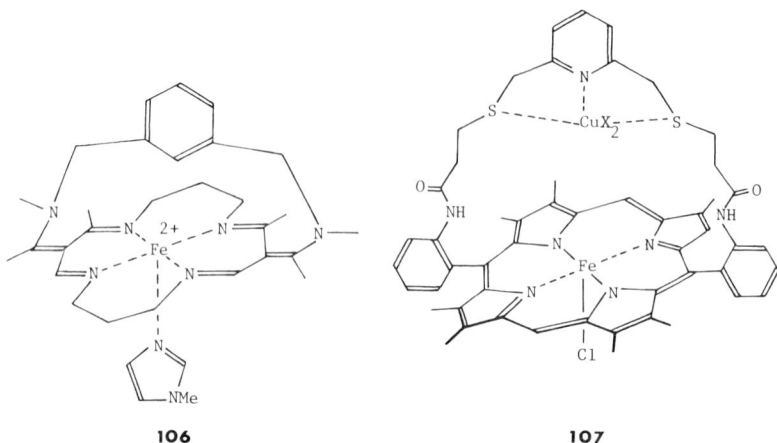

106   **107**

Bridged porphyrins can also be used as models for natural binuclear metal complexes if the bridge contains a ligand system. For example, the bridged porphyrin system **107** has been examined as a possible model for the Fe(III)–Cu(II) system believed to be present in cytochrome c.[52] Model **107** did not, however, show the magnetic coupling between the metal atoms that is characteristic of the natural system.

B. Capped Porphyrins

The capped porphyrins **100** present more extreme hindrance to autoxidation than the bridged porphyrins **99,** and in addition the cavity size is more closely controlled to prevent binding of ligands other than oxygen or carbon monoxide to the capped face. A number of systems of this type **(108),** prepared by an ingenious adaptation of the well-known tetraarylporphyrin synthesis, have been studied.[1] The critical step in the synthesis, the formation of the porphyrin ring system (see Scheme 7), was carried out under high-dilution conditions to maximize the yield. The capped porphyrin **108** undergoes irreversible complexation on exposure to carbon monoxide and N-methylimidazole, but the capped porphyrin **110,** with a greater cap-to-metal distance, binds carbon monoxide reversibly over several cycles of exchange with oxygen, thus behaving as a model for the natural hemoprotein systems. It was noted that the cap impedes both oxygen and carbon monoxide binding, as compared with simple unimpeded porphyrin systems, and the cap also conferred greater stability on the dioxygen complex than the bridge of the analogous bridged system **(109).**[2] In fact, it was noted that the bridge in **109** was not adequate

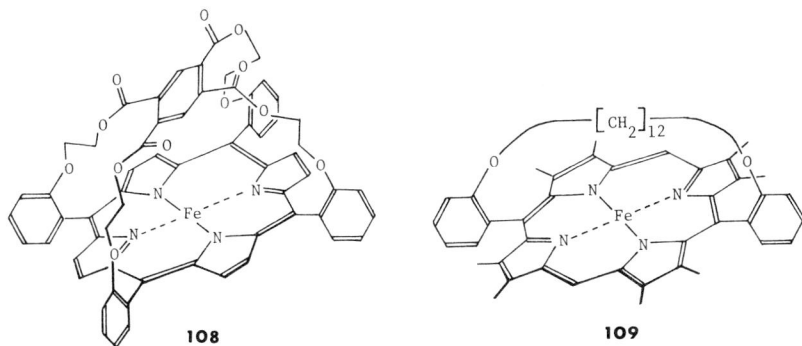

108 **109**

either to prevent the formation of the μ-oxo dimer or even to prevent the attachment of a second molecule of a nitrogen ligand, such as pyridine, with the formation of a hexacoordinate Fe(II) complex.

The oxygen affinities of the iron and cobalt complexes of the capped porphyrins have been studied in detail,[92] and it was shown that the natural hemoproteins have considerably greater oxygen affinities. This is apparently a consequence of the steric strain associated with the formation of a six-coordinate dioxygen adduct by the capped systems. The binding of a variety of ligands to Co(II) and Fe(II) complexes of the capped porphyrin

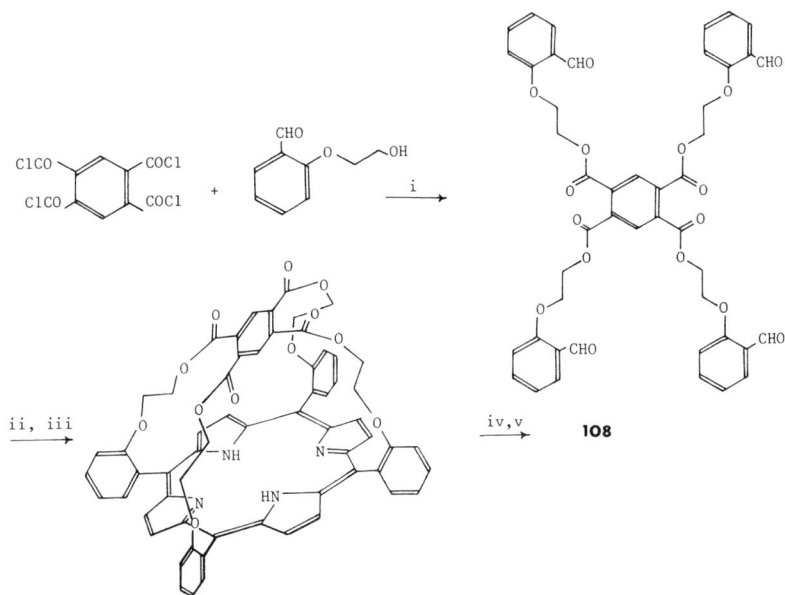

108

SCHEME 7. Reagents: *i*, Et₃N/THF; *ii*, [pyrrole], CH₃CH₂CO₂H, O₂; *iii*, [dichlorodicyanoquinone]; *iv*, FeCl₃; *v*, chromous acetoacetate.

has also been studied.[44] In general, only a single ligand can be bound to the free face of the porphyrin system, but if the separation between metal and cap is increased, as in a homolog **(110)** of the system **108,** it is possible to bind two small ligands, such as propylamine, with the usual hexacoordinate geometry **(111).** Larger ligands, such as N-methylimidazole and pyridine, also form pseudo-six-coordinate complexes with **110** that can add dioxygen to form pseudo-seven-coordinate species. The structure of these novel species in unknown.

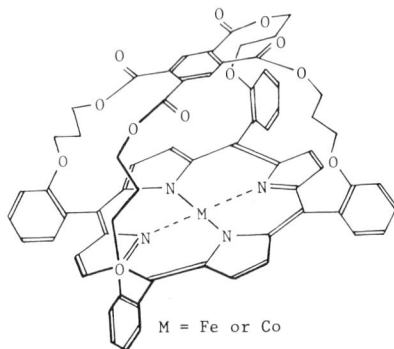

M = Fe or Co

110

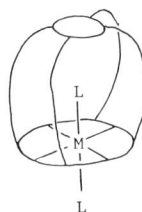

111

C. Face-to-Face Porphyrins

Face-to-face porphyrins **(101)** serve as models for the hemoproteins if they are regarded as bridged systems or as models for binuclear systems

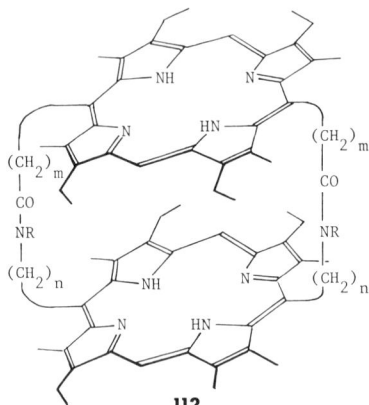

112

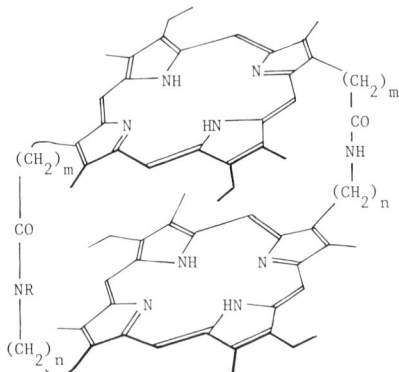

113

m = n = 1	m = n = 2	m = n = 2,	m = n = 1
m = 1, n = 2	m = 2, n = 1	m = 2, n = 1,	m = 1, n = 2

such as cytochrome c. The linkage between the two porphyrins making up such a system can be between two meso positions (e.g., **112**)[21] or between substituents on the pyrrole rings (e.g., **113**).[22,53,140] In most cases only two linkages have been used [see the bis(crown ether) systems **83**], but in one case a system **(114)** with bridges between all four pairs of meso positions has been reported.[68] The synthetic route used for the synthesis of this fully bridged (capped) system is of interest, and it is summarized in Scheme 8.

The face-to-face systems **113** were investigated as possible catalysts for the electrochemical reduction of oxygen to water, a reaction of importance in fuel cell technology. The bis-Co(II) complex of the system with the shortest distance between the two porphyrin rings (**112:** $m = n = 1$), when attached to the surface of a graphite electrode, was found to be an efficient catalyst for this reaction, whereas face-to-face porphyrins having

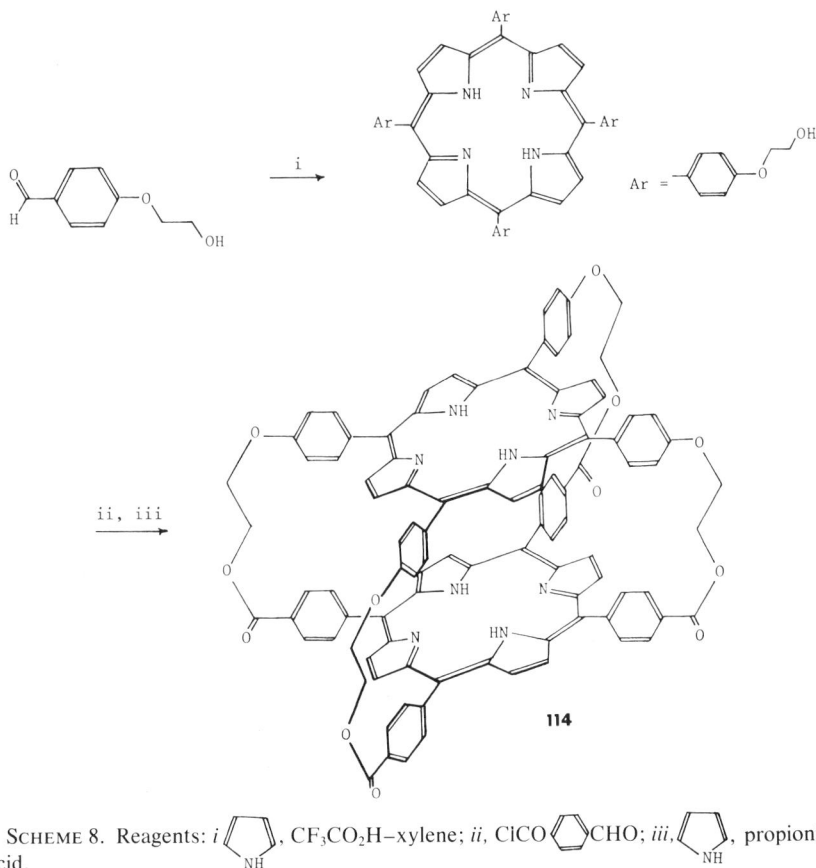

SCHEME 8. Reagents: *i* (pyrrole), CF_3CO_2H–xylene; *ii*, $ClCO$(benzene)CHO; *iii*, (pyrrole), propionic acid.

a greater separation were less efficient in catalyzing the required reaction and preventing the formation of hydrogen peroxide. It is interesting that face-to-face porphyrins, such as **113,** lacking a plane of symmetry that passes through the bridges can exist as diastereoisomers; nevertheless, a single major product was isolated and was believed to have the stereochemistry shown. Cofacial systems similar to **113** have also been used as hemoprotein models and, in particular, the Fe(II)–Cu(II) complexes **115** contain an active Fe(II) porphyrin bridged by an inert Cu(II) porphyrin.

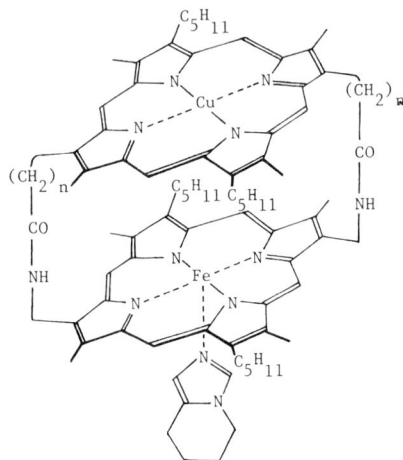

a n = 1; **b** n = 2

115

The latter presents extreme steric hindrance, and the dioxygen complexes of **115a** and **115b** are very stable, showing no signs of decomposition after 12 hr at room temperature. Furthermore, the rates of association of carbon monoxide with these highly hindered systems are reduced more than those of oxygen relative to an unhindered porphyrin system, but the reduction ratios (3 for **115b** and 4 for **115a**) are not as great as those shown by the natural myoglobin system, which is much less sterically hindered. The authors conclude[140] "that it would be a unique synthetic challenge to prepare heme models that match myoglobin's kinetic behaviour." Nevertheless, the success that has been achieved using various types of porphyrin cyclophanes as models for hemoproteins underlines the advantages of the stereochemical control afforded by the use of cyclophane systems. It is anticipated that this approach to guest–host chemistry will be developed more fully in the future.

REFERENCES

1. J. Almog, J. E. Baldwin, M. J. Crossley, J. F. DeBernardis, R. L. Dyer, J. R. Huff, and M. K. Peters, *Tetrahedron* **37,** 3589 (1981).
2. J. E. Baldwin, M. J. Crossley, T. Klose, E. A. O'Rear, and M. K. Peters, *Tetrahedron* **38,** 27 (1982).
3. A. R. Battersby and A. D. Hamilton, *J. Chem. Soc., Chem. Commun.* p. 117 (1980).
4. A. R. Battersby, S. G. Hartley, and M. D. Turnbull, *Tetrahedron Lett.* p. 3169 (1978).
5. H. F. Beckford, R. M. King, J. F. Stoddart, and R. F. Newton, *Tetrahedron Lett.* p. 171 (1978).
6. J. P. Behr and J. M. Lehn, *J. Am. Chem. Soc.* **98,** 1743 (1976).
7. M. L. Bender and M. Komiyama, "Cyclodextrin Chemistry." Springer-Verlag, Berlin and New York, 1977.
8. G. Borgen, J. Dale, K. Daasvatn, and J. Krane, *Acta Chem. Scand., Ser. B* **B34,** 249 (1980).
9. M. J. Bovill, D. J. Chadwick, M. R. Johnson, N. F. Jones, I. O. Sutherland, and R. F. Newton, *J. Chem. Soc., Chem. Commun.* p. 1065 (1979).
10. M. J. Bovill, D. J. Chadwick, I. O. Sutherland, and D. Watkin, *J. Chem. Soc., Perkin Trans. 2* p. 1529 (1980).
11. J. S. Bradshaw, S. L. Baxter, J. D. Lamb, R. M. Izatt, and J. J. Christensen, *J. Am. Chem. Soc.* **103,** 1821 (1981).
12. J. S. Bradshaw, S. L. Baxter, D. C. Scott, J. D. Lamb, R. M. Izatt, and J. J. Christensen, *Tetrahedron Lett.* p. 3383 (1979).
13. J. S. Bradshaw, G. E. Maas, R. M. Izatt, and J. J. Christensen, *Chem. Rev.* **79,** 37 (1979).
14. J. S. Bradshaw, G. E. Maas, R. M. Izatt, J. D. Lamb, and J. J. Christensen, *Tetrahedron Lett.* p. 635 (1979).
15. J. S. Bradshaw, G. E. Maas, J. D. Lamb, R. M. Izatt, and J. J. Christensen, *J. Am. Chem. Soc.* **102,** 467 (1980).
16. R. Breslow, P. Bovy, and C. L. Hersh, *J. Am. Chem. Soc.* **102,** 2116 (1980).
17. C. Brevard, J. P. Kintzinger, and J. M. Lehn, *Tetrahedron* **28,** 2447 (1972).
18. E. Buhleier, W. Wehner, and F. Vögtle, *Chem. Ber.* **112,** 546, 559 (1979).
19. J. P. Collman, *Acc. Chem. Res.* **10,** 265 (1977).
20. J. P. Collman, J. I. Brauman, K. M. Doxsee, T. R. Halbert, E. Bunnenberg, R. E. Linder, G. N. LaMar, J. Del Gaudio, G. Long, and K. Spartalian, *J. Am. Chem. Soc.* **102,** 4182 (1980).
21. J. P. Collman, A. O. Chong, G. B. Jameson, R. T. Oakley, E. Rose, E. R. Schmitton, and J. A. Ibers, *J. Am. Chem. Soc.* **103,** 516 (1981).
22. J. P. Collman, P. Denesevich, Y. Konai, M. Marrocco, C. Koval, and F. C. Anson, *J. Am. Chem. Soc.* **102,** 6027 (1980).
23. P. K. Coughlin, S. J. Lippard, A. E. Martin, and J. E. Bulkowski, *J. Am. Chem. Soc.* **102,** 7616 (1980).
24. D. J. Cram, *Acc. Chem. Res.* **11,** 8 (1978).
25. D. J. Cram and J. M. Cram, *Science* **183,** 803 (1974).
26. D. J. Cram, R. C. Helgeson, L. R. Sousa, J. M. Timko, M. Newcomb, P. Moreau, F. DeJong, G. W. Gokel, D. H. Hoffman, L. A. Domeier, S. C. Peacock, K. Madan, and L. Kaplan, *Pure Appl. Chem.* **43,** 327 (1975).

27. D. J. Cram, G. M. Lein, T. Kanada, R. C. Helgeson, C. B. Knobler, E. Maverick, and K. N. Trueblood, *J. Am. Chem. Soc.* **103**, 6228 (1981).
28. D. J. Cram, T. Kaneda, R. C. Helgeson, and G. M. Lein, *J. Am. Chem. Soc.* **101**, 6754 (1979).
29. D. J. Cram, T. Kaneda, G. M. Lein, and R. C. Helgeson, *J. Chem. Soc., Chem. Commun.* p. 948 (1979).
30. W. B. Cruse, O. Kennard, G. M. Sheldrick, A. D. Hamilton, S. G. Hartley, and A. R. Battersby, *J. Chem. Soc., Chem. Commun.* p. 700 (1980).
31. F. DeJong and D. N. Reinhoudt, *Adv. Phys. Org. Chem.* **17**, 279 (1980).
32. F. DeJong, D. N. Reinhoudt, and R. Huis, *Tetrahedron Lett.* p. 3985 (1977).
33. F. DeJong, D. N. Reinhoudt, C. J. Smit, and R. Huis, *Tetrahedron Lett.* p. 4783 (1976).
34. F. DeJong, D. N. Reinhoudt, and G. J. Torny, *Tetrahedron Lett.* p. 911 (1979).
35. J. G. DeVries and R. M. Kellogg, *J. Am. Chem. Soc.* **101**, 2759 (1979).
36. B. Dietrich, D. L. Fyles, T. M. Fyles, and J. M. Lehn, *Helv. Chim. Acta* **62**, 2763 (1979).
37. B. Dietrich, T. M. Fyles, J. M. Lehn, L. G. Pease, and D. L. Fyles, *J. Chem. Soc., Chem. Commun.* p. 934 (1978).
38. B. Dietrich, M. W. Hosseini, J. M. Lehn, and R. B. Sessions, *J. Am. Chem. Soc.* **103**, 1282 (1981).
39. M. Dobler and R. P. Phizackerley, *Acta Crystallogr., Sect. B* **B30**, 2746, 2748 (1974).
40. M. G. B. Drew, J. de O. Cabral, J. F. Cabral, F. S. Esho, and S. M. Nelson, *J. Chem. Soc., Chem. Commun.* p. 1033 (1979).
41. M. G. B. Drew, M. McCann, and S. M. Nelson, *J. Chem. Soc., Chem. Commun.* p. 481 (1979).
42. M. G. B. Drew, A. Rodgers, M. McCann, and S. M. Nelson, *J. Chem. Soc., Chem. Commun.* p. 415 (1978).
43. G. A. Ellestad, R. A. Leese, G. O. Morton, F. Barbatschi, W. E. Gore, W. J. McGahren, and I. M. Armitage, *J. Am. Chem. Soc.* **103**, 6522 (1981).
44. P. E. Ellis, J. E. Linard, T. Szymarski, R. D. Jones, J. R. Budge, and F. Basolo, *J. Am. Chem. Soc.* **102**, 1889 (1980).
45. D. E. Fenton, D. H. Cook, I. W. Nowell, and P. E. Walker, *J. Chem. Soc., Chem. Commun.* p. 279 (1978).
46. B. L. Feringa, *Tetrahedron Lett.* p. 1443 (1981).
47. K. Frensch, G. Oepen, and F. Vögtle, *Liebigs Ann. Chem.* p. 858 (1979).
48. K. Frensch and F. Vögtle, *Liebigs Ann. Chem.* p. 2121 (1979).
49. R. D. Gandour, D. A. Walker, A. Nayak, and G. R. Newkome, *J. Am. Chem. Soc.* **100**, 3608 (1978).
50. I. Goldberg, *Acta Crystallogr.* **31**, 2592 (1975).
51. E. Graf and J. M. Lehn, *J. Am. Chem. Soc.* **98**, 6403 (1976).
52. M. J. Gunter, L. N. Mander, K. S. Murray, and P. E. Clark, *J. Am. Chem. Soc.* **103**, 6784 (1981).
53. M. H. Hatada, A. Tulkinsky, and C. K. Chang, *J. Am. Chem. Soc.* **102**, 7115 (1980).
54. R. C. Helgeson, T. L. Tarnowski, and D. J. Cram, *J. Org. Chem.* **44**, 2538 (1979).
55. R. C. Helgeson, T. L. Tarnowski, J. M. Timko, and D. J. Cram, *J. Am. Chem. Soc.* **99**, 6411 (1977).
56. R. C. Helgeson, J. M. Timko, and D. J. Cram, *J. Am. Chem. Soc.* **96**, 7380 (1974).
57. N. Herron and D. H. Busch, *J. Am. Chem. Soc.* **103**, 1236 (1981).
58. L. C. Hodgkinson, M. R. Johnson, S. J. Leigh, N. Spencer, I. O. Sutherland, and R. F. Newton, *J. Chem. Soc., Perkin. Trans. 1* p. 2193 (1979).

59. L. C. Hodgkinson, S. J. Leigh, and I. O. Sutherland, *J. Chem. Soc., Chem. Commun.* pp. 639, 640 (1976).

60. L. C. Hodgkinson and I. O. Sutherland, *J. Chem. Soc., Perkin Trans. 1* p. 1908 (1979).

61. R. M. Izatt and J. J. Christensen, eds., "Synthetic Multidentate Macrocyclic Compounds." Academic Press, New York, 1978.

62. R. M. Izatt, N. E. Izatt, B. E. Rossiter, J. J. Christensen, and B. L. Haymore, *Science* **199,** 994 (1978).

63. G. B. Jameson, F. S. Molinaro, J. A. Ibers, J. P. Collman, J. I. Braumann, E. Rose, and K. S. Suslick, *J. Am. Chem. Soc.* **102,** 3224 (1980).

64. M. R. Johnson, I. O. Sutherland, and R. F. Newton, *J. Chem. Soc., Chem. Commun.* p. 309 (1979).

65. N. F. Jones, A. Kumar, and I. O. Sutherland, *J. Chem. Soc., Chem. Commun.* p. 990 (1981).

66. R. D. Jones, D. A. Summerville, and F. Basolo, *Chem. Rev.* **79,** 139 (1979).

67. P. Jouin, C. B. Troostwijk, and R. M. Kellogg, *J. Am. Chem. Soc.* **103,** 2091 (1981).

68. N. E. Kagan, D. Mauzerell, and R. B. Merrifield, *J. Am. Chem. Soc.* **99,** 5484 (1977).

69. J. R. Kalman and D. H. Williams, *J. Am. Chem. Soc.* **102,** 897 (1980).

70. R. M. Kellogg, T. J. Van Bergen, H. Van Doren, D. Hedstrand, J. Kooi, W. H. Kruizinga, and C. B. Troostwijk, *J. Org. Chem.* **45,** 2854 (1980).

71. J. P. Kintzinger, F. Kotzyba-Hibert, J. M. Lehn, A. Pagelot, and K. Saigo, *J. Chem. Soc., Chem. Commun.* p. 833 (1981).

72. K. E. Koenig, R. C. Helgeson, and D. J. Cram, *J. Am. Chem. Soc.* **98,** 4018 (1976).

73. K. E. Koenig, G. M. Lein, P. Stuckler, T. Kaneda, and D. J. Cram, *J. Am. Chem. Soc.* **101,** 3553 (1979).

74. F. Kotzyba-Hibert, J. M. Lehn, and K. Saigo, *J. Am. Chem. Soc.* **103,** 4226 (1981).

75. F. Kotzyba-Hibert, J. M. Lehn, and P. Vierling, *Tetrahedron Lett.* **21,** 941 (1980).

76. J. Krane and O. Aune, *Acta Chem. Scand., Ser. B* **B34,** 397 (1980).

77. D. A. Laidler and J. F. Stoddart, *J. Chem. Soc., Chem. Commun.* p. 979 (1976).

78. D. A. Laidler and J. F. Stoddart, *J. Chem. Soc., Chem. Commun.* p. 481 (1977).

79. D. A. Laidler and J. F. Stoddart, *Tetrahedron Lett.* p. 453 (1979).

80. D. A. Laidler, J. F. Stoddart, and J. B. Wolstenholme, *Tetrahedron Lett.* p. 465 (1979).

81. J. M. Lehn, *Struct. Bonding (Berlin)* **16,** 1 (1973).

82. J. M. Lehn, *Pure Appl. Chem.* **50,** 871 (1978).

83. J. M. Lehn, *Acc. Chem. Res.* **11,** 49 (1978).

84. J. M. Lehn, *Pure Appl. Chem.* **51,** 979 (1979).

85. J. M. Lehn, S. H. Pine, E. Watanabe, and A. K. Willard, *J. Am. Chem. Soc.* **99,** 6760 (1977).

86. J. M. Lehn and J. Simon, *Helv. Chim. Acta* **60,** 141 (1977).

87. J. M. Lehn, J. Simon, and A. Moradpour, *Helv. Chim. Acta* **61,** 2407 (1978).

88. J. M. Lehn, E. Sonveaux, and A. K. Willard, *J. Am. Chem. Soc.* **100,** 4914 (1978).

89. S. J. Leigh and I. O. Sutherland, *J. Chem. Soc., Chem. Commun.* p. 414 (1975).

90. S. J. Leigh and I. O. Sutherland, *J. Chem. Soc., Perkin. Trans. 1* p. 1090 (1979).

91. S. Lifson, A. T. Hagler, and P. Dauber, *J. Am. Chem. Soc.* **101,** 5111, and following papers (1979).

92. J. E. Linard, P. E. Ellis, J. R. Budge, R. D. Jones, and F. Basolo, *J. Am. Chem. Soc.* **102,** 1896 (1980).

93. W. J. McGahren, J. H. Martin, G. O. Morton, R. T. Hargreaves, R. A. Leese, F. M. Lovell, G. A. Ellestad, E. O'Brien, and J. S. E. Holker, *J. Am. Chem. Soc.* **102,** 1671 (1980).

94. M. A. McKervey and D. L. Mulholland, *J. Chem. Soc., Chem. Commun.* p. 438 (1977).

95. R. Mageswaran, S. Mageswaran, and I. O. Sutherland, *J. Chem. Soc., Chem. Commun.* p. 722 (1979).

96. A. E. Martin and J. E. Bulkowski, *J. Am. Chem. Soc.* **104,** 1434 (1982).

97. E. Maverick, L. Grossenbacher, and K. N. Trueblood, *Acta Crystallogr., Sect. B* **B35,** 2233 (1979).

98. J. C. Metcalfe, J. F. Stoddart, and G. Jones, *J. Am. Chem. Soc.* **99,** 8317 (1977).

99. B. Metz, J. Rosalky, and R. Weiss, *J. Chem. Soc., Chem. Commun.* p. 533 (1976).

100. S. S. Moore, T. L. Tarnowski, M. Newcomb, and D. J. Cram, *J. Am. Chem. Soc.* **99,** 6398 (1977).

101. O. Nagano, A. Kobayashi, and Y. Sasaki, *Bull. Chem. Soc. Jpn.* **51,** 790 (1978).

102. M. Newcomb, S. S. Moore, and D. J. Cram, *J. Am. Chem. Soc.* **99,** 6405 (1977).

103. M. Newcomb, J. M. Timko, D. M. Walba, and D. J. Cram, *J. Am. Chem. Soc.* **99,** 6392 (1977).

104. G. R. Newkome, T. Kawato, F. R. Fronczek, and W. H. Benton, *J. Org. Chem.* **45,** 5423 (1980).

105. G. R. Newkome, D. K. Kohli, F. R. Fronczek, B. J. Hales, E. E. Case, and G. Chian, *J. Am. Chem. Soc.* **102,** 7608 (1980).

106. G. R. Newkome, V. K. Majestic, and F. R. Fronczek, *Tetrahedron Lett.* pp. 3035, 3039 (1981).

107. G. R. Newkome, V. K. Majestic, F. R. Fronczek, and J. L. Atwood, *J. Am. Chem. Soc.* **101,** 1047 (1979).

108. G. R. Newkome, H. C. R. Taylor, F. R. Fronczek, T. J. DeLard, and D. K. Kohli, *J. Am. Chem. Soc.* **103,** 7376 (1981).

109. G. Oepen and F. Vögtle, *Liebigs Ann. Chem.* p. 1094 (1979).

110. D. P. J. Pearson, S. J. Leigh, and I. O. Sutherland, *J. Chem. Soc., Perkin. Trans. 1* p. 3113 (1979).

111. C. J. Pedersen, *J. Am. Chem. Soc.* **89,** 2495, 7017 (1967).

112. C. J. Pedersen and H. K. Frensdorf, *Angew. Chem., Int. Ed. Engl.* **11,** 16 (1972).

113. R. B. Pettman and J. F. Stoddart, *Tetrahedron Lett.* pp. 457, 461 (1979).

114. V. Prelog, *Pure Appl. Chem.* **50,** 893 (1978).

115. V. Prelog and D. Bedekovic, *Helv. Chim. Acta* **62,** 2285 (1979).

116. D. N. Reinhoudt and F. DeJong, *Prog. Macrocyclic Chem.* **1,** 157–217 (1979).

117. D. N. Reinhoudt, H. J. Den Hertog, Jr., and F. DeJong, *Tetrahedron Lett.* p. 2513 (1981).

118. L. Rossa and F. Vögtle, *Liebigs Ann. Chem.* p. 459 (1981).

119. F. P. Schmidtchen, *Angew. Chem., Int. Ed. Engl.* **16,** 720 (1977).

120. P. Seiler, M. Dobler, and J. D. Dunitz, *Acta Crystallogr., Sect. B* **B30,** 2744 (1974).

121. G. M. Sheldrick, P. G. Jones, O. Kennard, D. H. Williams, and G. A. Smith, *Nature (London)* **271,** 223 (1978).

122. S. Shinkai, T. Nakaji, Y. Nishida, T. Ogawe, and O. Manabe, *J. Am. Chem. Soc.* **102,** 5860 (1980).

123. S. Shinkai, T. Ogawa, T. Nakaji, Y. Kusano, and O. Manabe, *Tetrahedron Lett.* p. 4569 (1979).

124. J. F. Stoddart, *Chem. Soc. Rev.* **8,** 85 (1979).

125. K. Sugihara, K. Kamiya, M. Yamaguchi, T. Kaneda, and S. Misumi, *Tetrahedron Lett.* p. 1619 (1981).

126. F. Sztaricskai, C. M. Harris, A. Neszmelyi, and T. M. Harris, *J. Am. Chem. Soc.* **102,** 7093 (1980).

127. F. Sztaricskai, A. Neszmelyi, and R. Bognar, *Tetrahedron Lett.* p. 2983 (1980).
128. A. P. Thoma, A. Viviani Naver, K. H. Schellenberg, D. Bedekovic, E. Pretsch, V. Prelog, and W. Simon, *Helv. Chim. Acta* **62**, 2303 (1979).
129. J. M. Timko and D. J. Cram, *J. Am. Chem. Soc.* **96**, 7159 (1974).
130. J. M. Timko, R. C. Helgeson, M. Newcomb, G. W. Gokel, and D. J. Cram, *J. Am. Chem. Soc.* **96**, 7097 (1974).
131. J. M. Timko, S. S. Moore, D. M. Walba, P. C. Hiberty, and D. J. Cram, *J. Am. Chem. Soc.* **99**, 4207 (1977).
132. G. L. Trainor and R. Breslow, *J. Am. Chem. Soc.* **103**, 154 (1981).
133. T. G. Traylor, M. J. Mitchell, S. Tsuchiya, D. H. Campbell, D. V. Stynes, and N. Koga, *J. Am. Chem. Soc.* **103**, 5234 (1981).
134. K. N. Trueblood, C. B. Knobler, D. S. Lawrence, and R. V. Stevens, *J. Am. Chem. Soc.* **104**, 1355 (1982).
135. K. N. Trueblood, C. B. Knobler, E. Maverick, R. C. Helgeson, S. B. Brown, and D. J. Cram, *J. Am. Chem. Soc.* **103**, 5594 (1981).
136. B. Tummler, G. Maas, E. Weber, W. Wehner, and F. Vögtle, *J. Am. Chem. Soc.* **99**, 4683 (1977).
137. T. J. Van Bergen and R. M. Kellogg, *J. Chem. Soc., Chem. Commun.* p. 964 (1976).
138. T. J. Van Bergen and R. M. Kellogg, *J. Am. Chem. Soc.* **99**, 3882 (1977).
139. M. Van der Leij, H. J. Oosterink, R. H. Hall, and D. N. Reinhoudt, *Tetrahedron* **37**, 3661 (1981).
140. B. Ward, C. B. Wong, and C. K. Chang, *J. Am. Chem. Soc.* **103**, 5236 (1981).
141. N. Wester and F. Vögtle, *J. Chem. Res., Synop.* p. 400 (1978).
142. N. Wester and F. Vögtle, *Chem. Ber.,* **113**, 1487 (1980).
143. D. H. Williams, V. Rajanda, M. P. Williamson, and G. Bojesen, *Top. Antibiot. Chem.* **5**, 119 (1980).
144. M. P. Williamson and D. H. Williams, *J. Am. Chem. Soc.* **103**, 6580 (1981).
145. I. Yamashita, M. Fujii, T. Kaneda, S. Misumo, and Y. Otsubo, *Tetrahedron Lett.* p. 541 (1980).

Index

A

C

ORGANIC CHEMISTRY
A SERIES OF MONOGRAPHS

EDITOR

HARRY H. WASSERMAN

Department of Chemistry
Yale University
New Haven, Connecticut

ORGANIC CHEMISTRY

N

I

N

2